PRAISE FOR *GOD AND THE FOLLY OF FAITH*

"Stenger—a major powerhouse of honest thinking—has produced another massive exercise in logic. He shows why science is trustworthy as humanity's life-support. And, in contrast, he shows why supernatural beliefs are silly and childish, unworthy of thinking adults."

—James A. Haught, editor of the *Charleston (WV) Gazette* and author of *2000 Years of Disbelief*

"A magisterial survey demonstrating that religion is simply not compatible with science. In fact, as Stenger proves here time and again, religion keeps getting in the way of science, often motivates people to distort scientific facts to serve their religious hopes and wishes, and remains attached to ideas and ways of thinking that science has long since proved invalid. If you know someone who gainsays any of these conclusions, give them this book."

—Dr. Richard Carrier, author of *Sense and Goodness without God*

"Elisabeth Kübler-Ross famously outlined the stages through which we pass as we come to grips with death: denial, anger, bargaining, and acceptance. Well, religion seems to be dying a lingering death, and in *God and the Folly of Faith* Victor Stenger shows how it is passing through the same stages: fundamentalists are in denial, apologists are in the anger phase, and liberal theologians are bargaining. Stenger's fair and deadly arguments will help religion to accept its inevitable demise."

—Robert M. Price, professor of biblical criticism, Center for Inquiry Institute, and author of *Beyond Born Again*

"Puts to rest the myth that science and religion can coexist peacefully. Thoughtfully researched and delightfully readable!"

—Craig A. James, author of *The Religion Virus*

GOD

and the

Folly of Faith

VICTOR J. STENGER

Foreword by Dan Barker

GOD

and the

Folly of Faith

The Incompatibility
of Science and Religion

 **Prometheus Books**

59 John Glenn Drive
Amherst, New York 14228–2119

Published 2012 by Prometheus Books

Cover image © 2012 Media Bakery, Inc.
Cover design by Nicole Sommer-Lecht

Inquiries should be addressed to
Prometheus Books
59 John Glenn Drive
Amherst, New York 14228–2119
VOICE: 716–691–0133
FAX: 716–691–0137
WWW.PROMETHEUSBOOKS.COM

16 15 14 13   5 4 3 2

Library of Congress Cataloging-in-Publication Data

Stenger, Victor J., 1935–
    God and the folly of faith : the incompatibility of science and religion / by Victor J. Stenger.
        p. cm.
    Includes bibliographical references and index.
    ISBN 978–1–61614–599–6 (pbk. : alk. paper)
    ISBN 978–1–61614–600–9 (ebook)
    1. Religion and science. 2. Christianity—Controversial literature. I. Title.

BL240.3.S737 2012
201'.65—dc23

2011050570

Printed in the United States of America on acid-free paper

Dedicated to my wife, Phylliss,
for her fifty years of love, companionship, and support.

Science flies us to the moon.
Religion flies us into buildings.

—Victor J. Stenger

That which can be asserted without evidence can be dismissed without evidence.

—Christopher Hitchens

CONTENTS

7

ACKNOWLEDGMENTS

Once again I am deeply indebted to Brent Meeker and Bob Zannelli for their meticulous reading of my various manuscripts and their many invaluable corrections and changes. They are longtime members of the discussion group avoid-L ("Atoms and the Void") that I created over a decade ago, now ably managed by Bob. Others from avoid-L who have helped on this book include Greg Bart, Martin Bier, Yonatan Fishman, Don McGee, and Christopher Savage. I have also received valuable assistance from religion scholar Hector Avalos, ancient historian Richard Carrier, physicist Taner Edis, and writer Andrew Zak Williams. Keith Augustine was of great help with my discussion of near death experiences. Sean Carroll and Alexander Vilenkin graciously replied to my e-mail queries on cosmology. I am grateful to Peter Montgomery, senior fellow at People for the American Way, for tracking down some references for me. Also, I wish to express my appreciation to the prominent free thinkers Dan Barker and Michael Shermer for their continual encouragement and support.

Beside these friends and supporters, I am fortunate to have a loving family that continually provides me with inspiration and encouragement.

Finally, I wish to add that I am forever grateful to the late Christopher Hitchens for his inspiration, support, and friendship. He was the greatest writer, intellect, and gentleman I personally have ever known.

FOREWORD

Man, once surrendering his reason, has no remaining guard against absurdities the most monstrous, and like a ship without rudder, is the sport of every wind. With such persons, gullibility, which they call faith, takes the helm from the hand of reason and the mind becomes a wreck.

—Thomas Jefferson[1]

Madeline Kara Neumann was a fun-loving, eleven-year-old girl who liked to wear her straight, brown hair in a ponytail. A photo that ran in a newspaper after she died shows her kneeling proudly over a work of art she was creating at an outdoor chalkfest in Wausau, Wisconsin. Since Kara, as she was called, was completely under the care and authority of her parents, she was not allowed to choose her own religion, nor was she free to decide to take herself to the doctor. She died on Easter Sunday 2008, after suffering days of ghastly pain from undiagnosed but easily treatable diabetes. Her devout Christian parents had refused to take her to the hospital, believing that prayer alone can heal the sick.

Kara's father, who once studied to be a Pentecostal minister, testified that he neither wanted nor expected his daughter to die. Believing himself to be a good parent, he had faith that God would heal Kara, as promised in the Bible:

> Is any among you sick? Let him call for the elders of the church; and let them pray over him, anointing him with oil in the name of the Lord: and the prayer of faith shall save him that is sick, and the Lord shall raise him up. (James 5:14–15)

During the trial after her death, Kara's father was resolute. "If I go to the doctor," Dale Neumann said, "I am putting the doctor before God. I am not believing what he [God] said he would do."

15

Instead of acting like normal, prudent parents, who would naturally seek real help when it was obvious their daughter was in serious condition, the family contacted Unleavened Bread Ministries. The founder of that fundamentalist church, David Eells, has written: "Jesus never sent anyone to a doctor or a hospital. Jesus offered healing by one means only! Healing was by faith." A statement posted later on that church's website by Eells reveals that the Neumanns, believing the scripture verse cited above, "contacted one of our elders to ask that I call them to pray for their daughter. That elder got in touch with me Saturday evening and I called the Neumanns." As a direct result of her "loving" parents' faith-based inaction, Kara Neumann died the next morning, the day Christians celebrate their Lord Jesus' resurrection and victory over death.

The Neumanns, who were convicted of second-degree reckless homicide (and who remain unrepentant), are right about one thing: the Bible is very emphatic that faith will heal the sick. Jesus said, "All things, whatsoever ye shall ask in prayer, believing, ye shall receive" (Matt. 21:22).[2] Nothing could be more clear: he said all things. He didn't say "maybe," or "if I feel like it," or "if you are specially chosen." You will receive all things if you believe. Coupled with James 5:15 (above), who could blame the Neumanns for thinking their prayers would be answered by the creator of the universe who loves them and promised to hear their prayers and who has the power to perform miracles and heal illness?[3]

Here are a few of many other biblical passages that support the Neumanns' faith:

Ask, and it will be given you; search, and you will find; knock, and the door will be opened for you. For everyone who asks receives, and everyone who searches finds, and for everyone who knocks, the door will be opened. (Matt. 7:7–8; Jesus speaking)

If ye then, being evil, know how to give good gifts unto your children, how much more shall your Father which is in heaven give good things to them that ask him? (Matt. 7:11; Jesus spoke these words immediately before giving the Golden Rule)

> If two of you shall agree on earth as touching any thing that they shall ask, it shall be done for them of my Father which is in heaven. (Matt. 18:19; Jesus talking)

> And whatsoever we ask, we receive of him, because we keep his commandments, and do those things that are pleasing in his sight. And this is his commandment, that we should believe on the name of his Son Jesus Christ. (1 John 3:22)[4]

These and many other verses clearly teach that believers should expect miracles in their lives. Christ and his followers reportedly performed many healings, resurrections, and other miraculous deeds, and Jesus specifically promised that all Christians would be able to do the same, and more:

> Very truly, I tell you, the one who believes in me will also do the works that I do and, in fact, will do greater works than these, because I am going to the Father. I will do whatever you ask in my name, so that the Father may be glorified in the Son. If in my name you ask me for anything, I will do it. (John 14:12–14)

There is no room for misinterpretation here: if believers ask for anything in the name of Jesus, "I will do it," he assures them. Jesus healed the sick, and so can you.

To demonstrate such power, Jesus actually cursed and withered a fig tree because, being out of season, it had no fruit when he was hungry. When his disciples marveled at this petulant miracle, he replied:

> Have faith in God. Truly I tell you, if you say to this mountain, "Be taken up and thrown into the sea," and if you do not doubt in your heart, but believe that what you say will come to pass, it will be done for you. So I tell you, whatever you ask for in prayer, believe that you have received it, and it will be yours. (Mark 11:22–24)[5]

If the Bible is true, then why did Kara die? Why were her devout believing parents' prayers not answered? Why have no mountains been thrown into the sea by faith? Why have no miracles ever been proved?

The obvious answer is that the Bible is wrong. As Vic Stenger shows us in this book, anyone who respects observation over belief can see that wishful thinking does not equal truth. In the real world, faith healing is faith killing. So, after millennia of observing the patent folly and the perfect failure of faith, why do people still believe in "absurdities the most monstrous"?

***

Jews and Muslims have a slightly different idea of faith than Christians do (*aman* and *iman*, respectively, from the same root). In the Hebrew and Islamic scriptures, faith is more like trust. It is similar to how the word is used in the phrase "full faith and credit of the United States." For Jews and Muslims, faith is a "guarantee" or "promissory note." Christians might sometimes understand it in that secondary sense, and even a few atheists may use the word (unadvisedly, in my opinion) in a natural sense, as when expressing confidence in the character of one's father, or trusting that the scientists and engineers did a good job creating the plane one is riding in. "That is not faith," Dr. Stenger writes, "but trust. The term *faith* should be reserved for unfounded beliefs. Such faith is foolish."

Stenger is right. Christian faith is not trust. When the New Testament came along, "faith" and "belief" (from the Greek *pistis*) took on a more focused and more dangerous meaning. Although the Hebrew scriptures and the Koran offer descriptions of what we should have faith in, they do not provide a definition of what faith is. Only Christianity tries to elevate belief into a virtue (or gift) in and of itself. Faith has been reified. It has become something to have and to hold, a quality of character, an evidence of goodness, a divine endowment, or an act of the will that changes an outcome. Christians don't always agree on exactly what faith is, but they all claim it can change outcomes or thoughts. Thomas Aquinas defined faith as "the act of the intellect when it assents to divine truth under the influence of the will moved by God through grace."[6] Martin Luther shifted the emphasis, claiming that "faith is God's work in us," which "changes our hearts, our spirits, our thoughts, and all our powers," a "living, bold trust in God's grace," which brings a "confidence and knowledge of God's grace."[7] John Calvin had a more passive definition of faith: "A firm and certain knowledge of God's benevolence towards us, founded upon the truth of the freely given promise in Christ, both revealed to our minds and sealed upon our hearts through the Holy Spirit."[8]

Here is the actual New Testament definition of faith, which I memorized as a child in Sunday school: Now faith is the substance of things hoped for, the evidence of things not seen (Heb. 11:1). That is the famous King James Version, but it turns out that the words "substance" and "evidence" are not so evidently substantial in their meaning. As we saw with Aquinas, Luther, and Calvin, there is room for interpretation. Some English translations replace the word "substance" with "assurance," "realization," or "proof." (The Amplified Bible suggests "title deed.") Others replace the word "evidence" with "conviction" or "certainty," but they all come down to the actual confident possession of something that you don't have or see. The Bible simply assumes that the "things not seen" are existing eternal entities, as opposed to merely mundane physical observations: "The things which are seen are temporal, but the things which are not seen are eternal" (2 Cor. 4:18).

Of course, if faith were simply the "evidence of things not seen," it would be like a scientific hypothesis under testing. The highly probable but still postulated existence of quarks, for example, is based on "evidence of things not seen." (The evidence is indeed seen, but the quarks are not.) However, no scientist says that quarks exist "by faith." A probability, even if very high, is neither a substance nor a certainty. The only substance or certainty is observation.

But looking at the Bible itself, in its entirety and in context, it is clear to see that the biblical "evidence of things not seen" is not merely a tentative hypothesis in the scientific sense, falsifiable and subject to disconfirmation. It is a "substance of things hoped for," a "certainty." It can perform miracles, after all, a claim never made by any scientific postulation.

Notice that the Bible does not say faith is equal to knowledge. Just five verses after defining faith, the author of the book of Hebrews tells us that we can't really know if God exists before we decide to believe in him: "And without faith it is impossible to please God, because anyone who comes to him must believe that he exists and that he rewards those who earnestly seek him" (Heb. 11:6).

How's that for circular reasoning? You can't know by observation or evidence if God exists: you have to believe it. That is like saying: "Why should you believe in God? Because if you believe, then you will believe." (And if you don't believe, you won't be pleasing God, and you don't want to risk that, do you?) In fact, you can't have a God unless you first believe, according to that verse.

This does not mean that the existence of God is not testable in principle. If there is a God as defined by the Christian writings, then he should be observable. Responding to believers who claim faith equals evidence, Dr. Stenger writes: "The parallel between science and religion, that both are based on data (experience) and theory (interpretation), is strained. Science takes its data and forms theories (that is, models) that can be tested against other data. When religion does that at all, it always fails the test." If Kara Neumann had been miraculously healed, that would count as evidence in favor of faith. Her death, therefore, should count as evidence against it.

***

I can think of no better person to champion this point than Dr. Victor Stenger. Considered one of the "new atheists," along with Richard Dawkins (biology), Daniel Dennett (philosophy), Sam Harris (neuroscience), and Christopher Hitchens (the late journalist and literary critic), Vic Stenger (physics) has established himself as the authoritative debunker of bogus claims that the existence of God can be proved (or at least justified) by science. He even goes a step further, claiming that science has disproved God. I appreciate the fact that Stenger is the "real thing," the "go-to guy" physicist who knows what he is talking about and who can handily swat away the theistic mosquitoes buzzing around pretending to understand the science (and only managing to misrepresent it). But what I love most about Stenger's books is their unflinching and uncompromising attitude. I wish I had had this book you are holding back in the early 1980s when I was going through my wrenching deconversion from a sincere believing evangelical minister to an even more sincere disbelieving rational atheist. At its core, my struggle was a stark battle between faith and reason.[9] In the end, I had to "choose this day" whom I would serve, and I ended up choosing reason. I fell in love with learning and gained a refreshing respect for the abilities of the human mind to do the hard work and think things through, to follow the facts wherever they lead, even if they took me away from what was comfortable and precious to me. I was forced into the corner of asking, "Do I want truth, or do I want God?" Because you can't have both. This did not happen, however, until I went through the predictable wringer of attempted integration, a transitional period of trying to rationalize my faith by pretending it was just another kind of science, desperately trying to harmonize belief and reason. This book you are holding would have vastly accelerated that process for me. I now know that faith is a cop-out. It is the lazy way out. If the only way you can accept an assertion is by faith, you are admitting that it cannot be accepted on its own merits. As I write in my book *Godless* (telling my preacher-to-atheist story), what the world needs is a faithectomy. And there is no better surgeon for the task than Vic Stenger.

During a broadcast of *Freethought Radio*,[10] discussing Dr. Stenger's previous book, *The Fallacy of Fine-Tuning*, Annie Laurie Gaylor and I asked Vic Stenger about the relationship between faith and science:

Stenger: There is nothing wrong with postulating things that we haven't detected yet, as long as they are, in principle, detectable. The neutrino is an example. It took about thirty years between the time when the neutrino was predicted and when it was finally observed. But it has to be, at least in principle, observable.

Gaylor: What about the idea of a god? Is a god detectable?

Stenger: I claim in my book, *God: The Failed Hypothesis*, that the god that most people worship is in principle detectable, and should have been detected by now, because that god plays such an important role in the universe. He would have to deliberately hide himself. Of course, he could do that, but that would be kind of an unjust God, kind of an immoral God, because that would mean that God is up there and he's only allowing those people who grovel at his feet, and so on, to spend eternity with him, and if you're a disbeliever who's based your beliefs on the evidence, how could he hold that against you? That would be immoral and unjust for God to hold it against you for just trusting your own intellect.

Gaylor: Even then he's saying, "You gotta die before you can see me, and you can't come back and prove it."

Barker: Well, you guys don't understand. I used to be a preacher and I understand perfectly, because you guys are spiritually blind. You see, God wants to test our faith, and if God made himself evident to us, there would be no faith involved. It would just be a fact of science. So, he's really trying to test what kind of character and obedience and faith you have by deliberately hiding himself from you.[11] Can you fit that into a physical model?

Stenger: I don't know. I don't think so. But you can argue that, again, such a god that requires belief in the absence of evidence is not a very nice guy. That would not be a benevolent God.

On an earlier broadcast,[12] discussing his book *The New Atheism: Taking a Stand for Science and Reason*, Dr. Stenger gave a great analogy about seeking evidence, and also told us a little about his personal motivations:

Stenger: People often quote Carl Sagan, although he is not the only one to have said it, that "the absence of evidence is not evidence of absence." But I claim absence of evidence can be evidence for absence when the evidence should be there. I live near Rocky Mountain National Park, and if somebody said there were elephants in Rocky Mountain National Park, and you looked and you didn't see any, I think that would be a pretty good case for saying they don't exist in Rocky Mountain National Park. We would have surely discovered the evidence for them by now.

Gaylor: I notice that you had been raised Catholic. When did you leave that institution?

Stenger: Oh, I think when I was a teenager I began to have my doubts. I got interested in science. That is always a good step in that direction. I started reading about evolution, and even though the [Catholic] Church accepts evolution, most of the people around me didn't believe it. So, I drifted away. But I didn't become a really active atheist until really just in the last twenty or so years.

Barker: When you were younger, did you actually believe in transubstantiation?

Stenger: Sure. Sure, you believe whatever you are told when you are a kid.

Barker: But it still tasted like bread, didn't it?

Stenger: Right, right. Well, you know, like I tell people, that the Catholic Church is trying to modernize and produce a communion wafer with half the fat and a third fewer calories. They call it "I Can't Believe It's Not Jesus."

Barker: I notice that you say in your book that the world is actually worse off as a result of faith. What do you mean by that?

Stenger: The problem is that people think faith is something to be admired. In fact, faith means you believe in something which you have no evidence for. When that kind of attitude is present, it means that you're not making decisions—whether it's at a political level or at a personal level—you're not making decisions about your life, about the world, based on good rational, reasonable evaluation of the evidence but on preconceived ideas that you have no basis for. . . . Faith is folly. Faith is bad. It is not something to respect. We should be fighting its negative influence on the world.

Stenger has coined a now popular phrase, which became one of the favorite virtual billboard slogans of the Freedom From Religion Foundation: "Science flies us to the moon. Religion flies us into buildings."[13] He is convinced that "we must act for the sake of the betterment of humankind and the future of our planet. Based on the favorable signs that young people are increasingly abandoning religion, I have great hope that perhaps in another generation America will have joined Europe and the rest of the developed world in casting off the rusty chains of ancient superstition that stand as an impediment to science and progress. I just hope it's not too late."

***

For Kara Neumann, it is already too late. She might have grown up to think for herself, to discard her parents' faith, perhaps to work in science, or perhaps to find other ways to improve life on this planet. We will never know. Her lost life is a casualty of Christianity. But we do know that the danger is still there for others, as long as faith continues to be revered in society.

Dan Barker

PREFACE

Divine revelation, not reason, is the source of all truth.

—Tertullian (died 225)

The knowledge exists by which universal happiness can be secured; the chief obstacle to its utilization for that purpose is the teaching of religion. Religion prevents our children from having a rational education; religion prevents us from removing the fundamental causes of war; religion prevents us from teaching the ethic of scientific co-operation in place of the old fierce doctrines of sin and punishment. It is possible that mankind is on the threshold of a golden age; but if so, it will be necessary first to slay the dragon that guards the door, and this dragon is religion.

—Bertrand Russell[1]

Faith is belief in the absence of supportive evidence and even in the light of contrary evidence. No one disputes that religion is based on faith. Some theologians, Christian apologists, and even a few secular scholars claim that science is also based on faith. They argue that science takes it on faith that the world is rational and that Nature can be ordered in an intelligible way.

However, science makes no such assumption on faith. It analyzes observations by applying certain methodological rules and formulates models to describe those observations. It justifies that process by its practical success, not by any logical deduction derived from dubious metaphysical assumptions. We must distinguish faith from *trust*. Science has earned our trust by its proven success. Religion has destroyed our trust by its repeated failure.

Using the empirical method, science has eliminated smallpox, flown men to the moon, and discovered DNA. If science did not work, we wouldn't do it. Relying on faith, religion has brought us inquisitions, holy wars, and intolerance. Religion does not work, but we still do it.

Science and religion are fundamentally incompatible because of their unequivocally opposed epistemologies—the separate assumptions they make concerning what we can know about the world. Every human alive is aware of a world that seems to exist outside the body, the world of sensory experience we call the *natural*. Science is the systematic study of the observations made of the natural world with our senses and scientific instruments, and the application to human needs of the knowledge obtained.

By contrast, all major religions teach that humans possess an additional "inner" sense that allows us to access a realm lying beyond the visible world—a divine, transcendent reality we call the *supernatural*. If it does not involve the transcendent, it is not religion. Religion is a set of practices intended to communicate with that invisible world so that we can either cause forces from it to affect things here or at least apply insights gained from it to human needs.

The working hypothesis of science is that careful observation is our only reliable source of knowledge about the world. *Natural theology* accepts empirical science and views it as a means to learn about God's creation. But religion in general goes much further than science in giving credence to additional sources of knowledge such as scriptures, revelation, and spiritual experiences.

No doubt science has its limits. It is hard to imagine using science to distinguish between what expert critics decide is good or bad art, poetry, or music, although computers can do a credible job of producing works that the expert judges often can't distinguish from the "real thing."

However, the fact that science is limited doesn't mean that religion or any alternative system of thought can or does provide insight into what lies beyond those limits. For example, science cannot yet show precisely how the universe and life originated naturally, although many plausible scenarios exist. But the fact that science does not at present have a definitive answer to this question does not mean that ancient creation myths such as those in Genesis have any substance, any chance of eventually being verified.

The scientific community in general goes along with the notion that science has nothing to say about the supernatural because the methods of science as they are currently practiced exclude supernatural causes. However, if we truly possess an inner sense telling us about an unobservable reality that matters to us and influences our lives, then we should be able to observe the effects of that reality by scientific means.

If someone's inner sense were to warn of an impending earthquake unpredicted by science, which then occurred on schedule, we would have evidence for this extrasensory source of knowledge. Claims of "divine prophecies" have been made throughout history, but not one has been conclusively confirmed. In just one of countless examples of the same nature, the claimed prophecy that the Messiah would be born in Bethlehem was fulfilled after the fact by the Gospel writers inventing implausible, inconsistent, and historically disprovable scenarios by which Jesus could have been born there.[2]

So far we see no evidence that the feelings people experience when they perceive themselves to be in touch with the supernatural correspond to anything outside their heads, and we have no reason to rely on those feelings when they occur. However, if such evidence or reason should show up, then scientists will have to consider it whether they like it or not.

We cannot sweep under the rug the many serious problems brought about by the scientific revolution and the exponential burst in humanity's ability to exploit Earth's resources made possible by the accompanying technology. There would be no problems with overpopulation, pollution, global warming, or the threat of nuclear holocaust if science had not made them possible. The growing distrust of science found now in America can be understood by observing the disgraceful examples of scientists employed by oil, food, tobacco, and pharmaceutical companies who have contributed to the unnecessary deaths of millions by allowing products to be marketed that these scientists knew full well were unsafe.

But does anyone want to return to the prescientific age when human life was nasty, brutish, and short? Even fire was once a new technology. Unsafe products are more than overshadowed by miracle drugs, foods, and technologies that have made all our lives immeasurably better than those of humans in the not-too-distant past. At least in developed countries, women now rarely die in childbirth and most children grow to adulthood. This was not the case even just a few generations ago. Unlike our ancestors, we lead long, fulfilling lives largely free of pain and drudgery. The aged are so numerous that they are becoming a social problem. All this is the result of scientific developments.

We can solve the problems brought about by the misuse of science only by better use of science and more rational behavior on the part of scientists, politicians, corporations, and citizens in all walks of life. And religion, as it is currently practiced, with its continued focus on closed thinking and ancient mythology, is not doing much to support the goal of a better, safer world. In fact, as we will see, religion is hindering our attempts to attain that goal.

Many theistic-leaning historians and sociologists claim that religion contributed importantly to the development of science and so the two can live in relative harmony. They argue that treating the historical relationship between science and religion as a "war" is an oversimplification and inaccurate representation of the actual facts, at least up to the Enlightenment. Not until the seventeenth century was a distinction made between science (or "natural philosophy") and religion; many great scientists of the past merged the two in their thinking. While this may have been true in the past, it is not true now. Science and religion today mix like oil and water. Even scientists who are religious keep religion out of their work.

No doubt the great theologians of the past had few problems with science, seeing it as another way to learn more about the majesty of the Creator. Similarly, liberal theologians today fully accept the discoveries of science. Nevertheless, we will see that the theologies of all ages still promote a worldview that is antithetical to that of science. The differences between science and religion are not merely matters of different points of view that might be harmonized with some effort. They are forever irreconcilable.

Nevertheless, a common misunderstanding needs to be corrected. The conflict between science and religion should not be regarded as a conflict between reason and unreason. From the time of the ancient Greeks, reason has been a tool to gain insight into the nature of a postulated divine reality. The great Catholic theologians, notably Augustine of Hippo (died 430) and Thomas Aquinas (died 1274), applied reason in much of what they wrote—and they wrote a lot.

The distinction between theology and science is in the objects on which to apply reason. Nothing can be learned from reason acting alone. A logical argument contains no information not already embedded in its premises. Reason and logic must be supplemented by additional hypotheses about the nature of reality and the sources of our knowledge about that reality. In the case of science, that source is solely observation. In the case of theology, that source is primarily faith, with some observation thrown in as long as it does not conflict with faith. Theology is faith-plus-reason, with some observation allowed. Science is observation-plus-reason, with no faith allowed.

Now, that is not to say that theology always was and still is "reasonable" in every respect. For example, Saint Augustine (with a lot of help from Saint Paul) is primarily responsible for Christianity's obsession with sex, which is surely unreasonable.

Irreconcilable differences arising from the differing viewpoints and methodology of science and religion include the origin of the universe and its physical parameters, the origin of complexity, the concepts of holism versus reductionism, the nature of mind and consciousness, and the source of morality. I will document instances where religious authors have ignored or distorted science in an attempt to justify their unquestioned premise of a divine reality. But the folly of faith is even deeper than its history of factual error and misrepresentation.

Suppose a new religion was invented and it taught dogmatically that all that existed were material atoms that interacted exactly as described by physics, that Darwin was right and all life arose from the same primitive matter, and that the universe was not created but was part of an infinite and eternal multiverse. In other words, the new religion was based totally on the best scientific models of today. This religion would nonetheless be incompatible with science. The new religionists would still "believe" in an unchanging dogma while scientists would be open to change tomorrow, next week, or whenever new evidence is found.[3]

And this should provide a rebuttal to the oft-heard claim that "science is a religion." Religions are characterized by hardened beliefs that, when changed, result in a new spinoff sect while the old one continues with perhaps depleted membership. By contrast, from the beginning science has been one continuous flow of new knowledge and progress as old ideas are cast off and new ones take their place.

Today science and religion find themselves in serious conflict. Even moderate believers do not fully accept Darwinian evolution. Although they claim to see no conflict between their faith and evolution, they insist that God still controlled the development of life so humans would evolve, which is not at all what the theory of evolution says. In another example, greedy corporate interests and unscrupulous politicians are exploiting the antiscience attitudes embedded in popular religion to suppress scientific results on issues of global importance, such as overpopulation and environmental degradation, which threaten the generations of humanity that will follow ours.

Those who rely on observation and reason to provide an understanding of the world must stop viewing as harmless those who rely instead on superstition and the mythologies in ancient texts passed down from the childhood of our species. For the sake of the future of humanity, we must fight to expunge the fantasies of faith from human thinking.

As we will see in chapter 14, while other authors have documented the role of extreme conservatives in opposing scientific findings, none have emphasized the connection with religion. Others have chronicled the role of religion in the conservative movement, but have not emphasized the connection with antiscience. Here we complete the triad: religion, antiscience, and extreme conservatism.

This is not to say that antiscience does not exist on the liberal end of the political spectrum. It does, of course, and was particularly rampant in the sixties and seventies. But the extreme Left possesses little power in America today, while conservatives wield huge resources that give them influence far exceeding their actual numbers, making the extreme Right the far greater threat.

This book is a call for scientists and other rationalists to join together to put a stop to those who insist they have some sacred right to decide what kind of society the rest of us must live in—for the sake of the future of the planet and the betterment of humankind. Hopefully, in perhaps another generation, America will have joined Europe and the rest of the developed world in shucking off the rusty chains of ancient superstition that stand as an impediment to science and progress.

CHAPTER 1
INTRODUCTION

A reasonable way to interpret the long history of the conflict between scientific and religious models is to see these institutions as competing for the same ground, rather than operating in different domains.

—Gili S. Drori et al.[1]

The notion that science and religion have been long at war with each other is widespread but, as we will see, is somewhat of an oversimplification. The warfare model is largely the consequence of two influential nineteenth-century books: *A History of the Conflict between Religion and Science* by English-born American chemist John William Draper (died 1882),[2] and *A History of the Warfare of Science with Theology in Christendom* by the cofounder and first president of Cornell University, Andrew Dickson White.[3]

Draper had reacted angrily to proclamations from Rome asserting papal infallibility and claiming that revealed doctrine took precedence over the human sciences. He wrote that since coming to power in the fourth century, the Catholic Church had displayed "a bitter and mortal animosity" toward science and had its hands "steeped in blood."[4]

White's attack on religion was much broader, not limited to the Catholic Church, but like Draper's, it was motivated at least partially by ideology. At secular Cornell, White wished to create "an asylum for *Science*—where truth shall be sought for truth's sake, not stretched or cut exactly to fit Revealed Religion."[5] His book was largely in reaction to attacks from the religious community for his refusal to impose religious tests on students and faculty.[6] Nevertheless, White's efforts at Cornell helped lead to the conversion of the great private universities in America and Europe from the church-centered institutions they were originally to the secular ones they are today.

31

The second volume of White's tome documents the long history of meddling by religion in medicine: the legends of supernatural intervention in causing and curing disease, including miracles and satanic influences; the resistance against dissection and other anatomical studies; opposition to surgery, inoculation, sanitation, and the use of anesthetics; and demonic possession. While we still have faith healers and faith healing cults, these are not part of my concern in this book, which is the *current* intellectual battleground of theology and science.

Modern historical scholars, some with ideological motives of their own, have severely criticized the accuracy of Draper's and White's accounts, saying they oversimplified what was a far more complex relationship.[7] Historian John Hedley Brooke asserts that Draper's and White's arguments are "deeply flawed." He objects to their assumption of a dichotomy between nature and supernature, which he says oversimplifies the theologies of the past. He writes, "If a supernatural power was envisaged as working *through*, as distinct from *interfering with*, nature, the antithesis would partially collapse." Or, he says, another way to put it is, "an explanation in terms of secondary causes need not exclude reference to primary causes."[8]

In fact, a dichotomy does exist between nature and supernature. Later I will elaborate on the distinction between primary and secondary causes, but Brooke's mistake here is to assume, without some kind of evidence or rationale, that the mere fact that primary causes are theoretically possible means that they actually have a substantial likelihood of existing. Time and again we will run into this line of reasoning by religious apologists. Just because science cannot prove Zeus does not exist, we can't conclude he does.

The strongest indictment of Draper and White that I have seen is in The Great Courses lectures by chemist and historian Lawrence M. Principe, whose strong proreligion bias comes out no matter how hard he tries to hide it and to appear even-handed.[9] According to Principe, Draper's book is "one long, vitriolic, anti-Catholic diatribe." As for White, Principe says he "did not share the rabidity of Draper and did not sell as well," but he also uses "fallacious arguments and suspect or bogus sources."[10]

Let's take a look at one example that casts doubt on Principe's impartiality. He claims, without reference, that White said, "Earth's sphericity was officially opposed by the Church." I have looked through White's book, however, and find no claim regarding an official Church doctrine on the shape of Earth. White refers to certain figures in the early Church, such as Lactantius (died ca. 320) and John Chrysostom (died 407), who mainly distrusted science of any sort. But in contrast to such figures, White notes, "Clement of Alexandria [died ca. 215] and Origen [died 254] had even supported [sphericity]" and "Ambrose [died ca. 340] and Augustine [died 430] had tolerated it." Furthermore, White adds, "Eminent authorities in later ages, like Albert the Great [died 1280], St. Thomas Aquinas [died 1274], Dante [died 1321], and Vincent of Beauvais [died ca. 1200], felt obliged to accept the doctrine of the earth's sphericity."[11] On this point at least, Principe was attacking White for an error that White did not make.

In short, some historians have not been particularly careful or accurate in their criticisms of Draper and White.

Most discussions on the history of the interaction between science and religion focus on Europe, and, indeed, my main concern will be science and Christianity. However, it must be remembered that while Western Europe languished in the Dark Ages, science flourished for over seven hundred years during the golden age of the Islamic empire. In a recent wonderful book, *The House of Wisdom: How Arabic Science Saved Ancient Knowledge and Gave Us the Renaissance*, the distinguished Anglo-Iraqi physicist Jim Al-Khalili chronicles the contributions to human knowledge made by the great scholars of that period, about which I will have more to say.[12] Certainly there was little or no conflict between science and Islam during that period, when the international language of science was Arabic, the language of the Qur'an.

Nevertheless, while history cannot be neglected because of its effect on the present, the incompatibility between science and religion that we see today arises primarily from current conflicts, not from ancient history. So let me focus here on those.

In his 1999 book, *Rocks of Ages*, the late renowned paleontologist Stephen Jay Gould proposed that science and religion are "non-overlapping magisteria" (NOMA).[13] He argued that the two knowledge systems deal with different aspects of life. Science, Gould wrote, is concerned with describing the "outer" world of our senses, while religion deals with the "inner" world of morality and meaning. NOMA recalls the position enunciated by Galileo when he ran into trouble with the Church for teaching that Earth goes around the sun. Galileo is often quoted as saying, "The Holy Spirit's intention is to teach us how one goes to heaven, not how the heavens go," although it is generally assumed that he was in turn quoting Cardinal Cesare Baronius (died 1607).

Many scientists—believers and nonbelievers—have adopted the NOMA position. "Believing" scientists, that is, those who believe in God, compartmentalize their thinking by not incorporating into their religious thinking the "doubt everything" position they were trained to take in their professions.

A prime example is geneticist Francis Collins, who administered the Human Genome Project and at this writing directs the National Institutes of Health. His 2006 book, *The Language of God: A Scientist Presents Evidence for Belief*,[14] was a bestseller. As we will see in more detail later, his so-called evidence is not, as you might have thought from the title, based on his deep knowledge of DNA. Rather, it follows from his own inner feeling that the world is a moral place and only God could have made it that way. Nowhere in this book does Collins come close to applying to this notion the critical skills exhibited in his outstanding scientific career.

Unlike Descartes, Newton, Kepler, and many of the great founders of the post-Islamic scientific revolution (Galileo is a prominent exception), modern-day believing scientists such as Collins do not incorporate God into their science. This even includes those scientists who happen to also be members of holy orders, such as the Belgian Catholic priest Georges-Henri Lemaître, who proposed the big bang in 1927 but, as we will see in chapter 7, urged Pope Pius XII not to claim it as infallible proof that God exists.

Most nonbelieving scientists want to just do their research and stay out of any fights over religion. That makes the NOMA approach appealing because it allows these scientists to not worry much about what religion is or how it affects our social and political world. In my view, though, these scientists are shirking their responsibility by conceding the realms of morality and public policy to the irrationality and brutality of faith.

Neuroscientist and bestselling author Sam Harris observes, "The scientific community is predominantly secular and liberal—and the concessions that scientists have made to religious dogmatism have been breathtaking."[15]

He tells of attending a conference in the fall of 2006 at the Salk Institute in La Jolla, California, called "Beyond Belief: Science, Religion, Reason, and Survival." The other attendees included some of the leading figures in science. Harris remarks: "While at Salk I witnessed scientists giving voice to some of the most dishonest religious apologies I have ever heard. It is one thing to be told that the pope is a peerless champion of reason, that his opposition to embryonic stem-cell research is both morally principled and completely uncontaminated by religious dogmatism; it is quite another to be told this by a Stanford physician who sits in the President's Council on Bioethics."[16]

We will see later how the US National Academy of Sciences, along with several scientific societies and proscience organizations such as the National Center for Science Education, have compromised their principles in order to stay on good terms with religion. Even the prestigious science magazine *Nature* has adopted Gould's NOMA, editorializing that problems arise between science and religion only when they "stray onto each other's territories and stir up trouble."[17]

However, Gould's proposal and these views from the top tiers of science do not describe the actual roles science and religion play in society. Traditional religions are based on the belief in divinely inspired scriptures and other revelations, and they *do* try to tell us what "is" based on those beliefs. In doing so, they have proved to be almost universally incorrect.

Now, clever theologians will say that I am using science as my standard of what is correct and incorrect. Of course scriptures could be correct, but then we have to believe (as many fundamentalists do) that God is pulling the wool over our eyes, planting phony evidence that carbon-dated fossils, geological formations, and galaxies are older than the six thousand years since Creation implied in the Bible. The scientific descriptions of the world we observe with our senses and instruments aren't necessarily correct just because they are science; they simply work better than those found in scriptures. And if religion doesn't work in the sphere of nature, why should we expect it to work in the moral or other spheres?

Nothing prevents science from concerning itself with issues of morality and purpose. If these questions involve observable phenomena, such as human behavior, they can be analyzed with the rational methods of science. In his

2011 book, *The Moral Landscape*, Harris argues that science has an important role to play in analyzing moral questions and that it can be used to help
develop objective moral truths. That doesn't mean it has the final answers, but
science should be allowed to participate in the dialogue. Science is more than
making measurements and models; it is about applying empirical reasoning
to every aspect of life.

Many historians, scientists, and philosophers claim that, while a tension
exists between science and religion, an essential harmony between the two can
be maintained. Ian Barbour promoted this view in his 1997 book, *Religion and
Science: Historical and Contemporary Issues*. Barbour has both a PhD in physics
and a Bachelor of Divinity degree. In 1999 he won the lucrative Templeton
Prize for Progress in Religion, which is given annually to someone who has
advanced the reconciliation of science and religion.[18] Barbour's work will be
referred to often in the present book.

Also, in a recent book, *Science and Spirituality: Making Room for Faith in the
Age of Science*, philosopher Michael Ruse argues: "The basic, most important
claims of the Christian religion lie beyond the scope of science. They do not
and could not conflict with science for they live in realms where science does
not go."[19] But, once again, the fact that science cannot reject all conceivable
worlds cannot be used to argue for their existence. Furthermore, many fundamentalist Christian claims do not lie beyond the scope of science, they conflict
with it: the virgin birth, miracles, prophecies, revelations, the resurrection,
are just a few of these.

The John Templeton Foundation is behind much of the current effort to
reconcile science and faith. Financier John Templeton's legacy provides $70
million a year in grants to support research on "subjects ranging from complexity, evolution, and infinity to creativity, forgiveness, love, and free will."[20]
The foundation also provided support for another scholar, William Grassie,
who has argued for the essential harmony of science and religion. I will also
refer frequently to his 2010 book, *The New Sciences of Religion*.[21]

Barbour, Grassie, and others have interpreted historical events as evidence
for, though not in complete harmony with, a positive relationship between
science and religion where each has contributed constructively to the other.
They have argued, for example, that Puritanism in England contributed sig-

nificantly to the scientific revolution with its revolution against authority. So, they say, did Calvinist theology, in which people serve God not by shutting themselves away in a monastery or convent but by doing useful work. This is called the *Protestant ethic*.[22]

Science flourished in England after the Royal Society of London for Improving Natural Knowledge was chartered by King Charles II in 1660, the year he was restored to the throne. The society was formed from a group of royalists called the Oxford Circle, the members of which holed up in Oxford during the English Civil War. Spending their time dissecting human and animal cadavers, they established many anatomical facts, most notably that the brain is the primary organ of thought and that the heart is a pump that operates under the control of signals from the brain. The group, led by physician Thomas Willis (died 1675), included the great architect Christopher Wren (died 1723), the great chemist Robert Boyle (died 1691), and the great physicist Robert Hooke (died 1703). Willis was pretty great himself.[23]

The Puritans believed that God was revealed in the study of nature, and they gave strong encouragement to scientific work. However, most English scientists, such as those in the Oxford Circle, were actually Anglicans who saw in natural laws an analogy with the rule of law in society. Furthermore, everyone began to realize how technology was a source of control over nature with the resulting enhancement of economic and political power.[24]

There can be no dispute that the scientific revolution occurred in an atmosphere in which religious and scientific ideas were deeply intertwined. But religion still held the upper hand. In a lengthy essay titled "Puritanism, Separatism, and Science," historian Charles Webster concludes, "No direction or energy toward science was undertaken without the assurance of Christian conscience, and no conceptual move was risked without confidence in its consistency with the Protestant idea of providence."[25]

It is difficult to extract precise causes of the scientific revolution from the complex history of seventeenth-century Europe except to say that it happened there and no place else. China had made significant advances in technology but failed to develop science. And while science and learning flourished for a time in the Islamic world, there, too, a culture of scientific development failed to endure.

Barbour argues that the decline in science in the Islamic world was the result of the tight control of higher education by religious authorities. Although Barbour doesn't admit it, the same can be said of Christendom until the Reformation. Similarly, government authorities controlled education in China. From this perspective, it was the new openness in Europe that made science possible.[26]

However, Europe would not have been closed to independent thinking in the first place if it weren't for the Catholic Church. Science had flourished in pagan Greece and Rome, and, as we have seen, in medieval Islam. Now, I am not claiming that the Roman Empire declined because of the growth of Christianity. It declined because of the depravity of its leaders and people and from invasions from outside. Church-based leaders and social institutions were there to pick up the pieces, producing an authoritarian society that brutally suppressed the slightest traces of freethinking.

I will say more about Islam later.

The totality of evidence indicates that, on the whole, over the millennia the Christian religion was more of a hindrance than a help to the development of science. Surely it is no coincidence that the onset of the Dark Ages coincided with the rise of Christianity. It was only with the revolts against established ecclesiastic authorities in the Renaissance and Reformation that new avenues of thought were finally opened up, allowing science to flourish.

And these new avenues of thought are what we really need to explore. My position is that artistic and social activities with no significant political ramifications are far less important when considering the compatibility of science and religion than are intellectual matters. Scientific thinking is not dissonant with church art, music, and charitable work, or with the church's function of providing a structure where people can meet to enjoy one another's company and help one another. However, as Harris says, "Science and religion—being antithetical ways of thinking about the same reality—will never come to terms."[27] So long as religious people do not attempt to force their beliefs on others, they are mostly only harming themselves by the folly of their faith. But when religious notions dominate the political scene, as they do in Muslim countries and to some extent in America today, the world is in big trouble.

Barbour lists examples where he claims religion, and Christianity in particular, has had a positive influence on science:[28]

1. The conviction that nature is intelligible contributed to the rational component of science. Monotheism combined the Greek view of orderliness and regularity with the biblical view of God as lawgiver.

2. The Greeks had claimed that everything could be derived from first principles. Theists believe that God created the universe by an act of his own will. He didn't have to. So the facts of nature cannot be derived from first principles but must be learned by observation and experiment.

3. The Bible provides an affirmative view of nature. Creation implies the basic goodness of the world, or else God would not have made it.

None of these claims are very convincing, however. Without Christian monotheism the Greek (and Roman) view would not have been suppressed for a thousand years. And it's really stretching things to attribute empiricism to a belief in the Creation. Furthermore, monotheists were hardly the first people to imagine a created universe or to have an affirmative view of nature.

In honesty, Barbour must ask why the development of science in the Middle Ages, prior to the scientific revolution, was so meager—given that Greek ideas were prevalent in Europe by that time. He attributes this lack of progress to the dominance of the Catholic Church. Again, it is surely no coincidence that the scientific revolution occurred just after the Renaissance and Reformation challenged Church dominance. Still, Barbour concludes, "Many historians of science [not most?] have acknowledged the importance of the Western religious tradition in molding assumptions about nature that were congenial to the scientific enterprise."[29]

I will begin my narrative in the next chapter by going back to the very origins of science and religion as best we know them and tracing their history through the Greeks, Romans, and early Christianity. I will describe how in the Middle Ages much of Greek and Roman science and philosophy was lost in Europe but preserved and developed to new heights in the Islamic empire. We will see how this knowledge gradually crept back into Europe as theologians such as Augustine of Hippo and Thomas Aquinas developed rational theologies that incorporated the philosophies of Plato and Aristotle and translated texts became available.

When the Roman Catholic Church founded the first universities in Europe, Aristotle became the prime authority. Scholars used his logic as well as his science and philosophy to forge an amalgam of Greek and Christian thought that became known as Aristotelian *Scholasticism*. While the value of reason and observation was recognized, these were generally viewed as inferior to revelation since they were the products of imperfect human activity, whereas revelation came directly from God. The Renaissance and Reformation defied the authority of the Church, and a new science blossomed in which revelation and authority were replaced as final arbiters of truth by observation and measurement. Significantly, the scientific revolution occurred *outside* the church-dominated universities, which remained steeped in Scholasticism. Today, our secular universities lead the way in science while students at many church-connected universities and colleges are being taught creationism and other pseudosciences, along with mind-numbing biblical apologetics.

Nevertheless, a clean break between science and religion did not take place immediately at the start of the scientific revolution. All of the great pioneers of science—Copernicus, Galileo, Kepler, and Newton—were believers, although they hardly had a choice in the matter. Open nonbelief was nonexistent in the West at that time. Except for Galileo, these extraordinary figures incorporated their beliefs into their science. Galileo was the only one of the great founders of the new science who tried to separate science from religion.

In the brief period in the eighteenth century called the Enlightenment, thinkers in Europe and America began to distinguish science and philosophy from theology. Deism flourished and atheism became intellectually respectable, at least in France, as we will see.

The great bulk of humanity did not go along with atheism, however. Christianity found a way to incorporate science within its own system with the notion of *natural theology*. In natural theology, human scientific observations and theories are seen as a way to learn more about the majesty of the Creator who had made the natural world and its laws in the first place.

This was quite a reasonable position at the time. After all, prior to the mid-nineteenth century, science had no natural explanation for the complexity we see around us, especially in living things. When geologists showed that Earth was much older than implied in the Bible, and Darwin provided both the evidence and the theory for how life evolved without the need for God, the foundations of religious belief began to crumble.

This resulted in a very specific conflict between science and religion that has lasted to the current day, with the most recent battles being over the intelligent design brand of creationism. While the Catholic Church and moderate Christians have claimed to have no problem with evolution, their own words demonstrate that they do not accept unguided Darwinian evolution. Instead, they subscribe to a form of God-guided evolution that is just another form of intelligent design. We will have more to say about this in chapter 4.

The new physics of the twentieth century—relativity, quantum mechanics, and relativistic quantum field theory—have not struck many nerves with everyday religious believers since they are comprehended by only a tiny fraction of the public. In fact, these theories and the data that support them are monumentally misunderstood, misrepresented, and misused by many who naively write on these subjects without the years of study necessary to have any depth of knowledge.

This is especially the case with quantum mechanics, which has been made to look mysterious and weird, even by physicists who know better but think they can spark student and public interest, and sell their popular-level books, with overblown rhetoric.

While not technically theistic, modern quantum spiritualists and pseudoscientists should be included as part of the antiscience movement that is associated with religions and the transcendental worldview. Many members of this community assert that quantum mechanics tells us we can make our own reality just by thinking we can, and that it puts our minds in tune with a cosmic consciousness that pervades the universe. This claim results from a total misunderstanding of the wave-particle duality in which an object has the properties of a particle when you measure particle properties and the properties of a wave when you measure wave properties. Well, *duh*. Do you expect an object to have a particle property when you measure a wave property and a wave property when you measure a particle property? Physical objects have both properties, and no act of human consciousness has anything to say about it.

The other, more forgivable misuse of quantum mechanics is that made by theologians who look for a way for God to act in the universe without violating the laws of physics. They think they can do this by appealing to Heisenberg's uncertainty principle of quantum mechanics that puts limits on what you can measure with precision. They imagine God poking his finger in to make particles change their motion without any physicist noticing. Sure, God can do that, but he would then be breaking a law of physics, which theologians say they are trying to avoid.

Theists and quantum spiritualists also claim that modern physics has eliminated the reductionism—the breaking down of the whole into parts—that has marked physics and indeed all of science from the time of Democritus. In fact, the opposite is true. After flirting for a while with holism in the crazy sixties, when even I had hair almost down to my shoulders, by the late seventies physics had returned to an even deeper reductionism than before with the standard model of particles and forces. The whole is still equal to the sum of its parts, just as the Greek atomists said. We will see that this is another place where science and religion profoundly disagree.

Once again, some scientists and science writers who should know better have been roped into joining with theologians to announce a grand new scientific principle called *emergence*. They point to the fact that nature has a hierarchy of levels of complexity ranging from elementary particles to human society. At each level we find a new scientific discipline—physics, chemistry, biology, and so on up to sociology and political science. The scholars at each level do not derive their models from particle physics but develop models for each discipline by applying their own unique methods. The principles they uncover "emerge" from the level below by what is called "bottom-up causality."

No one should expect particle physicists to answer every question. However, speculations are being widely bandied about that some emergent principles have the power to control entities at lower levels by way of "top-down causality." At the very top of the pyramid, of course, is God up in heaven, acting down on us particles below. In this model, emergence by bottom-up causality is considered trivial. Emergence by top-down causality is considered world-shaking. We will see what can be made of that.

On the cosmic scale, twentieth-century cosmology also has been distorted by theists as constituting evidence for a creation of the universe when, in fact, modern cosmology points in just the opposite direction. Some previous gaps in our understanding of the physics of the cosmos provided some temporary comfort for those seeking evidence for a creator. However, these gaps were decisively plugged with astronomical discoveries as the century progressed. Today, cosmologists can provide a variety of plausible, mathematically precise scenarios for an uncreated universe that violate no known laws of physics. Furthermore, we have every indication that, despite the well-confirmed big bang, the universe, defined as all there is, had no beginning and thus no creator. We will see that so-called proofs that the universe cannot be eternal are erroneous.

Many theist authors, combining a naive understanding of physics and cosmology with their preformed unscientific beliefs, have been trumpeting that the constants of physics are so delicately balanced that any deviation would make life impossible. From this they conclude that the physical constants could only have been fine-tuned by God. This claim also can be shown to be erroneous, as we will see in chapter 7.

Believing scientists and theologians have also said they see evidence for divine purpose in the universe. This claim is likewise not supported by the evidence.

The fundamental religious belief is that transcendent reality beyond matter exists. Evidence for this reality is supposed to be found in human experiences termed mystical or spiritual. Specifically, a large amount of data has been accumulated over the years, and published in journals and books, on near-death experiences (NDEs). These experiences occur in about 20 percent of people resuscitated from clinical death, or something close to it. These people return with a memory of light at the end of a tunnel that they are convinced was a glimpse of heaven. (Few ever glimpse hell). We will look carefully at the data and conclude it has more plausible natural explanations.

We will also evaluate the data on reincarnation and psychic phenomena. Many dramatic claims have been made for well over a century now that evidence for these wonders exists, but these claims have never been independently confirmed. This discussion will be brought up to date with a critique of a recent highly publicized claim of retroactive causality published in a peer-reviewed psychology journal.

At the current stage of scientific development, we can confidently say that no empirical or theoretical basis exists for assuming anything other than that we inhabit a universe made entirely of matter (and energy into which matter can be transformed, and vice versa). Please understand that this is not a dogmatic position. Of course we don't know everything, and never will. The essential point is that within our existing knowledge we do not have a credible reason for requiring anything transcendent to explain anything we experience or observe. All science is provisional, and if sufficient evidence that meets all the most rigorous scientific tests were to come along to demonstrate the existence of a world beyond matter and energy, then nonbelieving scientists will change their minds. We will challenge the wide array of current claims that scientific observations and theories are already pointing toward transcendence. We will see that these claims have no basis.

We will also see that other metaphors for the "stuff" of the universe, such as information, do not diminish the need for, and primacy of, matter.

The one major area where we do not yet have a plausible physical model that satisfies a consensus of experts in the field is the question of the nature of consciousness. We can now ascribe much more of human thinking processes to the material brain than ever imagined in the past, when the mind was universally believed to be composed of some immaterial, spiritual substance separate from the body. However, the door to some immaterial reality in human consciousness is still open a tiny crack, and we will have to await further developments to see whether it, too, closes upon further scientific investigation.

Another important issue where fundamental disagreement between science and religion exists concerns the source and nature of morality. Believers cannot see how our notions of good and evil can come from any source other than God. They are joined by many nonbelievers who think science has no right to say anything on the question. But scientists are investigating morality anyway and coming up with discoveries that few believers will like. While a primitive morality can be found in animals and early humans that evolved *biologically*, our modern ideas of morality more likely evolved *socially* as humans found ways to overcome some of their animal instincts by force of intellect. Not only did these developments allow people to live together in some semblance of order, they also allowed us to use the ability to act cooperatively

to obtain resources from the environment, to protect ourselves from preda-
tors and other natural dangers. The incompatibility between science and religion becomes especially striking on the question of the origin of morality and
ethical behavior.

While the viewpoints of science and traditional religious beliefs are irreconcilable, contemporary science-savvy theologians are seeking to develop a
model of a deity that fits in with science. However, as we will see, such a
model is necessarily more deistic than theistic, as it has little in common
with the God of Judaism, Christianity, and Islam, or with other ancient gods
such as those of Hinduism and other faiths. All of those Gods can be ruled
out beyond a reasonable doubt by the absence of evidence for their existence,
evidence that should be there but is not.

Finally, we will see why the incompatibility of science and religion is more
than just an intellectual debate among scholars. Faith is a folly. It requires
belief in a world beyond the senses with no basis in evidence for such a world
and no reason to believe in it other than the vain hope that something else is
out there. While a false belief may be comforting or even temporarily useful,
it is a dubious guide to life or for the foundation of a successful society.

While not all believers have an uncompromising faith, and many recognize the power and value of science, we will see that an influential minority
of American Christians see materialist science as an enemy that needs to
be "renewed" so that God is restored to his rightful place in the scheme of
things. Backed by the financial resources needed to get their opinions heard
and to help elect officials who will legislate their line, this minority wields
far more political power than its numbers justify. It has succeeded in watering
down or eliminating the teaching of evolution in most high schools. Holding
extremely conservative views that they justify theologically, the members of
this minority join with unscrupulous politicians to protect the shortsighted
economic interests of their financial backers. In this way they help thwart
government actions recommended by scientific consensus that are needed to
reduce the gradual destruction of the planet by the exponential growth of our
species and its increasingly wasteful use of Earth's finite resources.

CHAPTER 2

THE EARLIEST SKIRMISHES

If we go back to the beginning we shall find that ignorance and fear created the gods; that fancy, enthusiasm, or deceit adorned or disfigured them; that weakness worships them; that credulity preserves them; and that custom, respect, and tyranny support them in order to make the blindness of men serve its own interests.

—Paul-Henri Thiry, Baron d'Holbach (died 1789)[1]

The various modes of worship which prevailed in the Roman world were all considered by the people as equally true; by the philosopher as equally false; and by the magistrate as equally useful.

—Edward Gibbon[2]

ORIGINS OF RELIGION

Neanderthal burial sites indicate that hominids buried their dead ritually as far back as fifty thousand years ago. Graves have been found containing tools and flowers, suggesting a belief in some sort of afterlife. Bodies were buried in the fetal position, perhaps indicating a notion of rebirth.

Archaeologists recently discovered a stone python twenty feet long that was carved seventy thousand years ago in a cave in the Kalahari Desert in Botswana. The modern San people living in the area have a legend that humanity descended from a python. Near the figure, digging uncovered colored stone spearheads that were not accompanied by any other signs of human habitation. The spearheads were evidently brought to the cave from

47

hundreds of miles away and burned in some kind of ritual. A secret chamber was found behind the python where a shaman could remain hidden and speak as if his voice came from the snake.[3]

Not only does this discovery show that humans have been thinking abstractly for at least seventy thousand years, it also demonstrates that superstition has been used as a means of control from its very beginning. We can easily imagine a shaman discovering some black powder that, when tossed in a fire, caused an impressive explosion. Using this in a public ritual, the shaman could convince the members of his tribe that he had supernatural powers, which then justified his right to tell them what to do.

Now, you might be inclined to say that such shamans were not only the first priests but also the first scientists. After all, the shaman just described must have been exploring nature when he found the black powder. He must have been experimenting when he discovered its explosive properties. And during the ritual, he was applying his knowledge the way a modern bomb maker applies chemistry or nuclear physics. However, as we will see, science is more than bomb building. As historian David Lindberg puts it, "It is one thing to know *how* to do things, another to know *why* they behave the way they do."[4] An electrician does not have to know Maxwell's equations of electromagnetism to wire up a new house. And, as invaluable as an electrician is to modern society, we do not regard such a tradesman as a scientist. Science develops out of technology but goes beyond mere practical application toward a deeper understanding of the world.

By the time civilization arose, supernatural beliefs were rampant in every nation, with complex rituals, human or animal sacrifices carried out to appease the gods (and to please the priests), and great temples of worship constructed. While only symbolic sacrifices are carried out in the present age, belief in a transcendent, controlling reality, one beyond the phenomena that present themselves to our senses, has remained a dominant feature of human thinking. As we will see, while this conviction, along with the accompanying rituals, has impeded human progress over the ages, the great temples of worship that were constructed and the sacred music and art that were produced constitute some of the finest achievements of humankind. I am not unappreciative of the positive elements of religion; I am just convinced that the positive elements are far outweighed by the negative and that the positives of religion probably would happen on their own anyway without religion.

CAVEMAN LOGIC

Today, a vast literature exists attempting to explain humanity's continuing obsession with religion. See, for example, *Religion Explained: The Evolutionary Origins of Religious Thought*, by anthropologist Pascal Boyer,[5] and *In Gods We Trust: The Evolutionary Landscape of Religion*, by anthropologist Scott Atran.[6]

No doubt the full story of religious obsession is complex. However, for my purposes I need only mention one human quality that affects *all* of our thinking, not just our religious thinking. Many authors have commented on the tendency of humans to ascribe animate agency to natural phenomena and argue that this was a natural inheritance from our ancestor animals in the evolutionary process.

As philosopher Daniel Dennett explains:

> A system or organization within the brain . . . has evolved in much the same way our immune system or respiratory system . . . has evolved. Like many other natural wonders, the human mind is something of a bag of tricks, cobbled together over the eons by the foresightless process of evolution by natural selection. Driven by the demands of a dangerous world, it is deeply biased in favor of noticing the things that mattered most to the reproductive success of our ancestors.[7]

Boyer calls this bag of tricks "gadgets," and Dennett notes that some of the patterns look like religion.

As Dennett points out, even the simplest animals have what psychologist Justin Barrett calls a *hyperactive agent detection device*, or HADD.[8] For example, a clam will retreat its foot into its shell whenever any vibration or bump is sensed. Most such disturbances are harmless, but the clam's motto is "Better safe than sorry."[9]

Animals with greater mobility than the clam have developed the ability to detect unusual motions that might be made by a predator but more often are not. This tendency toward imagining invisible causes of events leads them to sometimes engage in ritual behavior that serves no necessary purpose. In a famous experiment conducted in 1948, psychologist B. F. Skinner showed that pigeons exhibit what he called "superstitious behavior" in which they carry out repeated, stereotyped patterns of conduct to get food even when those patterns are not required.[10]

Humans have inherited the hyperactive agent detection device. In *The Believing Brain*, psychologist and prominent skeptic Michael Shermer provides evidence from many different areas of human behavior—from politics and economics to religion and conspiracy theories—that support the formation and reinforcement of beliefs in patterns and agency that have limited or no evidentiary support.[11]

In his book *Caveman Logic*, psychologist Hank Davis posits the following scenario:

> One of your ancestors is walking through the forest and sees something on the path ahead. It might be a predator. Then again, it might be a random array of shapes and textures that amounts to nothing. If he believes it to be dangerous, he takes appropriate defensive steps. Perhaps he freezes or arms himself or flees. What's the best that can happen? He survives a lethal encounter and gets to live and function another day. What's the worst? A false positive. He finds himself with heart pounding, pulse racing, hiding behind a tree with a spear drawn for no good reason. It was only a pile of twigs on the path. He's wasted some effort and experiences a baseless fear. But he gets to go home, eat dinner, and snuggle with his mate.[12]

Davis adds, "Perceptual accuracy was not an agenda of natural selection. Survival and reproduction were."[13]

Since our brains have hardly evolved physically and biologically since caveman days, they retain this protective agency module that does us more harm than good in the modern age. We no longer have to be excessively alert when taking a walk in the woods, although a city street is another matter. In the meantime, we assign invisible agency and causality to phenomena that have no agents or causes. This leads to behaviors that are a waste of time and energy. To make matters worse, these behaviors are reinforced by widespread

social support—by churches in particular.[14] And, as we will see, not only religious believers but scientists as well are burdened by this anachronistic brain
module.

Davis has this amusing but cogent summary of the situation:

> There is a popular bumper sticker that addresses the problem directly. It says
> SHIT HAPPENS. These two words are all but incomprehensible to the majority
> of people. The sticker does not say I CAUSED SHIT TO HAPPEN. It does not say
> SHIT WAS DONE TO ME BY A VENGEFUL GOD. It simply says that . . . SHIT does
> happen from time to time.[15]

This is not just an account of human reactions to everyday experiences.
It also applies on the cosmic scale, where great philosophers, scientists, and
theologians—as well as the typical churchgoer—find it difficult to grasp how
anything could happen without cause. As I will amplify later, many of our
scientific explanations, from evolution to quantum physics to the origin of
the universe, are acausal, that is, they describe events that are not the sole
result of previous circumstances but are events that just happen. We call these
events accidental or spontaneous. Stuff happens. And, as we will see, while
some theologians and other authors have woven "God" into this picture, such
a god has little in common with the traditional God of Jews, Christians, and
Muslims.

THE FIRST SCIENTISTS

While the prehistoric shaman described above resembles the priests we have
witnessed throughout history up to the present, he can be distinguished from
those we have identified as the first scientists.[16] Shamans and priests attribute
supernatural causes to phenomena, whereas scientists claim they are wholly
natural. The meanings of the terms "natural" and "supernatural" will become
clear from usage as we proceed, as will "religion" and "science." I prefer to let
words take on their meanings from context rather than attempting to define
them with other words, which is always an imperfect procedure. Yet, words
were a key to the development of philosophy and science, especially words

written down with alphabetic symbols.[17] It was the highly efficient Greek alphabet and language that made it possible for philosophy and science to develop in ancient Greece.

Thales of Miletus (ca. 624–546 BCE) is regarded by many historians to have been the first scientist, as well as the first philosopher, in the Western tradition.[18] Miletus is a city in the region of Asia Minor (modern Turkey) called Ionia. Thales was the first member of a philosophical school known as the *Presocratics*, which is not so much a chronological term but one used to distinguish the members of the school from the most famous and influential ancient philosophers Socrates[19] (died 399 BCE), Plato (died 347 BCE), and Aristotle (died 322 BCE), as well as just about every other ancient thinker West and East. Indeed, as we will see, the Presocratics were unique, and their differences with everyone else mark the very essence of the eternal conflict between science and religion. Unfortunately, only fragments of their writings survive and we know about them mainly through Aristotle and his successor Theophrastus (ca. 371–287 BCE).

Thales is famous for predicting an eclipse of the sun in 585 BCE (on May 28, according to modern astronomy) that, according to Herodotus, stopped a battle between the Lydians and Medes and ended their five-year war. Thales's prediction, if it really occurred, would have been based on astronomical tables that he likely picked up during journeys to Egypt and Babylonia. In those early civilizations, astronomy was a precise art, but it was used for divination rather than as empirical data upon which to build a model of the cosmos. The early Egyptian and Babylonian astronomers and astrologers were like the shaman: discovering empirical knowledge about the natural world but giving it a supernatural interpretation that could be put to use in managing the social order.

The Egyptians and Babylonians, as well as other early civilizations, had also developed what we now call technology. While today technology is based not only on observations of the physical world but also on the application of scientific theories, it is not what is known as "pure" science. Technology is applied science, while pure science is motivated by a search for understanding that is independent of possible applications (although usually justified by such to funding sources).

Thales's innovation, which we now identify as pure science, was to explain observed phenomena with reference to visible forces rather than to imagined, invisible spirits—which were, in Thales's time, the Greek gods. For example, Thales explained earthquakes by hypothesizing that Earth floats on water and is rocked by waves. After all, we can see for ourselves that the land we live on is surrounded by water.

Everyone recognizes that water is a very important ingredient of nature, not some abstract or imagined entity. It appears in solid, liquid, and vapor forms, and so is capable of change. All living things require water. Thales proposed that water was the sole ingredient from which everything else was constructed. Although this was wrong, as was his explanation of earthquakes, it nevertheless represented a revolution in human thinking in which gods were no longer needed to understand the universe. This was the first break between religion and science, and it suggests why the two are often in conflict—they have utterly opposing views about the nature of the world. Even contemporary theologians, while paying lip service to science, do not see reality as scientists see it.

Thales's notion that everything could be reduced to an elementary substance epitomizes a major difference between natural and supernatural thinking. For millennia after Thales, the common belief was that matter was composed of four elements: fire, water, earth, and air. Although this specific belief was wrong, it reflected Thales's key insight: this was not an abstract concept; it was observed in the natural world.

Still, Thales's picture of water as the primary ingredient of things was not pure materialism by our current standards. He still viewed matter as alive and as containing an immortal soul. Thales and the Presocratic philosophers who followed made no distinction between living matter, nonliving matter, and the soul.

Succeeding Thales, Anaximander (ca. 610–546 BCE) became the first to write in prose with a work called *On Nature*, which has been lost but is referred to by Aristotle. Prior to this achievement, all writing was poetic storytelling—myths and legends designed to provide explanations for the many puzzling and mysterious experiences in the lives of early humans that cried out for explanation.

Anaximander made *apeiron*, a word that is uncertainly translated as "without limits, boundless, or infinite," the principle of all things. No one really knows what Anaximander really meant, but here again the significance is not the specific teaching but the introduction of a new kind of thinking—philosophical argument. Aristotle interpreted Anaximander's principle as having no origin. This is important, I think, because Christian theology holds that everything has an origin except God, who is the origin, the uncreated creator. This is how Christian believers answer the nonbeliever's taunt of "If God was the creator, who created God?" God just is, say the believers. But if God can be uncreated, why can't the universe?

Anaximander constructed a spherical model of the universe with Earth at the center, surrounded by three hollow wheels containing fire. Holes in the wheels through which the fire could be seen corresponded to the sun, moon, and stars. Earth did not fall because there was no preferred direction for it to fall, no force to attract or repel it. Although Anaximander had no notion of gravity, we can see here an intuition of equilibrium or symmetry.

Anaximines (died 528 BCE) was a third philosopher from Miletus who presaged several modern ideas. He proposed that air was the source of all things. Many cultures have held the view that air is the breath of life, the soul. After all, it leaves the body upon death. Anaximines claimed that when thinned, air becomes fire, when condensed it becomes cooler wind and clouds, and when further condensed it becomes water and earth. He provided experimental evidence. Blow on your finger with a wide-open mouth and your finger becomes hot, indicating that when the air is rarefied it is fire. Blow again, but this time with pursed lips, and your finger feels cold, showing that when air is condensed, it becomes colder. Try it. This is an empirical fact and we are talking real science here, even if the specific theory is primitive.

THE ATOMISTS

So we have for the first time in recorded history the notion that everything is composed of fundamental stuff—matter and nothing more. Following on the heels of the Milesians, Leucippus (died ca. 440 BCE) and Democritus (ca.

460–370 BCE) proposed that matter was composed of tiny particles that could not be further subdivided. They called them *atoms*, meaning "uncuttable." This brilliant intuition, based on no empirical data whatsoever, remains the view of physics to the present day—buttressed by the last two centuries of supporting evidence with not a shred of evidence against it.

In the nineteenth century, the ninety-plus elements of the chemical Periodic Table were regarded as the basic constituents of matter, since they could not be broken down further by either alchemy or chemistry. In the twentieth century, however, the elements were split into smaller parts by nuclear collisions that take place at energies thousands of times higher than can be generated with a Bunsen burner or electric spark. This showed that the chemical elements are not elementary after all. While the term "atoms" is still retained to refer to the physical bodies that constitute the chemical elements, they are no longer "uncuttable." Thus I will refer to them as "chemical atoms" to avoid confusion with irreducible atoms.

Early in the twentieth century, the chemical atom was shown to be composed of a tiny nucleus surrounded by a cloud of electrons. The nucleus was then found to be composed of protons and neutrons. And these protons and neutrons also turned out to be cuttable, although electrons are still irreducible.

In the 1970s, physicists produced what became known as the *standard model of particles and forces*, in which everything is reduced to quarks, leptons, and gauge bosons (don't worry about what they are exactly). In particular, just three elementary particles—the electron and two quarks called "up" and "down"—are all that are needed to describe all the matter of normal experience, including the sun, moon, planets, and all the stars and galaxies in the sky. The proton is made of two up quarks and one down quark: *uud*. The neutron is made of one up quark and two down quarks: *udd*. Every chemical atom in the universe is then composed simply of *u*'s, *d*'s, and *e*'s. Add the photon, the particle of light represented by the symbol γ, and you have everything visible to the naked eye.

Although atomic matter is everything that "matters" to most of us, it constitutes only 5 percent of the total of the mass of the universe. The rest of the mass of the universe is contained in still unidentified *dark matter* and *dark energy*.

The reduction of matter to more elementary levels that are themselves irreducible contrasts not only with the god-centered universe of traditional religions, it also clashes with the "new spirituality" that claims the universe is one irreducible whole.[20] Once again, there is no sign of compatibility between religion or spirituality on the one side and science on the other.

As mentioned, the views of the Presocratics sharply collided with those of Socrates, Plato, Aristotle, and just about everyone else. This disagreement, which is particularly exemplified by the views of the early atomists and their successors, will turn out to be the archetype of the timeless conceptual clash between science and religion. Science as an alternative to religion was thus born in ancient Greece, although it must be said that it would be millennia before the distinction between the two was clear in people's minds.

PYTHAGORAS

Pythagoras (ca. 570–495 BCE) is the final Presocratic thinker that I will mention. What little we know about him is from later scholars, but the ideas associated with his name have had significant influence on both science and religion. Certainly the Pythagorean theorem is widely used in science, although it had previously been discovered in Babylonia and India. Pythagoras's empirical discovery that musical notes can be represented numerically was an important step in the realization that mathematics can be used to describe observed physical phenomena.

Born on the island of Samos, Pythagoras may have traveled widely before settling in Croton in southern Italy. There he set up a religious brotherhood called the "semi-circle" that was also concerned with philosophy and politics. Eventually the brotherhood was suppressed when its enemies took control of Croton.

Unlike the common belief among Greeks at the time, Pythagoras seems to have advocated the immortality and transmigration of souls.

CREATION AND DESIGN

While the Presocratics were materialists, most still attributed some kind of divine power as the source and continuing governor of the material world. An exception was Anaxagoras (ca. 500–428 BCE), who taught that the world was originally a homogeneous mixture in which everything was indistinguishable. Then *nous*—mind, or intelligence—created a cosmic vortex that separated out the ingredients, although it did so only partially. This brings to mind the modern notion of order coming out of chaos. Anaxagoras was the first in history to distinguish mind from matter. If mind were not separate from matter, he reasoned, mind could not act upon matter.[21] I don't see this myself. A fully material hammer can act on a fully material nail. In any case, Anaxagoras introduced the duality between mind and matter that to this day remains embedded in much human thinking but is within a hair of being completely ruled out by science.

Despite this, contemporaries regarded Anaxagoras as an irreligious figure. He taught that the sun and moon were not divine beings but simply huge, inanimate objects. His supreme power, *nous*, was not overtly divine but a naturalistic concept more akin to human intelligence. Anaxagoras had to leave Athens after being found guilty by a court of impiety.[22]

The Presocratic era ends, as you might have guessed, with Socrates. While most of what we know about this pivotal figure in the history of thought comes from Plato, Socrates's discomfiture with science is most marked in *Memorabilia*, written by Xenophon (died 354 BCE). Therein Socrates is said to object to the attempt to intellectually reconstruct divinely created mechanisms, since that overreaches the nature of human beings.[23] Furthermore, Socrates asserted that there was a study infinitely more worthy than that of trees, stones, and even the stars—the study of the human mind.[24]

The atomists had introduced the notion that chance, or accident, actually has creative power. This anticipated Darwinian evolution and modern cosmology, as well as some modern theologies in which God makes use of chance in his creation.[25] Creation by chance competes with that of intelligent causation. While no one except the atomists really questioned the presence of intelligence underlying matter (see Anaxagoras above), Plato was the first to intellectualize what we now call the *argument from design*. This appears in his *Timaeus*, perhaps the most significant philosophical text of antiquity.

I will not present all the explicit arguments but will just note that they sound very much like the arguments for intelligent design that we hear today. For example, Plato notes that products of chance do not serve a manifest purpose, while designed objects do. And, living beings, Plato insisted, exhibit manifest purpose. The notion that the universe has purpose, called *teleology*, was a common one in antiquity and was especially promoted by Plato's student Aristotle. This is very much a religious concept that modern science, to the distress of many, has questioned.

Plato's creator god is an intrinsically good, divine craftsman called the *Demiurge*. In Plato's view, our world is a single, spherical, intelligent entity consisting of the four elements—earth, water, air, and fire—plus a soul.[26]

While Plato cannot be classified as a scientist, his metaphysics continue to have a strong influence on modern mathematicians and theoretical physicists. Plato was a *realist*, meaning that he believed in the existence of an external world. In this he holds the same view as the overwhelming majority of scientists and philosophers of science, as opposed to the doctrine called *idealism* in which everything is all in our heads, a concept that has resurfaced in recent years with *quantum spirituality*. The realist Plato taught that the "true reality" was not the objects we perceive through our senses but an ideal world of perfect mathematical *forms*. A circle or triangle drawn in the sand with a stick is an approximation to the form of a circle or a triangle, which is precisely defined mathematically. Our senses distort the true reality the way a lens distorts an image. For example, the planets have been observed since antiquity to wander about the sky (the word *planet* comes from the Greek word for "wanderer"), sometimes turning around and going back the other way for a while before turning back to continue their original paths. In Plato's theory of forms, the planets really move in perfect circles around Earth and the observed wandering is an optical illusion.

In the *Phaedo*, Plato maintains that the senses are useless for the acquisition of truth, that knowledge can only obtained through philosophical reflection.[27] This contrasts sharply with the most important principle of science as we know it today: our best knowledge of the physical world is obtained from what we observe or can derive from what we observe. The ancients debated the point. For example, Parmenides (died ca. 450 BCE) agreed with Plato, asserting that truth cannot be known through sensory perception but only through reason. Empedocles (died 435 BCE), on the other hand, defended the senses—although he admitted they were imperfect and must be employed with care. Anaxagoras (died ca. 450 BCE) argued that the senses offer "a glimpse of the obscure."[28] Even the atomists did not hold the modern scientific view that only by observation can we learn about the universe.

The eminent contemporary mathematician and cosmologist Roger Penrose has argued that mathematics has a "robustness that goes far beyond what any individual mathematician is capable of perceiving" and that this provides evidence for a Platonic world.[29] I have not seen any surveys, but my personal observations after more than fifty years of academic life leads me to guess that most mathematicians and theoretical physicists are unacknowledged Platonists,[30] while most experimental physicists, including myself, see no way of using the tools of our trade—observations and measurements or reflection and logic—to decide what is ultimate reality.

EPICURUS AND LUCRETIUS

Usually discussions on ancient Greek philosophy follow Plato with his even more influential student, Aristotle. However, a more continuous narrative contrasts Plato's focus on divine agency, which we can safely label as religious in nature, with the natural philosophy of the atomists, which we can safely label as scientific. Plato's *Demiurge* suggests the single creator god of the Abrahamic religions; however, it is not personal and does not interfere in human events, and it fits in better with deism than with theism. On the other hand, the atoms of Leucippus and Democritus are a reasonable facsimile to the quarks and leptons of the current standard model of particles and forces.

Atomism was elaborated by the Greek philosopher Epicurus (died 270 BCE), who lived just a generation after Plato. While some of his writing has survived,[31] his ideas were passed down by other authors, in particular the Roman poet Lucretius (ca. 99–55 BCE), whose epic poem *De Rerum Natura—On the Nature of Things*—has reached us complete.

In a 2011 article in the *New Yorker* titled "The Answer Man," literary scholar Stephen Greenblatt tells how Lucretius's poem was "attacked, ridiculed, burned, or ignored" as the Roman Empire collapsed, since it was "so incompatible with any cult of gods." Luckily, a copy survived.[32]

When Greenblatt was a student at Yale he stumbled upon a translation of *De Rerum Natura* in the university bookstore on sale for ten cents. (I similarly found a cheap paperback translation in a used bookstore). Greenblatt describes the core of Lucretius's poem as "a profound, therapeutic meditation on the fear of death." He movingly relates how it helped him come to grips with the family problems brought upon by his mother, who "brooded obsessively on the immanence of her end." Although she lived to almost ninety, "she had blighted much of her life" and cast a shadow on Greenblatt's own by her constant harping over death.[33]

Greenblatt quotes lines from *De Rerum Natura* translated by seventeenth-century poet John Dryden (died 1700), which I repeat in part:

> So when our mortal frame shall be disjoin'd
> The lifeless lump uncoupled from the mind,
> From sense of grief and pain we shall be free;
> We shall not feel, because we will not be.
>
> Nay, though our atoms should revolve by chance,
> And matter leap into the former dance;
> Though time our life and motion could restore.
> And make our bodies what they were before,
> What gain to us would all this bustle bring?
> The new-made man would be another thing.[34]

In a 2010 book titled *The Return of Lucretius to Renaissance Florence*, historian and philosopher Alison Brown writes how during the Renaissance, Lucretius was rediscovered and widely influenced Florentine thinkers on three "dangerous" themes: "Lucretius' attack on superstitious religion; his pre-Darwinian theory of evolution; and his atomism, with its theory of free will and the chance creation of the world."[35] This may have helped pave the way for the scientific revolution that followed.

Greenblatt tells how the Church tried to suppress *De Rerum Natura* while Niccolò Machiavelli made his own copy and Thomas More (died 1535) openly engaged in Epicurianism in his *Utopia*. Lucretian materialism can be found in Shakespeare, Donne, Bacon, and others. Newton declared himself an atomist. Thomas Jefferson owned at least five copies of *De Rerum Natura*, in Latin and translations into English, French, and Italian. He wrote to a correspondent, "I too am an Epicurean."[36]

Epicurus's universe was eternal and so lacked a creation, divine or otherwise. It was also boundless, with an unlimited number of atoms moving about in a void. There was no ruling mind or force and no life after death. Gods exist, but they are made of atoms like everything else.[37]

Epicurus rejected the idea that the gods punish the bad and reward the good. As philosopher David Sedley puts it, "Belief in divine creation brings with it, according to Epicurus, intolerable religious consequences, compelling us to assume that our own lives are under divine surveillance, and to live in terror of the threat this poses."[38] Atomism frees us from that threat by posing a universe whose contents are the products of accident. This does not mean gods do not exist. According to Epicurus, gods simply do not concern themselves with human beings.

As we have seen, atomism anticipated much of modern science, from Darwinian evolution to elementary particle physics. It also anticipated modern cosmology and the now commonly believed, although admittedly unproved, view that our universe is just one of many.[39] The atomists saw that if our world was but an accident, then many worlds could also arise by accident. This explained how one world, ours, happened to be congenial to human life. With an unlimited number of universes and chance, every permutation is allowed.[40] However, as we will see, another mechanism is suggested by both modern physics and biology: the evident capability of matter to assemble itself; that is, the natural growth of complexity from simplicity.

Epicurus and Lucretius also questioned the motivations for a supremely happy divine being to create a world in the first place:

> What novelty could have tempted hitherto tranquil beings, so late on, to desire a change in their earlier lifestyle? For those who are obliged to delight

in the new are plainly those who are troubled by the old. But where someone had no ill befall him up to now, because he had lived his life well, what could have ignited a passion for novelty in such a one?[41]

Besides, it is inconsistent for a perfectly good being to commit the impious, immoral act of bringing misery down on lesser beings. This disproof of a perfect God has been placed on a firm, logical foundation by contemporary philosophers Richard LaCroix[42] and Nicholas Everitt.[43]

The Roman philosopher and statesman Cicero (died 43 BCE) raised another interesting issue in his work *De Natura Deorum* (*On the Nature of the Gods*), asking "why the world-builders suddenly appeared on the scene after infinitely many centuries."[44] Saint Augustine of Hippo (died 430) provided sort of an answer: time did not begin until the creation. Another related major area of disagreement between science and theology today is whether the universe had a beginning or is eternal. We will see that, while there is no serious disagreement that *our* universe began with the big bang, nothing forbids, and indeed modern cosmology suggests, an eternal *multiverse* containing many universes besides our own. In chapter 7, a scenario will be discussed in which our universe "tunneled" from an earlier one.

In any case, the proposal that time did not exist until it was created is incoherent. Creation has the concept of time embedded in it. Something that did not exist "before" now exists "after." You can't have a creation if you have no time.

ARISTOTLE

While I have no trouble placing Plato in the religion camp, I see Aristotle with a foot in both camps. Certainly he was a scientist, having made observations of animals, especially sea life. His writings on physics are almost completely wrong, but this was not the last time a scientist got physics wrong. Here, unlike his biology, Aristotle does not seem to have relied on observations but on philosophical reasoning. Many of his claims, such as heavier bodies falling faster than lighter ones, could have been easily checked empirically, as Galileo

did centuries later. However, unlike his mentor Plato, who argued that ultimate reality was far removed from sensory experience, Aristotle gave primary
reality to the objects of sensory experience.[45]

Aristotle's greatest contribution was to provide the foundation for formal
logic that remains in use today in philosophy, mathematics, and science.

At the same time, as we will discuss further in the next chapter, Aristotle
had enormous influence on the theology of Thomas Aquinas (died 1274),
which set the standard for the Catholic Church. The great universities of medieval Europe, which were essentially run by the Church, taught Aristotelian
Scholasticism, in which Aristotle's word was virtually dogma. Scholasticism
continued to dominate European universities well into the scientific revolution, which, as we will see, developed outside of Church-controlled academia.

This does not mean to imply that Aristotle's notion of the divine much
resembles the Christian God. (In this book, as in my others, I will capitalize
God when referring to specific deities such as the Judeo-Christian-Islamic
God, while generic gods will appear in lower case). Aristotle and Plato shared
the conviction that god is the ultimate explanatory principle. However, his
supreme deity is an unmoved mover detached from the world and engaged in
self-contemplation, whose activity is "pure actuality." Everything else in the
world functions by striving, in its own way, to emulate that actuality.[46]

Aristotle identified two states of things: potentiality and actuality. A state
moves from potentiality to actuality by the action of one of four causes:

1. Material cause, or the elements *out of which* an object is created
2. Efficient cause, or the means *by which* it is created
3. Formal cause, or the expression of *what* it is
4. Final cause, or the end *for which* it is.

I will rely on the peer-reviewed *Internet Encyclopedia of Philosophy* to explain:

Take, for example, a bronze statue. Its material cause is the bronze itself. Its
efficient cause is the sculptor, insofar as he forces the bronze into shape. The
formal cause is the idea of the completed statue. The final cause is the idea
of the statue as it prompts the sculptor to act on the bronze. The final cause
tends to be the same as the formal cause, and both of these can be subsumed

by the efficient cause. Of the four, it is the formal and final which are the most important, and which most truly give the explanation of an object. The final end (purpose, or teleology) of a thing is realized in the full perfection of the object itself, not in our conception of it. Final cause is thus internal to the nature of the object itself, and not something we subjectively impose on it.[47]

Final cause was adopted by the Scholastics in Europe during the Middle Ages as an argument for a divine purpose behind all things, and this remains today a bone of contention between science and religion, since science sees no evidence for purpose (see chapter 8).

The one and perhaps only characteristic of Aristotle's god that Aquinas gave to the Christian God was his status as the ultimate first cause. Aquinas enumerated five ways that the existence of God can be "proven." The first two rest on the notion that the chains of causes and effects cannot be extended back indefinitely but must terminate in a first cause to which everyone gives the name of God. I will not repeat all the arguments, but note the caveman logic.

Furthermore, all Aquinas's proofs automatically fail for the simple reason that any logical argument already has its conclusion built into its premises. Barring supernatural revelation, premises must be based on observations.

However, I am not writing a treatise on theology, the logic of god, or for that matter, atheology, the logic of no god. My subject is the contrast of science and religion. That issue does arise in this case since the first cause argument rests on the assumption that everything must have a cause. Darwinian evolution and most interpretations of quantum mechanics suggest that this is *not* the case, that there is no first cause—a lot of stuff happens by accident, as the atomists proposed millennia ago. While Aristotle allowed a role for accident (or "luck") in the lives of humans and animals, he saw no place for it in the workings of the cosmos:

> Some say that although each animal is and comes by nature, the heaven has been formed in the way it is from luck and the fortuitous. Yet nothing whatsoever in the heaven appears to be the result of luck and disorder.[48]

Of course Aristotle was unaware of the cosmic background radiation composed of very low-energy photons left over from the big bang. These photons outnumber the atoms in the universe by a factor of a billion, and they are random to one part in one hundred thousand. In short, the universe is mostly random motion. Nearly everyone is unaware of this fact and, because we happen to live in a tiny pocket of complexity, wrongly assumes the universe is highly ordered.

EARLY CHRISTIAN PHILOSOPHY

In the second and third centuries, the Christian church developed its own intellectual tradition designed to defend the faith against learned opponents, a pursuit that continues today and is referred to as "apologetics." Greek philosophical methods were applied. Plato's teachings of divine providence, a monotheistic creator god, and the immortality of the soul fit in well with Christian beliefs, although, as we have seen, such a god is not personal.

However, there were theological objections to other Greek philosophers, notably Aristotle and, of course, Epicurus. Tertullian (ca. 155–230) denounced most philosophy as heresy. I have mentioned already, and we will discuss further in the next chapter, how the medieval Church came to grips with Aristotle, finding his teachings indispensable despite their many contradictions with Christian thought. Augustine popularized the concept that had been proposed before his time but afterward would be long adopted by the Church: philosophy should serve as the handmaiden of religion, put to use in the service of theology.[49]

By the thirteenth century, thanks mainly to Aquinas, the teachings of Aristotle and Greek scientific writings became incorporated into Christian theology.[50] Included in Aristotle's scientific notions was the concept that the sun revolved around Earth. Aristarchus of Samos (died ca. 230 BCE) and Seleucus of Seleucia (died ca. 190 BCE) are the only ancients on record who suggested that Earth revolved around the sun.[51] The heliocentric view was rejected by most of the greatest minds in ancient Greece as well as by Aristotle.[52]

However, Lucretius informs us that Epicureans held the view that there was no center of the universe and that Earth and its sun were just one planet-star system out of countless many, each star in the sky being a distant sun with its own planets.[53] This is precisely what astronomy says today.

There were many Epicurean mathematicians and scientists in the period between Aristotle and the rise of Christianity, but because they were perceived as being atheists, their writings were largely suppressed.[54] What all this means is that when Greco-Roman scientists stuck with the idea of Earth-centered cosmology, they did so knowing there was a contrary view, and it's likely that only their view was preserved because it agreed with Christian beliefs.

Religious historian Vern Bullough has also noted that Aristotle's teachings became fixed dogma in the universities, while science itself is continually changing and challenging old assumptions.[55] As a result, Christian theology and academic Scholasticism were unprepared for the scientific revolution that was triggered when Copernicus proposed that Earth moves about the sun, Galileo and Johannes Kepler (died 1630) provided the data, and Isaac Newton (died 1727) provided the theory that demolished the ancient belief that Earth was the center of the universe.[56] The new science they created transformed the world and made religion even more irrelevant.

CHAPTER 3

THE REBIRTH AND TRIUMPH OF SCIENCE

Our advancement in natural science is not dependent on our faith. All the problems of physical science are worked out by laborious examination, and strict induction.

—Benjamin Silliman (1842)[1]

Physical science, at the present day, investigates phenomena simply as they are in themselves. This, if not positively atheistic, must be a dangerous tendency. Whatever deliberately omits God from the universe, is closely allied to that which denies him.

—James Read Eckard (1860)[2]

ARABIC SCIENCE

Greek civilization did not remain confined to the Mediterranean but spread to Asia and North Africa with the conquests of Alexander the Great (died 323 BCE). Over the centuries that followed, Greek culture and language were assimilated over a vast region. It did not reach into the Arabian Peninsula, however, where Muhammad (died 632) united the tribes under the banner of a new monotheistic religion that could not have been more dissimilar to Greek teachings.

Thus it is one of the ironies of history that it was Islam and not Christendom, whose scriptures were written in Greek, that preserved for the world much of the written record of ancient Greek thought after much of that record was lost with the fall of Rome.

Muhammad's successors had rapidly conquered Byzantium and Persia, and within a century had gained control of most of North Africa and Spain. The conquering Arabs came into contact with the educated elite in these lands and, even after being converted to Islam, this elite continued the Hellenistic tradition of serious intellectual activity. Within 250 years, many of the scientific works of the classical tradition as we know them today had been translated into Arabic.[3]

However, there were some major exclusions, such as the Archimedes Codex, which includes almost everything we have of what Archimedes wrote. This work was abandoned by Christians and largely ignored by Muslims. Scholars only recently recovered a big piece of his work by using a particle accelerator(!) to read a book containing some of the writings, written over with hymns to God by some Christian who scraped off the ink.[4]

While Islamic religious scholars expressed concern with foreign ideas, all thinkers in the major urban centers, including Jews and Christians, still enjoyed considerable intellectual freedom in the empire. Scientific work was recognized for its practical value, especially in medicine, chemistry, and astronomy. As was the case elsewhere, astronomy was essential for the dating of religious events. Islam had a need for precise astronomical measurements that went well beyond the crude charts used in astrology, which essentially was rejected by Muslim scholars. Astronomy provided the means to precisely locate the direction of Mecca and to determine important dates on the Islamic lunar calendar, such as the beginning of Ramadan.

Islamic astronomers improved on the Earth-centered model of the solar system developed by Claudius Ptolemy (ca. 90–168), a Roman citizen living in Egypt. Muslims built the first observatories—buildings solely dedicated to astronomical study. The observatory at Maragha in northern Persia, constructed in the thirteenth century, had a dome with an aperture for observation and a library said to contain forty thousand volumes, probably an inflated number.[5]

In mathematics, Muslim scholars introduced Arabic numerals, based on the Hindu system. They also invented what they called "algebra." Although it used geometrical methods, it contributed to the eventual development in Western Europe of symbolic algebra, so important to science.[6]

In Spain, Ibn Rushd (died 1198), known as Averroes in Christendom, produced many works on virtually every scholarly subject from astronomy to Islamic jurisprudence. He also commented on Plato and Aristotle. His name is attached to the philosophical school known as *Averroism* that reconciles Aristotle with faith, both Islamic and Christian.

Islamic science began to decline as the Abbäsid Empire, centered in Baghdad, slowly broke up into smaller emirates and came to an end in 1258 with the sack of Baghdad by the Mongols. Baghdad, however, continued as the center of religious authority. About the same time in the West, Christians recaptured Spain, and the knowledge accumulated by the Muslims began to trickle back into Europe.

In chapter 1, I mentioned the extensive review of Arabic science by physicist Jim al-Khalili, in his book, *The House of Wisdom*.[7] I have followed al-Khalili's suggestion that "Arabic science" represents a more accurate description of the subject than "Islamic science," since the work we are describing was not done exclusively by Muslims, or by Arabs for that matter, but was largely written in Arabic.

Al-Khalili makes an important point:

In contrast to the Greek philosophers' abstract notions, Arabic scientists were grounded in something very close to the modern scientific method in their reliance on hard empirical evidence, experimentation, and testability of their theories. Many of them, for instance, dismissed astrology and alchemy as not being part of real science and being quite distinct from astronomy and chemistry.[8]

While the translation of ancient texts was of great importance to the advancement of science, al-Khalili emphasizes the original contributions made by Arabic-speaking scientists. In particular, he informs us that chemistry began in the Islamic world and is not, as often thought, an outgrowth of European alchemy.

The Arabic word for chemistry comes from the word kīmiyā, and alchemy is just the same word with the addition of the definite article *al* ("the"). The two were not distinguished in Europe until the mid-eighteenth century.

The scholar whom al-Khalili credits with being the founder of the science of chemistry is Jābir ibn Hayyān (ca. 721–815), known as Geber the Alchemist in the West. Although steeped in mysticism, Jābir transformed chemistry into an empirical science. He reportedly said, "The first essential in chemistry is that thou should perform practical work and conduct experiments, for he who performs not practical work nor makes experiments will never attain to the least degree of mastery."[9]

Jābir perfected many chemical techniques, such as crystallization, distillation, evaporation, and calcination. Other important industrial processes developed about the same time include the making of paper and soap, although these had been invented elsewhere. Medieval European scholars were forced to copy texts onto much more expensive parchment and so produced far fewer books than Arabic scholars. And Europeans viewed bathing as unhealthy while Muslims considered it a sacred duty.

EUROPE AWAKENS

The thousand-year period from 500 to almost 1500 has been referred to as the Dark Ages, when ancient learning was almost completely lost in Western Europe. Less pejoratively, historians call this period the Middle Ages, or Medieval Ages. The era was not completely dark. Although little science or classical philosophy was taught, monasteries maintained at least the tradition of learning. While they focused on religious matters, some science seeped through.[10]

The Muslims had captured Spain in 711 and built a great capital at Cordoba. There, Christians and Jews participated freely with Muslims in scholarly and other activities, even in government, and learning flourished. The library in Cordoba was reputed to have four hundred thousand volumes, also probably an exaggerated number.[11]

In the eighth and ninth centuries, Charlemagne united much of the territory of modern France, Germany, Belgium, and Holland. He was literate himself and encouraged a renewal of scholarship. Again the emphasis was on sacred needs, with astronomy applied to timekeeping for monastic rituals and establishing the dates for religious holidays such as Easter. Nevertheless, important ancient works were translated into Latin in the monastery and cathedral schools that Charlemagne created.[12]

While learning was deemphasized after Charlemagne, history can supply the names of several scholars in the following centuries who pursued the rational study of nature as a tool for learning about the Creator. For example, Thierry of Chartres (died ca. 1156) managed to merge Platonic, Aristotelian, and Stoic natural philosophy with the account of Creation in Genesis.[13]

The full return of scholarship to Western Europe did not occur until the Christians regained control of Spain. Cordoba fell in 1236, but the reconquest was not completed until the capture of Granada by Queen Isabella of Castile and her husband King Ferdinand II of Aragon in 1492. Coincidentally, this happened the same year Columbus, whom they had commissioned, landed in the New World. Thus the year 1492 marks the end of the Dark Ages in more ways than one.

In the meantime, the golden age of Arabic science also came to an end. What were the reasons, since it had made such an auspicious start 700 years before? Al-Khalili lists several factors. He downplays the common claim of Western scholars that it was primarily the result of growing Islamic orthodoxy in which Greek philosophy was attacked as being anti-Islamic.[14] This dispute was purely philosophical and did not affect other disciplines such as mathematics, medicine, and astronomy. He also rejects the notion that the golden age came to an abrupt end with the sack of Baghdad by the Mongols in 1258, since by that time there were a number of other centers of Arabic scholarship.

One rather intriguing suggestion made by al-Khalili is that the rise of the printing press in Europe was a major impetus for science, while the cursive nature of Arabic script was very difficult to typecast. Furthermore, in the Islamic world, calligraphy was a major art form, and reducing it to a mechanical process was strongly resisted.

SCIENCE IN ISLAM TODAY

Although out of sequence with the historical timeline being followed here, this seems the best place to contrast the golden age of Arabic science with what we find in the Islamic world today. That contrast could not be more marked. There are more than a billion Muslims in the world today in over fifty-seven nations. These states spend less than 1 percent of their GDPs (gross domestic products) on research and development, five times below that of the developed world. They have fewer than 10 scientists, engineers, and technicians per 1,000 of the population compared to an average of 4 per 140 in the developed world. Except for Turkey and Iran, the scientific output in terms of cited journal articles is negligible.[15]

However, as physicist Taner Edis points out in his enlightening 2007 book, *An Illusion of Harmony: Science and Religion in Islam*,[16] many of the conservative Islamic thinkers who enjoy great influence in the Islamic world today refuse to accept any division or discordance between science and Islam, where the Qur'an contains all revealed truth. As al-Khalili notes, "Thousands of elaborately designed Islamic websites [purport] to prove that the Qur'an predicts the Big Bang, black holes, quantum mechanics, and even the notion of relativistic time dilation."[17] He adds, "Science is attacked therefore on the grounds that it seeks to explain natural phenomena without recourse to spiritual or metaphysics causes, but rather in terms of natural or material causes."[18]

The incompatibility of modern science and faith could not be better illustrated than in the world of Islam where everything is centered on God.

THE FIRST EUROPEAN UNIVERSITIES

In 1100, small urban schools existed in Europe that in the following century grew into the first universities. Initially these were not fixed locations but consisted of corporations of teachers who moved around from city to city. Gradually they settled down, with the first university in a fixed setting appearing in Bologna in 1150, followed by Paris in around 1200, and Oxford in 1220. These became the models for future institutions. In the fourteenth century the typical medieval university had about eight hundred students, although Oxford had more than a thousand and Paris over two thousand.[19]

Students at these universities were under the protection of the Church, with certain privileges of clergy even if they were not training to assume the cloth. However, their masters were not as bound to ecclesiastic regulations as we might be led to think. Within limits, scholars debated and criticized almost every philosophical and theological doctrine, and teachers of the natural sciences generally were not restricted by Church authority. Nevertheless, a fairly uniform curriculum was taught in all the universities, with emphasis on logic and other teachings of Aristotle eventually becoming compulsory. Christian theology was not forgotten, of course, but was integrated with Greek and Arabic learning.[20]

This freedom was not totally without interference. In 1210, a decree issued by a council of bishops in Paris forbade the teaching of Aristotle's natural philosophy at the University of Paris because it was conceived as promoting *pantheism*, the doctrine that identifies God with the universe. This regulation was affirmed by Pope Gregory IX (died 1241), who ordered it enforced until "examined and purged of all suspected error." The commission was never carried out, and eventually, after Gregory's death, Aristotle's works were accepted without censorship and by 1255 had become the principle ingredient of university teaching.[21] Ironically, Aristotelian dogma joined Christian dogma in impeding the development of science.

Of course, conflict existed between the two dogmas. For one, Aristotle said the universe was eternal, with no beginning and no end, in obvious contradiction to the fundamental Christian belief in a divine creation. Furthermore, Aristotle's prime mover was eternally unchanging and not capable of intervening in the world—hardly consistent with the Christian God who plays such an important role in the running of the universe. Many believers today think Jesus decides even the outcome of football games and their tennis shots. In one more important example, Aristotle did not believe the soul was immortal since it could not have an existence independent of matter.[22] All this was swept under the rug in medieval universities.

Modern cosmology today views the universe as probably eternal, having no beginning and no end. Theologians object, pointing to the big bang as evidence that the universe had a beginning, arguing that the universe cannot be eternal because then it would take an infinite time to reach the present.[23] However, while the big bang is the origin of *our* universe, nothing forbids the existence of a prior universe or, indeed, many other universes. Indeed, current cosmological models imply a vast, eternal multiverse in which our universe is just one member.

As for the claim that we would never reach the present if the universe were eternal, this assumes the universe had a beginning an infinite time ago. In fact, the eternal universe of science had no beginning, and the time interval to any moment in the past, no matter how distant, is finite.

In chapter 7, we will expand on the fundamental cosmological differences between science and Christian theology that form part of the case that the two are irretrievably incompatible. Here we will see that this difference goes back to the Middle Ages. It wasn't settled then, and it won't be settled now.

SCIENCE AS THE HANDMAIDEN OF RELIGION

As we saw in chapter 2, Saint Augustine proposed that philosophy be used as a "handmaiden of religion." This well describes the relationship between church and science that developed in the Middle Ages. While, as mentioned, there was much in Aristotle's teachings that conflicted mightily with Christian dogma, Aristotle's logic was too useful and his science was about the only science anyone knew at the time. So scholars sought ways to put Aristotelian philosophy to work on behalf of Christendom.[24]

The first chancellor of Oxford, Robert Grosseteste (died 1253), began the effort at reconciliation by bringing in Neoplatonic ideas that were more compatible with Church teachings, particularly the notion that the universe emanates from God. This program was continued by Roger Bacon (ca. 1220–1292), who argued that science contributes to the understanding of God's creation, helps to establish the religious calendar, and prolongs life. He asserted that the seeming conflicts between philosophy and Christian belief were due to faulty reasoning, since philosophy was given by God and thus cannot conflict with faith.[25] Bacon disputed the theory that disease was caused by demonic possession, and he was generally regarded as being ahead of his time (at least ahead of Christendom) in promoting empirical science.

A Dominican friar known as Albert the Great (died 1280) provided a comprehensive interpretation of Aristotlelianism while managing to maintain the priority of Christian doctrines obtained by way of revelation. He regarded philosophy and theology as separate domains.

Albert the Great's student, another Dominican named Thomas Aquinas, became the central theologian of the Catholic Church and is still recognized as such today. We have already discussed, in chapter 2, how Aquinas used Aristotle's notion of first cause to provide what he and many since have thought were unassailable proofs of the existence of God. Few philosophers or theologians today regard these proofs as airtight. The theologians advocating these "proofs" had assumed what they set out to prove.

Aquinas, like his predecessors, argued that philosophy could not be inconsistent with what is revealed to us by faith. All throughout the medieval period, we see that the assumption that God exists went unchallenged. There were some nonconformists, such as Siger of Brabant (ca. 1240–1280) who, while professing faith (perhaps for his own good), pointed out that properly conducted philosophy can lead to conclusions that contradict theology.[26] Accused of heresy during a crackdown on Aristotlelianism in 1277, Siger fled to Italy where he died mysteriously. Also fleeing with Siger was Boethius of Dacia (ca. 1240–1290), who defended Aristotle's notion of an eternal universe.

Although Siger and Boethius were Averroists, agreeing that Church doctrine took precedence over philosophy, they were nonetheless condemned by authorities. In 1277, the Bishop of Paris, Étienne Tempier, issued a list of 219 forbidden propositions, including the eternity of the world, denial of personal immortality, and denial of free will.[27]

Early in the twentieth century, French physicist, philosopher, and historian Pierre Duhem speculated that the 1277 attack on entrenched Aristotlelianism opened up the path for modern science, especially in physics. However, historian David Lindberg points out that in 1277 no such orthodoxy existed, at least not yet. As he puts it, "The condemnations were a ringing declaration of the subordination of philosophy to theology."[28]

Even Thomas Aquinas was tainted by his flirtation with Aristotle. But by 1323, Aquinas had been canonized as a saint, and in 1325, all the articles of condemnation from 1277 that were applicable to Aquinas's teachings were revoked by the then bishop of Paris. Aristotlelianism became firmly entrenched and served as the foundation of intellectual effort of any kind.

With the work of John Duns Scotus (ca. 1266–1308) and William of Ockham (ca. 1285–1347), the ability of philosophy to address articles of faith was seriously questioned, leaving faith alone as the source of religious truth. The predominance of observation and experimentation was still centuries away.[29]

THE MEDIEVAL WORLDVIEW

Ian Barbour regards Thomas Aquinas as the most influential medieval scholar, whose synthesis of Aristotelian philosophy and Christian theology dominated Western thinking until the seventeenth century and still holds sway in the Catholic Church today. Barbour summarizes the medieval worldview elucidated by Aquinas in terms of five categories:[30]

1. *Methods*

Both Greek and medieval science were primarily deductive: Knowledge was to be obtained by contemplation. Observation was not ignored but was regarded as imperfect. The human intellect was believed capable of grasping the true essence of the world. Barbour argues that when Aquinas and his followers asserted that God is rational, they indirectly contributed to the rise of science since it is based on the concept of a universe ruled by law.

2. *View of Nature*

According to Aquinas, nature was a hierarchy: God, planets, angels, men, women, animals, and plants. (An eighth level probably should have been added: inanimate objects.) No distinction was made between science and magic, sorcery, and astrology. Nature was believed to be *static*, with species created in their present forms and with no changes except those mandated by God.

3. *The Context of Theology*

In the medieval world, God was known through observation, reason, and revelation. According to Aquinas, the existence of God could be established by observational evidence and rational argument. The observed world provides support for intelligent design. Reason tells us that there must have been a first cause, which everyone identifies with God. Still, these sources are secondary to the revealed truths, such as the Trinity and the Incarnation, that cannot be accessed by observation or reason. The Bible is authoritative, but only as interpreted by the Church.

4. *Concept of God*

Aquinas's God is more than Aristotle's prime mover. Aquinas's God is personal and concerned with humanity. Indeed, Aquinas considers the possibility of an infinite sequence of causes, with God as the necessary being for all existence. While God is the primary cause, he works through secondary causes, in particular, natural processes. Barbour sees this concept of nature as also contributing to the science that eventually emerged in the seventeenth century. But, as we have seen, the Greeks differentiated the natural from the supernatural long before Aquinas did.

5. *Humans Are Central*

While God is the supreme member of the hierarchy of being, Aquinas places humanity at what Barbour calls the "center of the cosmic drama." Humans are special compared to other creatures, with bodies and immortal souls. They are part of a divine plan leading ultimately to God.

AQUINAS REBUFFED

As we will see, all five of these views were challenged by the new science that began with Copernicus and Galileo. Observation is central to the new science, and its superiority in obtaining knowledge of the physical universe is demonstrated by its success and the utter failure of alternatives, such as revelation and pure reason. This does not speak well for the capabilities of these alternatives to tell us anything about "ultimate reality."

Today, the notion that the universe is governed by natural laws analogous to the Ten Commandments remains in the thinking of most scientists, including the majority who are atheists. However, this view is unfounded. What are called natural laws are simply human inventions—ingredients of the models that scientists introduce to describe observations. Far from aiding in the advancement of science, the notion of universal law that became deeply embedded in our thinking from medieval theology has held science back to this day.

Evolution and modern cosmology not only exemplify the ubiquity of change but also the absence of hierarchy, in direct contradiction to the medieval view. Humans share the same genetics with bacteria and the same physics with inanimate objects such as rocks. Magic, sorcery, and astrology are not confirmed as separate forces of nature.

Modern science finds no evidence to support revelation as a source of information, no sign of intelligent design, and no need for everything to have a cause. It is likely that the multiverse is eternal, with no beginning, no end, and no need of a creator. The Bible is so filled with violence, contradictions, and downright errors that it provides no reliable source for the nature of reality or morality. God is not needed to help explain any human observation. Our universe and probably other universes are so vast in space and time that humanity is beyond insignificance.

While it is possible that a deist god could be behind everything as the primary cause and could use natural processes as the secondary causes of events (no miracles), we will see in detail later how science has no need to assume any primary cause beyond nature itself.

Yet, despite these well-established facts, billions of people today, including many theologians and religious scholars such as Barbour, still cling to medieval religious concepts. The Catholic Church is defined by Aquinas's theology. Protestant churches follow the same basic principles, differing from Catholics mainly in their emphasis on the Bible as the ultimate authority rather than the pope and ecclesiastic traditions. At least popes have the power to modernize, which they occasionally do, albeit far too slowly. Protestants are frozen in their ancient myths.

Likewise, Islam and Judaism today do not differ widely from medieval beliefs, and both had their own influential teachers from the same period. While science developed within this framework, organized religion has remained mired in medieval thinking to this very day. Where Christian theology has tried to keep up, the result is a modernized deism that bears little resemblance to traditional Christianity.[31] It has had no impact on what is preached from the pulpit or taught in Bible study classes.

Nevertheless, the medieval era was not the Dark Age that it is often made out to be. A visit to one of the great Gothic cathedrals of Europe, such as Chartres or Salisbury, will quickly disabuse anyone of this misapprehension. Built for the glory of God, such cathedrals testify to the glory of human genius. And while religion has been the source of much human misery with its wars, crusades, and inquisitions, it has inspired some of humankind's greatest achievements. No doubt many Christian thinkers during the centuries before Copernicus participated in the intellectual activity leading to the development of natural philosophy. However, suggestions that the scientific revolution was somehow a direct consequence of Christian thinking[32] are refuted by the historical facts.

Much genius, such as that of Darwin and Einstein, was not inspired by thoughts of a world beyond the senses. Who can say that magnificent buildings would not have been built in a world without God? Certainly we have built many great structures in our secular age.

THE NEW SCIENCE

Science as we know it today is generally regarded as having been triggered by Nicolaus Copernicus, who in 1543 published his model of the solar system that placed the sun near the center in a book called *De Revolutionibus Orbium* (*On the Revolutions of the Heavenly Spheres*). Ptolemy's Earth-centered model was a complicated conglomeration of circles upon circles called *epicycles*. Copernicus regarded his own system as simpler, although, as it turned out, it did not place the sun exactly at the center either, in order to more accurately fit astronomical data. At first, the Copernican model did no better than the Ptolemaic model in describing accurate observations; it was just simpler, and it eventually did better when the data got better.

Perhaps the most significant new conception here, contributed to by both Ptolemy and Copernicus, was that Aristotle was wrong to use philosophical reasoning to determine the nature of reality. Rather, one must make careful observations and then fit these observations to mathematical models. Theological considerations, such as apparent contradictions with the Bible, are to be considered separate from science. In the sixteenth-century perspective, however, God was still behind everything and was still needed as an ingredient of any scientific theory.

The revolutionary development in human thinking that we call the new science was punctuated by Johannes Kepler (died 1600). Having available the comprehensive data from the precise nontelescopic observations of Tycho Brahe (died 1601), Kepler discovered that the motion of the planets could be beautifully described by elliptical orbits with the sun at one of the two foci of each ellipse. When Newton proved that elliptical motion followed from his laws of motion and gravity, and his friend Edmond Halley (died 1742) used those laws to successfully predict the reappearance of the comet that bears his name seventy-six years later, well after his and Newton's deaths, the scientific method became triumphant. Everything in the history of science since then has been epilogue.

The eventual conquest of observation as the final arbiter of scientific truth had begun in earnest with Francis Bacon (died 1626). Bacon, an English lawyer, was an important Parliamentarian and aristocrat during the reigns of Elizabeth I and her successor James I. Bacon rose as high as Lord Chancellor before falling from power.

Bacon initiated the eventual discrediting of Aristotle's scientific methodology, in which knowledge is obtained primarily by reason and reference to authority. Bacon introduced the part of the method of scientific inquiry whereby we start with observations and use inductive reasoning to arrive at underlying principles. True to his beliefs, he died from pneumonia obtained while studying freezing as a means for preserving meat.

Barbour observes that Bacon neglected the theoretical side of science and omitted the role of creative imagination. He gives as an example of such creativity Galileo considering the motion of a body without air resistance, a conception we do not experience (or at least did not at the time of Galileo). I would add Galileo's notion of unaccelerated motion, which was not a common experience riding in a horse-drawn carriage on a cobblestone street. We will get to Galileo in a moment.

But first, let us mention another important figure who helped set the stage for the scientific revolution, René Descartes (died 1650). His Cartesian coordinate system and analytic geometry became basic tools that are still used in physics. He was able to reconcile God with the new mechanistic science that was just starting to develop, with his help, by making a clear distinction between body and soul, or, as we would now put it, matter and mind.

Prior to the seventeenth century, it was universally believed that the movements of bodies were the result of spiritual action, with God as the prime mover, as taught by Aristotle and Aquinas. Descartes replaced *spiritual* causality with *mechanical* causality, in which inanimate bodies and animals were machines moved around by natural forces. The one exception was the human being.

Descartes proposed that humans possess a distinct immaterial, immortal mind, or soul, that controls their movements. This duality became a fundamental ingredient of Christianity, as well as fitting in well with just about every other religious belief in the world. And it is a major place where religion and modern science, as I will later argue in detail, irreconcilably disagree.

Finally, let us discuss another thinker of the seventeenth century who has had some influence on modern thought, although he was largely disputed or ignored in his lifetime. This is the Jewish, Dutch philosopher Baruch Spinoza (died 1677).

Spinoza formulated the doctrine known as pantheism, mentioned above, that associates god with an impersonal cosmic order. When a rabbi asked Einstein if he believed in God, the great physicist was able to escape disapproval by saying he believed in "Spinoza's god." The rabbi may have been unaware that Spinoza had been excommunicated (a rare event in Judaism) from his synagogue in Amsterdam for his teachings. The Catholic Church banned all of Spinoza's books, and many were burned by Protestant reformers.

Einstein and many other avowed atheistic physicists of our era, notably the celebrated cosmologist Stephen Hawking, use the term "god" metaphorically in the Spinozan sense, to the enormous confusion of laypeople who have never heard of Spinoza or of pantheism and are thereby led to falsely think that Einstein was, and Hawking is, a believer. It would be better for all if nonbelieving scientists avoided using this metaphor. More recently, Hawking called heaven a "fairy story."[33]

GALILEO

Galileo is generally recognized as the first true scientist of the new science because he placed observation over authority except on purely theological matters. As we have seen, several Islamic scholars had done the same, so Galileo was not the first to emphasize the importance of observations. But he took the ball and ran with it.

The Florentine scholar improved the previously invented telescope, increasing its magnification factor from three to twenty, and turned it on the heavens. He observed four satellites of Jupiter and inferred that they rotated around that planet rather than around Earth, providing support for Copernicus's heliocentric conjecture. He saw sunspots and mountains on the moon, rejecting the common belief that the celestial bodies, being part of heaven, were perfect spheres.

By watching a swinging lamp in church, Galileo deduced the principle that the period of a simple pendulum depends only on its length. He noted that all heavy objects fall with the same acceleration, neglecting air resistance.

He was the first to make a careful distinction between velocity, the rate of change of position, and acceleration—the rate of change of velocity. When asked the reasonable question, "If the Earth moves, why do we not notice it?" Galileo gave two responses. First, he insisted that his observations of celestial bodies provided evidence that Earth moves and thus our intuition that we can detect our motion must be wrong. Second, he explained that when we sense our motion we are observing our acceleration, not our velocity. This has become known as the *principle of Galilean relativity:* velocity is relative. There is no such thing as absolute motion.

When we measure the velocity of an object, it is always with respect to some particular observer's "frame of reference." For example, suppose I watch you drive by in a car and measure the car's velocity to be 80 kilometers per hour. This is the velocity of the car with respect to my frame of reference. However, you are sitting in the car, so the car is at rest, zero speed, in your frame of reference.

The story of Galileo's conflict with the Catholic Church is widely known and also widely misunderstood. Many authors who agree with me that science and religion are incompatible have used the Galileo affair as a prime example. I expect some readers will expect me to take the same approach. However, I must bow to the more expert conclusions of most contemporary historians that the story is more complicated and that Galileo brought much of his trouble on himself.

Not that Church authorities were blameless, but the facts are that Galileo's crime was not so much his teaching that Earth moved about the sun; his crime was more about disobedience. This was before free speech was accepted as a basic human right. In 1616, Galileo had been ordered to cease teaching the Copernican picture as a true representation of reality. He was instructed to instead teach it only as a convenient mathematical formalism. This was not unreasonable, because the strongest evidence in support of the Copernican model, such as the moons of Jupiter and the imperfection of the celestial bodies, had not yet been observed.

In 1632, Galileo published *Dialogue on the Two Chief World Systems*, in which he argued that the tides provide evidence, incorrectly as it turned out, that Earth moves. Worse, Galileo used a foolish character, foolishly named Simplicio, to express some arguments that had been advanced by Pope Urban VIII, a longtime friend and supporter of Galileo. The pope must have said, "We are not pleased."

Although the Inquisition had worked out what we today call a "plea agreement" that would have left Galileo with little more than a slap on the wrist, the Pope intervened in his case, and in 1633 Galileo was tried by the Inquisition, found guilty, and sentenced to house arrest. Although he was forbidden to write further, his *Discourse on the Two New Sciences*, which laid the foundation of mechanics that would be built upon by Newton, was published in Holland in 1638.

The Catholic Church was, in fact, almost ready to adopt the Copernican model, as better observations began to show the model's superiority as a calculational tool—a fact that was not immediately evident in Copernicus's time. As we have seen, both Islam and Christianity relied on astronomy to determine important dates in their religious calendars, so the Church was happy to have a better method.

However, with the Protestant Reformation, the Church backed off from its flexibility on cosmology, feeling pressure to maintain its traditional conservative doctrines. *Dialogue* was placed on the Church's prohibited list and not removed until 1822. In 1992, a commission appointed by Pope John Paul II acknowledged, "Church officials had erred in condemning Galileo."[34]

Martin Luther (died 1546) opposed the Copernican model because it contradicted the Bible. The other great reformer, John Calvin (died 1564), is often quoted as saying, "Who will venture to place the authority of Copernicus above that of the Holy Spirit?" However, this quotation had not been found in any of Calvin's work, and it is not clear what Calvin's position was on the new astronomy.[35]

As the Reformation developed, Protestant countries, notably England, became more intellectually free than Catholic countries and provided the environment for Newton and others to proceed with a revolution in human thought. After all, the Reformation demonstrated that authority could be questioned, even the authority of the Roman Church, which was claimed to follow in an unbroken chain from Peter, Christ's first Vicar of God. Surely it is no coincidence that Luther and Calvin were contemporaries of Copernicus.

Galileo always insisted that his teachings did not conflict with those of the Church. He is often quoted as saying, "The Holy Spirit's intention is to teach us how one goes to heaven, not how the heavens go," although it is generally assumed that he was in turn quoting Cardinal Cesare Baronius (died 1607). The new science opened the door to what is known as *natural theology*, in which God's majesty is revealed through the wonders of nature. It also led to deism, which reached full flower briefly in the eighteenth century during the brief period known as the Enlightenment, or the Age of Reason. In Enlightenment deism, an impersonal creator sets the universe in motion and interferes no further with it as it proceeds deterministically according to the laws of mechanics set down by the creator.

The new cosmology removed humanity from the center of the universe and challenged the teachings of both Aristotle and the Church. Still, in the medieval view humans were special in that they had not only a material body but also an immaterial soul.

As for the Galileo affair, historian Thomas Dixon says it was not a clash between science and religion, as it is usually remembered, but rather a dispute about who was authorized to produce and disseminate knowledge. Galileo's claim that a sole individual could arrive at knowledge by his own observations and reasoning was considered presumptuous and a direct threat to the authority of the Church.[36]

Moreover, Galileo promoted a philosophical view now defined as *scientific realism*, which has become the dominant perspective of most scientists ever since. According to this doctrine, scientific observations and inferences about the unobserved represent a true glimpse of ultimate reality. Galileo insisted that Earth really moved around the sun.

The Church did not object to scientific models such as those of Copernicus and Kepler as being useful tools. But it insisted that science was just in the business of making predictions of observable phenomena and that one should leave questions of ultimate truth to be determined by scripture as interpreted through the revelations and traditions of the Church.[37]

The Catholic Church would have been perfectly happy if Galileo had, like earlier astronomers, just followed the lead of the Lutheran theologian, Andreas Osiander (died 1552), who oversaw the publication of *De Revolutionibus Orbium Coelestium* (*On the Revolutions of the Heavenly Spheres*) and added an unsigned preface that said:

> It is the duty of an astronomer to compose the history of the celestial motions through careful and expert study. Then he must conceive and devise the causes of these motions or hypotheses about them. Since he cannot in any way attain to the true causes, he will adopt whatever suppositions enable the motions to be computed correctly from the principles of geometry for the future as well as the past. . . . These hypotheses need not be true nor even probable. On the contrary, if they provide a calculus consistent with the observations, that is enough.[38]

This is what we call *instrumentalism*. Today, scientific realism is the common belief among most scientists—not because they have thought about it but because they would rather not. Philosophers of science, however, have made a convincing case that our scientific models and their ingredients—while not arbitrary, culturally dependent narratives, as claimed by postmodernists—are formulated with a great deal of subjectivity and need not precisely describe "true reality."

NEWTON'S LAWS

Galileo, and to some extent Descartes, laid the foundation of mechanics. Isaac Newton carried the program to its conclusion. Although many improvements to Newton's methods and observations have since been elaborated, these basically decorate the Newtonian edifice.

Newton's greatest achievements are so simple that they can be easily understood by anyone with a high school education—and that is what makes them so great. They show just how comprehensible the universe really is. Science writers love to extol the "mysteries of the universe." That sells books. But in today's physics and cosmology, there are no current mysteries that we require a new revolution in human thinking to solve—just unanswered questions on finer details.

Newton's masterwork, published in 1687, was titled *Philosophiae Naturalis Principia Mathematica* (*The Mathematical Principles of Natural Philosophy*), or referred to as just *Principia*. The book introduces the laws of motion, the law of universal gravitation, and a derivation of Kepler's laws of planetary motion. I will discuss each in turn, using my own words rather than Newton's for simplicity and to bring them up to date.

The Laws of Motion

1. A body at rest will remain at rest and a body in uniform motion will remain in uniform motion unless acted on by an external force.

We readily observe that bodies at rest do not start moving spontaneously. They require something to get them moving, an external agent we call a *force*.

I will precisely define force below. As we saw previously, Galileo had introduced the principle of relativity, which says that velocity is relative. There is no difference between being at rest and being in motion at constant velocity, that is, at both constant speed and constant direction, which we call *uniform motion*. It follows that a body moving at constant velocity will not change that velocity unless acted on by an eternal force. Thus, a force is needed to produce a change in velocity—an acceleration.

Now, Newton actually put it more generally. He introduced a "quantity of motion" we today call *momentum*. Also called *linear momentum*, it is defined as the product of the mass and velocity of a body. I will define *mass* below. The first law more fundamentally says that, in the absence of an external force, the momentum of a body does not change. That is, the mass and velocity could change, as in a rocket that expels mass as it accelerates, as long as the product remains the same.

2. Force is defined as the time rate of change of momentum.

I promised you I would precisely define force. The second law does that for us.

3. For every action there is an equal and opposite reaction.

More quantitatively, if you kick a brick wall, it kicks back at you with an equal and opposite force. You can feel that force, especially if you take off your shoe.

In another example, the force of gravity on your body is called your weight. It is pressing down on the floor where you are standing. The floor is pressing back up on you with an equal and opposite force, keeping you from crashing through to the room below.

Today we can derive the first and third laws from a single principle: *conservation of momentum*. The momentum of a body remains constant unless acted on by an external force, which I have already noted is another way to state the first law. To understand the third law, consider the example of firing a rifle while standing on ice wearing ice skates. You feel a recoil as the bullet goes flying off with high momentum in one direction, which must be balanced by your body recoiling in the opposite direction with the same momentum. Since your mass is much greater than that of the bullet, you will recoil with a speed proportionally lower than that of the bullet.

An operational definition of mass, based on this observation, can be given as follows. Place two objects, such as wooden blocks, on a frictionless surface, with a compressed spring separating them. Release the spring and measure the recoil speeds of the objects. The ratio of the masses of the objects is then defined as inversely proportional to the ratio of their respective speeds. That is, the higher the mass, the lower the recoil.

The Law of Universal Gravity

Newton's great breakthrough here was to realize that the moon fell to Earth at the same rate as an apple. The only difference, other than their relative distances, is that the moon has an orbital velocity that keeps it from hitting Earth. Imagine firing a projectile from the top of a tall building with a velocity parallel to the ground. The projectile will follow a parabolic path and hit Earth some distance from the building. Now picture firing the projectile parallel to the ground at such a high speed that, before it lands, the ground falls away because of the curvature of Earth. This is the case with the moon and any other orbiting body, including the planets. The planets fall around the sun because of the same universal force by which the apple falls to Earth.

The universality of gravity was an enormous insight. Until it was discovered, everyone assumed a different set of laws for Earth and the heavens. This was the first example of the unification of forces, which continues to the present day.

Newton was able to infer from the relative accelerations of the apple and the moon, and their relative distances from the center of Earth, that this universal force fell off as the square of the distance between two bodies. He also was able to explain Galileo's observation that objects fell with the same acceleration, independent of their masses, by assuming that the force was proportional to the masses of the bodies.

More precisely, Newton's law of gravity gives the force between two particles as proportional to the product of their masses and inversely proportional to the square of the distance between them. The value of the proportionality constant G, called Newton's constant, was not known until Henry Cavendish (died 1810) actually measured the gravitational force between two bodies in the laboratory. This is called the "experiment that weighed Earth," since Newton's law of gravity could then be used to calculate the mass of Earth.

Now, since the law of gravity is defined for point particles, what right do we have to use it for calculating our own weight as we stand on the surface of a large planet? Newton put to use the mathematical tool called calculus, which he invented independently from the German philosopher Gottfried Leibniz (died 1646). Newton proved that the force between two extensive bodies is the same, as if the mass of each body is concentrated at a certain point in each, called the center of mass. Treating Earth as an approximate sphere, that point is the center of Earth.

Kepler's Laws of Planetary Motion

Newton's contemporary and less brilliant (but still very smart) rival, Robert Hooke (died 1703), had also inferred the inverse square law. One day Hooke, Edmond Halley, and Christopher Wren were sitting around trying to figure out what planetary orbits would be with an inverse square law of gravity. Halley said he would ask Newton. He did so and Newton said, "An ellipse." Halley asked how he knew this and Newton replied, "I have proved it."[39]

Again this turns out to be easy to understand, once someone tells you the principle. Let us look at the explanation found in today's elementary physics textbooks.

Linear momentum is accompanied by another quantity called *angular momentum*. Although this is not the general case, the angular momentum of a body moving in a circle is the linear momentum of the body multiplied by the radius of the circle.

Just as linear momentum is conserved when there are no external forces on a body, angular momentum is conserved when there are no external *torques*. A torque is the twisting force you apply when you use a screwdriver. Angular momentum conservation keeps a bicycle from falling over when it is moving and enables a figure skater to speed up her spin by pulling in her arms.

The planets conserve angular momentum to a good approximation since their attraction to bodies other than the sun is of second order. From conservation of angular momentum one can show that a line from a planet to the sun will sweep out equal areas in equal times, defining an ellipse.

HALLEY'S COMET

Edmond Halley was a good friend of Newton's and an accomplished scientist in his own right with a lifetime of achievement. Today, however, he is best remembered for the comet bearing his name.

After learning that Newton had proved Kepler's laws, Halley saw to it that *Principia* was published, at his own expense. In 1705, Halley published his own *Synopsis of the Astronomy of Comets*, in which he used Newton's laws, along with historical tables, to determine that a comet that had appeared in 1682 and two comets seen in 1531 had the same orbits and were probably the same body. In the 1531 case, one sighting must have occurred when the comet was heading toward the sun; the second sighting must have happened when the comet came back after going around the sun. Halley then predicted that the same object would return in seventy-six years, in 1758. An amateur astronomer first observed the return of Halley's comet on December 25, 1758.

THE NEWTONIAN WORLD MACHINE

The wonder of Newtonian mechanics is that it enables scientists, and even high school physics students, to predict the future with precision—an ability that no prophet of any religion nor any modern-day psychic has ever demonstrated. Given the mass, initial position, initial velocity of a body, and the net force acting on it, the position and velocity of the body can be calculated at any future time with, in principle, unlimited accuracy that depends only on the accuracy of the initial measurements. And since, in the materialist worldview, the physical universe is composed of nothing more than bodies with mass, the initial conditions of the universe and the laws of physics determine everything that happens in the natural world. This is called the *clockwork universe*, or the *Newtonian world machine*.

As we will see in chapter 6, Heisenberg's uncertainty principle of quantum mechanics, introduced by the German physicist Werner Heisenberg in 1927, showed that the accuracy of such predictions is, in fact, limited in principle—not just by measurement errors. However, in the 240 years between the publication of *Principia* and the discovery of the uncertainty principle, it was assumed that everything that happens in the natural world has been predetermined since the creation.

Newton himself did not go so far as to reject God. In fact, he wrote extensively about the Bible, in particular the prophecies of an eventual paradise on Earth when Christ returns. He was not, however, an orthodox Christian. Although he held the chair in mathematics at Trinity College in Cambridge (the same chair held by Stephen Hawking today), Newton was an Arian and rejected the Trinity. He also spent a lifetime's effort on alchemy, which was motivated by his belief that mechanical action alone is not sufficient to account for living organisms and that God directs organisms' behavior through an animating force.[40]

Newton saw a need for God to determine the initial conditions and set the solar system in motion. He believed that the coplanar orbits of the planets and their unidirectional motion could not be explained naturally. He thought that these orbits would collapse together because of the mutual gravitational attraction of the planets, so it was necessary for God to step in continuously to adjust their orbits.[41]

By including God as a part of scientific theories, Newton was applying what we now know as the "God of the gaps" argument, in which God is introduced to account for some observation that the science of the time, or sometimes just a science-challenged author (not Newton, of course), cannot explain. The God-of-the-gaps argument inevitably fails since we have no way of knowing that science will never find the missing explanation. In order to defeat the argument, the atheist merely has to produce a plausible explanation consistent with existing knowledge. No proof is needed; the theist has the burden of proving that no explanation will ever be found—an almost, but not quite, impossible task.

And, as it turned out, the scientists and mathematicians in the generations following Newton were able to fill in the gaps in Newton's understanding.

In 1734, the Swedish scientist and philosopher Emanuel Swedenborg (died 1772) proposed the *nebular hypothesis* that is still used today to describe the origin of the solar system. According to this picture, a massive ball of gas condenses down into smaller clumps by gravitational attraction.

The philosopher Immanuel Kant (died 1804) and the French mathematician and astronomer Pierre-Simon, Marquis de Laplace (died 1827) argued that the rotation of the gas will cause it to flatten into a disk, and the clumps that became planets and other smaller objects would all rotate in the same direction and on the same plane around the central clump that became the sun. While this model has many problems that are still not worked out in detail, most likely it is basically correct.

Laplace is remembered for his encounter with Napoleon, in which he presented the emperor with a copy of his work. Napoleon had heard the book contained no mention of God, and so said to Laplace, "M. Laplace, they tell me you have written this large book on the system of the universe, and have never mentioned its Creator." Laplace answered, "*Je n'avais pas besoin de cette hypothèse-là.*" ("I had no need of that hypothesis.")[42] I mark this as the place in modern times where science and religion began to part company.

Laplace and others elucidated the view that became known as the previously mentioned clockwork universe, or Newtonian world machine. In this model, nature is a complete mechanical system determined by the laws of physics. The present state of the universe is the effect of its previous state and the cause of its following state.

Galileo, Newton, Laplace, and the physicists who followed were explicitly *reductionist*. They divided matter into independent bodies that moved around subject to certain mathematical principles. The movements of such independent bodies were affected by colliding with other bodies.

Reductionism provided support for the ancient atomic model of Leucippus and Democritus. While the technology of Newton's day was far from providing the tools needed to uncover the nature of atoms, the general picture of a world composed of fundamental particles finally would be established in the 1970s with the standard model of particles and forces.

The notion that everything is composed of matter and nothing more did not take hold immediately after Newton; however, it was waiting in the wings. Richard C. Vitzthum has provided a very useful in-depth review of the literature on materialism from ancient Greece to the present day.[43]

THE ENLIGHTENMENT

The scientific revolution helped trigger what is called the Enlightenment. The story is a complex one. Let us follow and expand upon Barbour's summary of the three ideas that affect the interaction of science and religion.[44]

1. Nature As a Deterministic Mechanism

We have just seen how Laplace and others interpreted Newton's laws to imply that the natural universe is a vast machine in which everything that happens is determined by what happened before. We found in the discussion of Descartes that humans were exempted from this scheme: They were special creations of God, having souls that enabled them to decide, with God's guidance, on their own actions.

During the eighteenth century, this notion was challenged. In 1746, French physician Julien Offray de La Mettrie (died 1759) published *Histoire naturelle d'lâme* (*A Natural History of the Soul*), in which he argued that since a physical phenomenon such as a fever affected mental activity, our minds are simply extensions of our bodies. Thus there was no soul or afterlife. The book was condemned and burned by court order and La Mettrie fled France for the Netherlands.[45]

In 1748, La Mettrie published *L'homme Machine* (*Man Machine*), which rejected the dualism of matter and mind and argued that the human body is a purely material machine.[46] This got him kicked out of the Netherlands.[47] He became court physician to Frederick II of Prussia and died in Potsdam in 1751.

France became the center of this unprecedented atheistic philosophy. Denis Diderot (died 1784) edited the thirty-five volume *Encyclopédie*, which contained a history of everything that was known at the time and included contributions from Americans Benjamin Franklin, Thomas Jefferson, and Benjamin Rush along with thousands of Diderot's own essays and those of dozens of European scholars. Diderot disregarded all dogma, religious and secular. Both state and church were threatened, but *Encyclopédie* proved too popular to suppress.

Paul-Henri Thiry, Baron d'Holbach, whom I quoted in the epigraph of chapter 2, sponsored Diderot and his encyclopedia and wrote hundreds of its entries. Multilingual, d'Holbach translated many enlightenment works into French. His best-known work, *The System of Nature; or, Laws of the Moral and Physical World*, was published anonymously, as were all his works. Uncompromisingly atheistic, it was smuggled into Holland for printing and then smuggled back into France, where authorities attempted, though not too successfully, to suppress it.[48]

D'Holbach hosted one of the salons for which the Paris of the day was famous. He held dinners on Sundays and Thursdays in his townhouse on the rue Royale Saint-Roch, today known as the rue des Moulins, a short walk from the Louvre. His guest list included all the most prominent political and intellectual figures in Europe. The discussions were open and friendly with disagreements kept on an intellectual plane. Many of the arguments are still heard today with little recognition of the far-reaching thoughts that were generated at these historic gatherings. Historian Philipp Blom has made d'Holbach's salon come alive again in *A Wicked Company: The Forgotten Radicalism of the European Enlightenment*.[49]

The basic thrust of the Enlightenment, championed by La Mettrie, Diderot, and d'Holbach, was that reason could be applied to every aspect of human life. For example, social engineering was anticipated to accompany mechanical engineering, thus enabling the perfection of humanity, which was being retarded by existing institutions.

Barbour is quick to reject this notion, however, quoting historian William Cecil Dampier as saying that assuming mechanics was capable of giving an exhaustive account of events was "a natural exaggeration of the power of the new knowledge which had impressed the minds of men with its range and scope, before they realized its necessary limits."[50] Today science has a better understanding of it limits, but all attempts to find evidence for a nonmaterial component in human body or mind have so far failed. Mechanics, updated to include the prefix "quantum," remains consistent with all available data.

2. The God of Deism

Barbour sees three states in the development of Enlightenment deism in the eighteenth century. First, reason was used to develop a *rational religion* that was simply an alternate route to the same basic truths. In the second stage, natural theology was seen as a substitute for revelation. Then, in the third stage, deism fell out of favor because the notion of a cosmic designer who did not act in the universe was viewed as too impersonal, too remote. Eventually this led to the reaction known as *romanticism*, in which Nature was viewed "not as an impersonal machine but as a living companion, a source of warmth, vitality, and joy, of healing and restoring power."[51]

A religious revival took place in England with John Wesley (died 1791), whose Methodism carried a message of "spiritual rebirth." Wesley wrote a five-volume work called *Survey of the Wisdom of God in Creation* that supported Copernican astronomy but was less enthusiastic about Newtonianism because it was used to justify deism.

In America, the early nineteenth century saw the Enlightenment deism of the major founding fathers, and much of the new science, fade from view as itinerant preachers traveled the country calling on sinners to repent and to accept Christ as their personal savior. This continues to this day, with the well-financed efforts of fundamentalist and evangelical Christians to undermine science. They agree with me that science and religion are incompatible. However, since they "know" the truth, they conclude that wherever science disagrees with that truth, science must be wrong.

THREE PHILOSOPHICAL VIEWS

The seventeenth and eighteenth centuries also saw significant developments in philosophy that supplemented the new science and gave new insights into theology by asking the usual philosopher's question: What does it all really mean? I will just mention three philosophers, whose views most directly address the science-religion interface: John Locke (died 1704), David Hume (died 1776), and Immanuel Kant. We have already briefly discussed Kant and the nebular hypothesis for the origin of the solar system.

John Locke

Locke came earlier than the other two and his influence was primarily political. He is regarded as the founder of liberalism and had a profound effect on the American Revolution. However, his ideas relating to science and religion are important to mention. Locke's classic work is *An Essay Concerning Human Understanding*, which he wrote over a twenty-year period starting in 1671.[52]

Locke rejected the doctrine of innate ideas and principles that at the time were believed to be necessary to gain knowledge of religious truths, morality, and natural law. Such truths were to be obtained by reason. According to Locke, ideas occur in the mind as the result of reflection and sensation. Contradicting Descartes, Locke accepted what is called the *empiricist's axiom*: that there is nothing in the intellect that was not previously obtained by the senses. He also disagreed with Descartes—and Aristotle—by promoting the atomist view that the world is composed of atoms and the void. Aristotle had reasoned that a void was impossible.

Locke objected to authority of any kind, sacred or secular, and promoted reason over superstition. Still, he believed life had a divine purpose. He had little use for Aristotle and was one of the founders of the English Royal Society. The society provided the institutional structure that enabled the new anti-Aristotelian science to blossom in England outside the church-controlled (Anglican, in this case) universities.

David Hume

The philosophy of David Hume comes closer to representing the sense of modern science than that of any thinker before him. Hume's major works are *A Treatise of Human Nature* (1739–1740), the *Enquiries Concerning Human Understanding* (1748), and *Concerning the Principles of Morals* (1751), as well as the posthumously published *Dialogues Concerning Natural Religion* (1779).

Hume argued forcefully against metaphysics and the existence of innate ideas, following Locke in saying that human knowledge is obtained from experience. The laws of nature, according to Hume, are not built into the universe; they are human expectations based on observations. In this he challenged the

Enlightenment belief that reason can be used to obtain truth and questioned that science can ever obtain certain knowledge. He also challenged the argument from design and was accused of skepticism and atheism, although he was more of an agnostic, saying that God's existence cannot be proved or disproved. Here he was willing to reserve judgment, awaiting the evidence.[53]

Hume's most important insight relating to the theme of this book is his rejection of our ability to infer causation from the observation of a series of events. Just because one event follows another, Hume reasoned, this does not mean that the event was determined by the previous event.

Here Hume makes a clear break from the caveman logic discussed in chapter 2. As we have seen, this is a major area of incompatibility between science and religion. Religion uses caveman logic to see divine agency, that is, causality, behind everything that happens in the universe. Biology, physics, and cosmology, on the other hand, have shown that observed phenomena occur without cause. Chance, under no control—human or divine—is not simply a minor player deciding the outcome of events; it is the major player.

Barbour finds Hume's criticism of religion less cogent today, claiming that religious belief is "not based on rational argument but on historical revelation or on moral and religious experience."[54] This hardly agrees with the fact that when believers are asked to provide a reason for their belief, most typically point to the world around them and say, "How can all of this be an accident?" That is, they use the argument from design.[55] And many theologians continue to provide what they regard as reasonable arguments for the existence of God and the truth of Christianity.[56]

Immanuel Kant

When Kant read Hume's critique of causality, he said it awakened him from his "dogmatic slumber." This wakening motivated him to find a way other than observation and reason to arrive at knowledge of the world. He came up with the notion that the human mind is not the *tabula rasa*, the blank slate, that Locke and Hume proposed. Rather, it possesses certain *a priori* knowledge by which it organizes the data sent to it by the senses.

Kant gives the examples of space and time and, in particular, Euclidean geometry. Although not directly observed, we express all our observations in terms of space and time. He saw no alternative. Furthermore, Euclid was able to deduce the theorems of geometry from a few simple axioms that could be seen to apply to the observable universe.

However, one of the principles of Euclidean geometry, that parallel lines never meet, turned out to be simply a definition. By relaxing that rule, other geometries are possible. The simplest example is the geometry of the surface of a sphere, where parallel line of longitude meet at the poles. Non-Euclidean geometry was used by Einstein in his general theory of relativity.

Furthermore, space and time are human inventions, defined operationally by what one measures with a meter stick and a clock. Space and time are used in our mathematical models, but those models can all be rewritten by means of what is known as a *Fourier transform* (see chapter 6) so that the independent variables are momentum and energy or wavelength and frequency.

Kant also introduced what Barbour calls the *independence hypothesis* in the competition between science and religion. Each has its own realm. Science occupies the realm of the senses, and religion does not have to keep defending itself by seeking gaps in scientific explanations or arguments from design. The realm of religion is the moral life and its relation to ultimate reality.

As we saw in chapter 1, the separated compartmentalization of science and religion was reintroduced in 1999 by the famed paleontologist Stephen Jay Gould in *Rocks of Ages*.[57] Gould defined science and religion as "two non-overlapping magisteria" (NOMA). Religion was to concern itself just with matters of morals and ultimate meanings, while science would deal solely with the world of the senses.

Kant's and Gould's division of labor has been viewed favorably by the minority of scientists who are believers (see chapter 12). I know many personally and find that, as I have mentioned, they are able to compartmentalize their own thinking, leaving their science at the church door on Sunday and leaving God at home when they return to work on Monday morning. These individuals are sincere and able to perform their tasks just as competently as any atheist colleague. They simply avoid applying the same critical thinking tools they use on the job to their faith, and avoid letting their faith intrude on their science.

However, as science and religion are actually practiced, they overlap considerably. Religion often intrudes on science's recognized turf in asserting that the universe was divinely created. Science intrudes on religion's self-claimed turf by subjecting human morality to observational analysis. But even if science and religion were independent, that would not make them compatible.

DID CHRISTIANITY BEGET SCIENCE?

As I have alluded to previously, a number of contemporary authors have asserted that Christianity provided a unique intellectual framework that led to the scientific revolution.[58] For example, in *The Politically Incorrect Guide to the Bible*, self-styled "beach philosopher" Robert Hutchinson writes:

> It was the rational theology of both the Catholic Middle Ages and the Protestant Reformation—inspired by the explicit and implicit truths revealed in the Jewish Bible—that led to the discoveries of modern science.[59]

In a similar vein, sociologist Alvin Schmidt wrote in his book *How Christianity Changed the World*:

> Belief in the rationality of God not only led to the inductive method but also led to the conclusion that the universe is governed rationally by discoverable laws.[60]

Historian Richard Carrier has vigorously disputed these claims in an essay titled "Christianity Was Not Responsible for Modern Science."[61] From the evidence he infers that "Christianity fully dominated the whole of the Western world from the fifth to the fifteenth century, and yet in all those thousand years there was no Scientific Revolution."[62]

As Carrier points out, the notion that the universe is rational was already highly developed in pagan antiquity. He concludes, "Had Christianity not interrupted the intellectual advance of mankind and put the progress of science on hold for a thousand years, the Scientific Revolution might have occurred a thousand years ago, and our technology today would be a thousand years more advanced."[63]

CHAPTER 4

DARWIN, DESIGN, AND DEITY

There is simple grandeur in this view of life . . . that from so simple an origin, through the selection of infinitesimal varieties, endless forms most beautiful and most wonderful have been evolved.

—Charles Darwin[1]

What is Darwinism? It is atheism.

—Charles Hodgea[2]

NATURAL THEOLOGY

As we have noted, the new science of Galileo and Newton was not immediately regarded as being inconsistent with religion. Certainly the two giants didn't think so. Or, if they did, they never said so. At their particular moment in history, the study of nature was viewed as a way to learn more about God, since God, in their view, after all, is the author of nature. By making observations and describing them in terms of "laws," scientists were reading another one of God's law books alongside the Bible. This was called *natural theology*. Indeed, the deists of the eighteenth century saw more magnificence in a God who created a world ruled by the beauty of natural law and who didn't have to step in, the way John Calvin had conceived, to grow every blade of grass and guide very falling leaf to the ground.

Natural Theology: or, Evidences of the Existence and Attributes of the Deity, Collected from the Appearances of Nature was the title of a highly influential book published in 1802 by the Archdeacon of Carlisle, William Paley (died 1805).[3]

Using the analogy of a watch, which is clearly an artifact, Paley compared it with the human eye and other complex biological systems and argued that they also exhibited incontrovertible proof of intelligent design. These systems all showed evidence of purpose not found in inanimate objects such as rocks. Humans stood at the top of this ladder of complexity, the pinnacle of creation. The argument seemed unassailable to many, and it remains convincing to many today, despite having been refuted by philosopher David Hume decades before Paley. How could all this beauty and functionality have happened without an intelligent creator? However, this was another God-of-the-gaps argument awaiting an original thinker to fill the gap. And that thinker was Charles Darwin.

NATURAL SELECTION

When Charles Darwin (died 1882) read Paley at Cambridge, he was greatly impressed. But after serving as a naturalist aboard HMS *Beagle* during a round-the-world survey expedition lasting from December 27, 1831, until October 2, 1836, and analyzing the vast wealth of data he had collected for another twenty-three years, Darwin produced a revolution in human thought that matched, and in many ways exceeded, even Newton's.

Newton's theories did not directly challenge the best arguments for the existence of God; Darwin's did. In Darwin's books, beginning with *On the Origin of Species by Means of Natural Selection, or the Preservation of Favored Races in the Struggle for Life*, published November 24, 1859, Darwin proposed the alternative to intelligent design that, while fully accepted by science, is still bitterly fought over today on the political front in the war between science and religion.[4] The Bible teaches that living species, or "kinds" as they are referred to therein, were created exactly as they now exist by God at a time estimated at about six thousand years ago. Darwin provided an alternative that agreed much better with the scientific facts.

British geologist Charles Lyell (died 1875) had suggested that Earth was far older than implied in Genesis. Darwin took volume 1 of Lyell's masterwork, *The Principles of Geology*,[5] with him on the *Beagle*, and he received volume 2 in South America. His observations in the months spent on land were not limited to studying plant and animal life but also included studies of rock formations, the structure of atolls, and other phenomena that tended to confirm Lyell's ideas.

In *Origin of Species*, Darwin presented the theory that populations change over long periods of time as small changes are selected out in the struggle for existence. While others, including Darwin's own grandfather Erasmus Darwin, had earlier considered the notion that life evolves, Charles provided the mechanism.

Perhaps the most important of the pre-Darwinian evolutionists was Jean-Baptiste Lamarck (died 1829). Although Lamarck was not widely recognized during his lifetime, Darwin acknowledged him for proposing that changes in both the organic and inorganic worlds were the result of natural law and not of direct divine intervention.

Lamarck's proposed mechanism for evolution was "the inheritance of acquired traits." In the theory known as *Lamarckism*, environmental changes result in individuals finding greater need for some organs and lesser need for others, causing the first to develop stronger and the second to wither away. These changes were then inherited in the next and future generations. Erasmus Darwin had similar ideas. For a while Lamarckian inheritance provided an alternative to Charles Darwin's mechanism for evolution. However, it was eventually shown that the characteristics that an individual acquires naturally during his lifetime are not inherited.

The innovation made by Darwin was the process of natural selection. Naturalist Alfred Russel Wallace had independently arrived at the same idea and communicated with Darwin in a friendly, collegial fashion.[6] Each praised and supported the other's work, and Darwin was finally stirred into action after having sat on his ideas for years without making them public. On July 1, 1858, Darwin and Wallace presented joint papers to the Linnean Society announcing their findings.

Although Wallace deserves recognition, the huge mass of data Darwin gathered and the thorough analysis he performed over decades rightly justifies his place as the primary discoverer of evolution by natural selection. While the two men announced their results simultaneously, Darwin had the idea a good twenty years before Wallace, and only Darwin's innate conservatism kept him from coming out sooner.

Evolution by natural selection is a remarkably slow process, at least for organisms with appreciable lifetimes. Artificial selection, or breeding, which humans have been carrying out with animals and plants for thousands of years, is far more rapid because it is planned rather than accidental. We are now beginning to modify genes themselves to produce more desirable organisms, a technique whose value and ethical implications continue to be hotly debated. Here we do have a Lamarckian process in which acquired traits are passed to the next generation because the changes are made directly to the gene, the agent of inheritance.

In 1871, Darwin followed *Origin of Species* with *The Descent of Man*, which showed how humans were also part of the evolutionary process and not the separate creation described in the Bible.[7] This, of course, became the source of the greatest controversy in the Darwinian scheme. It was one thing for ants and tulips to have evolved, but humans have always regarded themselves as special. Just look at the huge gap in intelligence between humans and every other species of life.

DISSENT AND DISPUTE

While most, but not all, churchmen objected to the notion that humans were just intelligent apes, some prominent contemporary scientists also expressed doubts about Darwin's theory. Lyell thought Darwin had overemphasized the role of selection. The biologist and paleontologist Richard Owen (died 1892) accepted evolution but also strongly opposed natural selection. Geologist Adam Sedgwick (died 1873), one of the young Darwin's mentors, was scathing in his opposition, saying that *Origin* was "utterly false." He wrote to a correspondent, "It repudiates all reasoning from final causes; and seems to shut the door on any view (however feeble) of the God of Nature as manifested in His works. From first to last it is a dish of rank materialism cleverly cooked and served up."[8]

It will be important to keep in mind as we go forward that the term *evo-*

lution does not automatically include natural selection. As we will see, many religious people will say they believe in evolution, but evolution guided by God. Darwinian evolution by natural selection, as the overwhelming majority of biologists now view it, is unguided. Although even some evolutionary biologists still use the term "design" in the context of natural selection, it is confusing. It is non-design, purely and simply.

To my physicist mind, the strongest argument against Darwinian evolution at the time, and one that Darwin himself worried considerably about, was made by one of the greatest physicists of the day, William Thomson, who later became Lord Kelvin. Thomson had pioneered the science of thermodynamics, introducing its first and second laws. Using thermodynamics, he calculated that the sun could not possibly have lasted long enough for Darwin's scheme to work. He also made various estimates of Earth's age, which ranged from 20 million to 40 million years. A devout believer, Thomson did not clearly separate his faith from science, and he concluded that evolution, which he did not question, must be speeded up by God.

As the story goes, in 1903, another great physicist, Ernest Rutherford, gave a lecture with the now Lord Kelvin in the audience and argued that the newly discovered nuclear energy could provide a longer lasting source of energy than the sun. Rutherford reported, "The old guy beamed." The actual tale is more complicated.[9] But make no mistake about it; evolution was inconsistent with the known physics of the nineteenth century. If nuclear energy or something equivalent did not exist, evolution would be impossible and thereby falsified.

In 1907, the age of Earth was estimated by radioactive dating. The best value of the Earth's age today is 4.55 billion years with about a 1 percent uncertainty. Also, it is now well established that the sun is powered by *nuclear fusion*, in which light nuclei, especially hydrogen, fuse into heavier nuclei with the release of energy. I am proud to have been involved in an underground experiment in Japan that in the late 1990s provided direct proof by taking "pictures" of the sun in neutrinos produced by the fusion reaction. The sun should last about another 5 billion years before its nuclear fuel burns out.

Darwin also had strong supporters, notably the biologist Thomas Huxley (died 1895), who became known as "Darwin's Bulldog" for the fervor of his advocacy of evolution. Huxley wrote: "Distinguished theologians lie about the cradle of every science as the strangled snakes beside that of Hercules; and history records that whenever science and orthodoxy have been fairly opposed, the latter has been forced to retire from the lists, bleeding and crushed if not annihilated; scotched if not slain."[10]

Huxley gained everlasting fame from his exchange with Bishop Samuel Wilberforce (died 1873) at a meeting of the British Association for the Advancement of Science in Oxford in 1860.

Legend has it that Wilberforce asked Huxley if he was descended from an ape on his mother's side or his father's side. Huxley reportedly replied something to the effect that he would rather be descended from an ape than a man who misused his great talents to ridicule a grave scientific discussion.[11]

Also present at the Oxford meeting was Robert FitzRoy, the captain of the *Beagle* during Darwin's adventure. The tragic story of Captain FitzRoy is beautifully told by Peter Nichols in *Evolution's Captain: The Dark Fate of the Man Who Sailed Charles Darwin around the World*.[12] FitzRoy's surveying work during two missions on the *Beagle* was justly celebrated on his return and proved critical, particularly his charts of the southern tip of South America, which made that route a practical way to reach the Orient.

However, FitzRoy's outstanding accomplishments were far overshadowed by those of Darwin, whose own account of the voyage, finished in 1837, became a bestseller, while FitzRoy's detailed reports were only of interest to a small group of experts. Although the two had worked well together for the duration of the voyage, sharing the same quarters, FitzRoy reacted furiously to Darwin's book for the lack of credit given to him and his crew, for which Darwin quickly apologized and corrected. More important, FitzRoy could not abide the theological implications that, while still far from explicit in Darwin's words, were evident to the intelligent but devoutly blinkered captain.

Darwin and FitzRoy increasingly grew apart. When *Origin* was finally published in 1859, FitzRoy came unhinged and told of the "acute pain" he felt for his role in making Darwin's discoveries and insights possible. He wrote Darwin, "I, at least, *cannot* find anything 'ennobling' in the thought of being descended from even the *most* ancient ape."[13]

After an up-and-down career that was diminished by an unstable mental disposition, FitzRoy became founding director of the British weather office and set up invaluable weather stations around Britain and northern France that greatly reduced the number of shipwrecks. He travelled by train to the British Advancement of Science meeting in Oxford to give a talk on British storms and was present during the confrontation between Wilberforce and Huxley. Standing up and waving a Bible, FitzRoy was shouted down by the audience.

After becoming the butt of jokes for the perfectly understandable inaccuracies of his weather forecasting, which still saved many lives, Robert FitzRoy became a broken man. On April 30, 1865, he kissed his daughter, Laura, and repaired to his dressing room where he slit his own throat.

Robert FitzRoy had spend his large inherited fortune supplementing the costs of his many ventures for the public good with little reimbursement and died with most of it drained away. Darwin contributed £100 to a fund set up for FitzRoy's family.

Darwin kept his own beliefs largely to himself out of love and respect for his devout wife, Emma, and the desire for a quiet, socially respectable life. At the time of *Origin*, he still seemed to see God behind it all but did not support Christian doctrine, especially the notion that only the faithful are saved while everyone else, no matter how virtuous, was consigned to eternal torment.[14]

As time went on, however, Darwin became increasingly doubtful of a beneficent deity. In his explorations he had seen close up the cruelty of nature. He famously gives the example of the ichneumon wasp that lays its eggs inside a caterpillar so that when the larvae hatch they eat their host alive.

However, to my knowledge Darwin never described himself as an atheist, preferring the designation "agnostic" that was invented by Huxley. Even today the simple observation of nature remains the best scientific argument against the existence of a beneficent God.

The scientific controversy that occurred with the publication of *Origin* was not, as is often thought, basically over the idea that species evolve, which disagrees with the common interpretation that the Bible's "kinds" are immutable. Only the most diehard fundamentalists then, as today, believed that life has remained unchanged since creation. The breeding of animals demonstrated the mechanism of artificial selection. The issue was, and still remains, the role of natural selection. Biologists raised legitimate questions, and there were many unsolved problems with the Darwin-Wallace proposal.

For example, some organic structures had no useful function and did not seem to be adaptive. Now we know that this is the way evolution works. A long-held misconception assumes that every part of every biological organ must have had some survival value to have become part of the whole organism. In fact, many parts are simply neutral, retained just because they do no damage and were never evolved away. Or, as we will see below, functions often change.

Recent excellent books by biologists Jerry Coyne[15] and Richard Dawkins[16] are the latest in a huge literature that amply makes the case for the basic Darwinian mechanism in which unguided, random mutations are selected out that either aid in survival or at least do not inhibit it. And, as will be elaborated later, it is not the survival of the individual organism that matters but the survival of the information stored in its DNA. If other physical processes play a role, such as the self-organization that occurs with inanimate objects such as crystals, they are minor except, most likely, in the origin of life itself on which biological evolution has little to say.[17]

DARWIN AND DESIGN

More relevant to my thesis are the philosophical and theological implications of Darwinism. The argument from design, as applied by Paley and others to the complex forms of life, was countered by providing a purely natural process that lacked design or purpose. Contrary to common sense, complexity can arise naturally from simplicity. Just as a snowflake arises spontaneously and uncaused from simple water vapor, so living organisms arise spontaneously and uncaused from simple collections of quarks and electrons.

Giving an apologetic spin on Darwinism, Ian Barbour refers to the "design" of the laws by which evolution occurs. He claims that evolution is not just a matter of blind chance but actually extends the rule of law to nature. Variations arise accidentally, Barbour admits, but they are preserved lawfully. However, it should be noted that the term "natural selection" is misleading, suggesting that some agency is in operation that decides which mutations will aid in an organism's survival. A better term is "environmental sifting."

Barbour says a new kind of law is introduced in evolution that incorporates chance: a "statistical law."[18] But is it proper to refer to a principle that arises out of chance a "law"? Let us consider another example of such a statistical law. In the radioactive decay of atomic nuclei, no known theory can predict when a particular nucleus will disintegrate. Assuming that only chance is in operation, the probability for the nucleus to decay in any given time interval is the same for all equal time intervals. For example, so long as a radioactive nucleus has not decayed, its probability of doing so in the next second is a fixed value characteristic of the type of nucleus. The time of its decay is as random as logically possible.

From this one can easily derive mathematically what is called the *exponential decay law* in which the number of nuclei in an initial sample of identical nuclei will drop off exponentially with time. This is precisely what has been observed for radioactive nuclei for a century now, providing strong evidence that nuclei decay randomly without a predetermined cause.

Calling such a statistical principle a "law" misleads the reader into thinking that some lawgiver or law-abiding process is behind it all. In fact, the conclusion is quite the opposite. When an observed phenomenon follows a statistical pattern that can be predicted from pure randomness, then the conclusion can be drawn that the phenomenon occurred spontaneously without any action by an outside agent.

And this is exactly the place where the greatest conflict between science and religion exists in biology. Although the Catholic Church and moderate Protestant churches claim they support evolution by natural selection, the fact is they do not. In a message to the Pontifical Academy of Sciences on October 22, 1996, Pope John Paul II refers to encyclical *Humani Generis* (1950) composed by Pope Pius XII as stating that "there was no opposition between evolution and the doctrine of the faith about man and his vocation, on condition that one did not lose sight of several indisputable points." Pope John Paul hedged considerably on his acceptance of evolution, implying it has not yet been validated and there is more than one hypothesis in the theory. And he made it very clear that mind or spiritual soul did not emerge from matter but is a creation of God.[19]

To reinforce this teaching, in his 2011 Easter Homily, Pope Benedict XVI said:

> It is not the case that in the expanding universe, at a late stage, in some tiny corner of the cosmos, there evolved randomly some species of living being capable of reasoning and of trying to find rationality within creation, or to bring rationality into it. If man were merely a random product of evolution in some place on the margins of the universe, then his life would make no sense or might even be a chance of nature. But no, Reason is there at the beginning: creative, divine Reason.[20]

A recent survey indicated that only 9 to 14 percent of Americans believe that evolution is *not* God-guided.[21] This is about the same percentage of Americans who do not belong to any church. This would indicate that virtually all Christians who accept that species evolve, contrary to the Bible they believe is the word of God, think evolution is God-guided. This is intelligent design, not Darwinism. How many American Christians believe in evolution as it is understood by science? The data indicate *none*.

As the Christian theologian and apologist Alvin Plantinga spins it:

> The claim that God created human beings *in his image* . . . is clearly consistent with evolution. . . . God could have caused the right mutation to arise at the right time. He could have preserved populations from perils of various sorts, and so on; in this way, by orchestrating the course of evolution, he could have ensured that there come to be creatures of the kind he intends.
>
> What is *not* consistent with Christian belief, however, is the claim that evolution and Darwinism are *unguided*—where I'll take that to include *being unplanned and unintended*. What is not consistent with Christian belief is the claim that no personal agent, not even God, has guided, planned, intended, orchestrated, or shaped the process. Yet precisely this claim is made by a large number of contemporary scientists and philosophers who write on this topic.[22]

Plantinga and other theists argue that even if there is no evidence for design, this does not prove that God isn't still behind it all. As philosopher Daniel Dennett has pointed out, this logic allows for other possibilities besides the creator God of the Abrahamic religions. There could be other gods. Prehistoric visitors from another galaxy could have fiddled with the DNA of earthly species. Here is his Superman scenario:

> Superman, son of Jor-El, also later known as Clark Kent, came from the planet Krypton about 530 million years ago and ignited the Cambrian explosion. Superman could have [quoting Plantinga] "caused the right mutation to arise at the right time. He could have preserved populations from perils of various sorts, and so on; in this way, by orchestrating the course of evolution, he could have ensured that there come to be creatures of the kind he intends."[23]

Plantinga calls this example "silly" and a "foolish proposition."[24] Dennett agrees, but he asks: "Is it *relatively* foolish? Compared to his [Plantinga's] alternative?"[25]

Obviously this becomes the philosophers' endless cyclical war of words. The only way to break out of such vicious cycles is with evidence. There is no evidence for design, but there could have been. If God or Superman interfered with the normal course of evolution, it should have resulted in some observable effect in the fossil record.

EVOLUTIONARY POLITICS

The United States is a remarkable anomaly on the question of the public acceptance of evolution. In a 2005 survey of thiry-four nations and their beliefs in evolution, only Muslim Turkey scored lower.[26] Unfortunately, this survey did not ask the key question: guided or unguided evolution? Perhaps the most telling response in the survey was to the question of whether evolution was "definitely true, probably true, probably false, definitely false." Only 14 percent of American adults thought that evolution is "definitely true." A third said it was "definitely false," compared to just 7 percent in Denmark, France, and Great Britain, to 15 percent in the Netherlands who said evolution was definitely false.

Objection to evolution is strong within moderate and orthodox Islam. In early 2011, a prominent British imam, Dr. Usama Hasan, a physicist and fellow of the Royal Astronomical Society, received death threats for giving a speech claiming that Islam was compatible with evolution. Despite issuing a retraction saying, "I seek Allah's forgiveness for my mistakes and apologise for any offence caused," he was fired from his position at the Masjid al-Tawhid mosque.[27] As we saw previously for Turkey, evolution is widely unaccepted in Islamic nations.[28]

After Darwin, many Christian churches accepted evolution and viewed it as another example of the grandeur of God. It made more sense, they claimed, for God to use the natural laws he had created than to constantly intervene in the running of the world. This was deism more than theism, but preachers swept that fact under the rug.

However, many objected to the notion that humans were cousins to apes and apricots, which denied the special nature of humanity implied in the Bible. Since God had come to Earth in the person of a man, it was degrading to both men and God to view humans as nothing more than less-hairy apes.

America has proved to be a unique battlefield in the conflict between creationism and evolution, starting with the Scopes "Monkey trial" in Dayton, Tennessee, in 1925 and carrying on through the Dover, Pennsylvania, courtroom clash in 2005. The issue, which has almost always centered on what to teach in schools, has been more political than scientific. We do not have an intellectual debate here; we have a power struggle over who should decide what is taught in class—academic eggheads or the pious majority.

The defendant in the 1925 trial, teacher John Scopes, was accused of discussing evolution in class, which had been forbidden by state law. The trial attracted wide attention and was broadcast nationally, with the prosecution led by three-time unsuccessful presidential candidate William Jennings Bryan and Scopes defended by the legendary attorney Clarence Darrow. Scopes was convicted and fined $100, which he never had to pay. He did break the law, after all. However, the press came down hard on Bryan.

The famous journalist H. L. Mencken had reported on the trial from Dayton. He wrote that the campaign against evolution "serves notice on the country that Neanderthal man is organizing in these forlorn backwaters of the land, led by a fanatic [Bryan], rid of sense and devoid of conscience."[29] Bryan died shortly afterward.

The play and film *Inherit the Wind* does a good job of depicting the mood of the time, although it takes a number of liberties with the facts.[30]

Laws forbidding the teaching of evolution remained on the books in several states until being overturned by the US Supreme Court in 1968. With the teaching of evolution legalized, a push came to teach biblical creationism as an alternate science, termed "creation science."[31] That movement began in the early twentieth century when a Seventh Day Adventist evangelist with a minimal college education, George McCready Price, wrote a number of books challenging the conclusions of geologists that Earth was much older than the six thouand years implied in the Bible.[32] Price wrote, "Do you know that the theory of evolution absolutely does away with God and with His Son Jesus Christ, and with His revealed Word, the Bible, and is largely responsible for the class struggle now endangering the world?"[33]

Price's ideas were scoffed at by professional geologists and paleontologists, but they caught on among fundamentalists.[34] Price was referred to by Bryan at the Scopes trial, although Price was in England at the time and did not testify. This brought Price some recognition but also led to a retort from the acerbic Darrow: "You mentioned Price because he is the only human being in the world so far as you know that signs his name as a geologist that believes like you do. . . . Every scientist in this country knows [he] is a mountebank and a pretender and not a geologist at all."[35]

With two world wars and a worldwide depression sandwiched between the two, little public attention was paid to evolution. But with the launch of Sputnik by the Soviet Union in 1957 and the suddenly awakened threat to US dominance, America became increasingly focused on science. The government generously funded the rewriting of science textbooks, including the incorporation of evolution. The flooding of classrooms with evolution textbooks produced a backlash among conservative Christians who began to take action to counter the trend.[36]

In 1961, in Louisiana, a civil engineering professor, Henry Morris, followed the lead of Price and attempted to put the biblical story on a scientific footing by claiming geologic evidence for a young Earth and worldwide flood.[37] While strongly criticized by the scientific community for inaccuracies, biased data selection, and placing religious beliefs ahead of science,[38] the book sold more than two hundred thousand copies. A law requiring the teaching of creation science in public schools in Arkansas was struck down in federal court in 1982.[39] A similar law in Louisiana was rejected by the US Supreme Court in 1987.

All the court rulings were based on the clearly overt purpose of the laws to teach religion disguised as science, in violation of the Establishment Clause of the First Amendment to the US Constitution, which says, "Congress shall make no law respecting an establishment of religion, or prohibiting the free exercise thereof." The US Supreme Court has issued several interpretations of this clause, but perhaps the most important was in the case of *Lemon v. Kurtzman* in 1972 when it established the so-called Lemon Test:[40]

1. The government's action must have a secular legislative purpose
2. The government's action must not have the primary effect of either advancing or inhibiting religion
3. The government's action must not result in an "excessive government entanglement" with religion.

However, the judge in the 1982 Arkansas ruling, William R. Overton, went further than necessary in declaring that creation science was not a science. This part of the ruling was disputed by the eminent philosopher of science Larry Laudan, later a colleague of mine at the University of Hawaii. Laudan pointed out that creation science had many of the properties that the court itself had used to define science, such as referring to empirical data. Creation science was science—just wrong science, since it disagreed with the empirical facts on matters such as the age of Earth.[41]

After their failure to enable public schools to teach creation science, creationists came up with a better plan. Led by biochemist Michael Behe and theologian William Dembski, a theory was proposed called *intelligent design* (ID). This theory raised no objections to the established scientific view of evolution over a vast time period of hundreds of millions of years. Behe and Dembski simply claimed that many life forms are too complex to have arisen solely by the Darwinian process of random mutations and natural selection and that this was evidence for intelligent design. Here we see the argument from design once again, although despite their indisputably religious motives, Behe and Dembski insisted they were not claiming any specific entity or deity as the source of design.

It took no time at all for the evolutionary science community to demonstrate the flaws in ID arguments. Nevertheless, supported by large sums of money provided by wealthy fundamentalist Christians, Behe, Dembski, and their collaborators enjoyed several years of media attention and attracted a huge following among believers, including a modest number of scientists who were not specialists in evolutionary biology. Neither were Behe or Dembski, although otherwise their academic credentials were sound—an improvement over earlier creationists.

Behe argued that many biological systems contain parts that cannot be reduced to simpler parts that still have functions and so had to have been designed.[42] Examples of what he called "irreducible complexity" included the eye, the blood-clotting cascade, and bacterial flagella.

Bacterial flagella have provided the prototype example. These are little whip-like molecular propellers composed of forty complex proteins that drive certain bacteria through water. Behe claimed that if a single part of the flagellum were missing, what remained would have no function and hence could not have evolved. However, evolutionary biologists have provided plausible scenarios for how bacterial flagella could have evolved.[43] Behe proved unaware of much of the earlier research in this field. What he failed to understand was that parts of organisms often evolve for one function but then combine with other parts to form a new, more complex system with another function.

Dembski also claimed that life is too complex to have arisen without intelligent design. His argument was based on the notions that the more complex a system, the greater the information it contains and that no natural process can result in an increase in information. He called this the *law of conservation of information*.[44]

In my 2003 book *Has Science Found God?* I showed that the law of conservation of information is provably incorrect.[45] Dembski had used the conventional definition of information introduced in 1948 by the father of information theory, Claude Shannon.[46] Shannon mathematically defined the information transferred in a communication channel as proportional to the decrease in the entropy between the transmitter and receiver. Entropy is a measure of disorder, so an entropy decrease, or information increase, represents an increase in the order of a system.

Here Shannon used the standard equation for entropy found in statistical mechanics textbooks. As is well known, the entropy of a system is not conserved in physics. It can increase or decrease depending on the interaction of the system with other systems. If a system is isolated, that is, does not interact with anything else, then the entropy of the system will remain the same or increase. This is the *second law of thermodynamics* that was discovered in the nineteenth century. Since information is defined as the *decrease* in entropy, information for an isolated system is either unchanged or lost; it can't increase.

However, a transmitter and receiver are two interacting systems. They are not individually isolated. So, the entropy lost by one system can be gained by the other. Or, equivalently, the information lost by one can be gained by the other. So a physical system, such as a biological organism or Earth itself, which gets energy from the sun, can become more ordered by purely natural processes.

Although Behe and Dembski claimed to be strictly scientific, the large movement that grew around them was clearly motivated by religion. Their work and the huge propaganda machine that promoted it in the media was largely financed by a Seattle organization called the Discovery Institute. In a 2004 book, *Creationism's Trojan Horse: The Wedge of Intelligent Design*, philosopher Barbara Forrest and biologist Paul Gross document how the institute's Center for the Renewal of Science and Culture (now called the Center for Science and Culture) sought nothing less than a scientific and cultural revolution by overthrowing "scientific materialism."[47]

The movement to renew science and culture was spearheaded by a retired criminal law professor, Phillip Johnson, who wrote a series of books denouncing evolution. Johnson recognized what many people still fail to grasp about the impact of Darwinian evolution on religion. As Johnson says in his book, *Defeating Darwinism by Opening Minds*, evolution "doesn't mean God-guided, gradual creation. It means unguided, purposeless change. The Darwinian theory doesn't say that God created slowly. It says that naturalistic evolution is the creator, and so God had nothing to do with it."[48] That's exactly what I have been trying to say.

Johnson attributes to naturalism many of the evils of the world, from homosexuality to genocide.[49] He proposed the "wedge strategy"[50] adopted by the Discovery Institute, a five-year plan to drive a "wedge" into the trunk of scientific materialism and split it at its weakest points. The primary goals of the wedge were:

- To defeat scientific materialism and its destructive moral, cultural, and political legacies
- To replace materialistic explanations with the theistic understanding that nature and human beings are created by God.

The method was to promote intelligent design theory and see it become the "dominant perspective" in all fields of science until it permeates "our religious, cultural, moral, and political life."[51]

The startup funds for the Center for the Renewal of Science and Culture were provided by the estate of Howard F. Ahmanson, whose company, the parent of Home Savings of America, had over $47 billion in assets in 1997. Ahmanson and his wife were associated with the movement called Christian Reconstruction, which seeks nothing less than to replace American democracy with a fundamentalist theocracy. This organization, according to one source, would require the death penalty for adulterers, homosexuals, witches, incorrigible children, and those who spread "false" religions.[52]

In 2005, intelligent design met its Waterloo in a Dover, Pennsylvania, courtroom. Across the country, legislatures and school boards had been considering mandating the teaching of ID in public school science classes as an "alternative" to Darwinian evolution. The Dover school board had adopted a policy requiring teachers to read a statement in class stating that intelligent design provides an alternative to evolution as an "explanation of the origin of life." This was challenged by a group of parents represented by the American Civil Liberties Union and other national organizations. Both sides called expert witnesses, including Michael Behe and sociologist Steve Fuller for the defense and Barbara Forrest and biologist Kenneth Miller for the plaintiffs.

Behe did not fare well. Under cross examination he was forced to admit, "There are no peer-reviewed articles by anyone advocating for intelligent design supported by pertinent experiments or calculations which provide detailed rigorous accounts of how intelligent design of any biological system occurred."[53]

Fuller, who is a highly controversial participant in disputes over the nature of science, argued for an "affirmative action" program for intelligent design.

Forrest gave testimony on the history of intelligent design and its connection with the Discovery Institute and the institute's wedge strategy as documented in her book mentioned above, *Creationism's Trojan Horse*, coauthored with Paul Gross.

Miller was asked what harm would be done if the board's statement were read in school classrooms. His answer: (1) "It falsely undermines the scientific status of evolutionary theory and gives students a false understanding of what theory actually means;" and (2) "As a person of faith who was blessed with two daughters, who raised both of my daughters in the church . . . had they been given an education in which they were explicitly or implicitly forced to choose between God and science, I would have been furious, because I want my children to keep their religious faith."[54]

Miller is one of the most effective spokespersons in support of Darwinian evolution not only because of his intricate knowledge of the science and gifted ability to explain it to laypeople, but also because he is a devout Catholic who sees no conflict with his beliefs.[55] While this may be positive for the goal of teaching evolution in schools, it is negative if it leads people to the wrong conclusion about the compatibility of Darwinism and faith. In fact, Miller's theology is more deism than theism. He postulates that God set up evolution as the means by which to achieve his ends and, at the same time, allow for human free will.[56] The Enlightenment deist god set the universe in motion according to its fully deterministic laws and had no further need to intervene. Miller and several Christian theologians seem to envision a new form of deism in which God "plays dice," allowing chance to operate, giving considerable freedom to the universe and humanity to evolve independently of God's direct involvement.[57]

The judge in the Dover trial was John E. Jones III, a conservative Republican and 2002 appointee of George W. Bush. On December 20, 2005, Jones issued a 139-page ruling concluding, "The Board's ID Policy violates the Establishment Clause." He criticized some members of the school board "who so staunchly and proudly touted their religious convictions in public" and "would time and time again lie to cover their tracks and disguise the real purpose behind the ID policy."[58] All eight members of the board who voted for the ID policy were defeated for re-election. The new board complied with an order to pay slightly over $1 million for legal fees and damages.

Recall that in the 1982 Arkansas case involving creation science, Judge Overton had ruled that creation science was religion and not science. We saw that philosophers of science disputed his definition of science. Likewise, Judge Jones in Dover felt compelled to rule that ID was not science. In both cases this was unnecessary, since applying the First Amendment was sufficient to overrule the Arkansas law and the Dover board's policy once it was proved that these did not have a secular purpose.

Here again, one of my philosopher colleagues, Bradley Monton of the University of Colorado, voiced his disagreement and argued that intelligent design is science, although probably incorrect science.[59] He notes that ID can be formulated as a testable hypothesis and says it should be discussed in science classes.

Monton is an avowed atheist, but those theists who rally behind the teaching of intelligent design creationism, as with those who want to see the Bible taught in schools, may get more than they bargained for if they have their way. Why shouldn't ID and the Bible be taught, as long as it is done with academic freedom and honesty? That would mean pointing out their errors, which for both cases are many.

Evangelical Christians seem to think that teaching the Bible will bring more people to Jesus. In fact, the opposite is likely to happen. A recent poll showed that the more people know about religion, the less likely they are to be religious.[60] The best way to become an atheist is to read the Bible from cover to cover—everything, not just the select passages taught in Sunday school or Bible study classes. More than one devout Christian preacher has become an atheist after studying the Bible.[61]

The same can be said about intelligent design theory. The more you know about it, the better you will understand why it is wrong.

Even today, evolution is usually treated with kid gloves in high schools. A 2010 study of 926 high school biology teachers by Pennsylvania State

University professors Michael Berkman and Eric Plutzer found that 17 percent of high school biology teachers do not cover human evolution at all in their classes, while 60 percent devote only one to five hours on the subject.[62] Many of these teachers said they wished to avoid confrontation with students and parents. The same authors found that 52 percent of Americans do not believe that "human beings developed from earlier species of animals.[63] Once again the great Christian propaganda machine is doing its best to push us back into the Dark Ages.

It should come as no surprise that, at least partially as the result of religious intolerance of critical thinking, American students score low in science and mathematics. In a 2009 international education test for fifteen-year-olds in sixty-five nations by the Organization for Economic Cooperation and Development, the United States ranked twenty-third in science, seventeenth in reading, and thirty-second in math. Comparing the US scores with those of two nations having similar cultures, Canada ranked sixth and Australia twelfth in science.

Of course, universities are another matter. Within the scholarly world there is little conflict over the teaching of evolution. There, academic freedom and integrity are enforced and evolution is intricately woven into every biological subject.

EVOLUTION THEOLOGY

The theory of evolution by natural selection had a great impact on human thinking in many areas.[64] In science, evolution reaffirmed the power of meticulous observations coupled with imaginative theoretical conjecture. It introduced the notion that chance plays a major role in the occurrence of events—before that idea arose in physics, although as noted in chapter 2, the ancient Greek atomists had recognized the creative power of chance.

In theology, Darwinian evolution drove the last nail into the coffin of biblical literalism, threw doubt upon revelation, undermined the argument from design and the God of the gaps, and appealed to human experience as a basis for morality. Theologians since have sought less traditional concepts of God as he acts through the evolutionary process. Of course, they did not give up on God. They are paid to take the existence of God for granted and to come up with models of deity that are consistent with our best knowledge as well as with traditional religious teachings. Drawing the alternate conclusion that God does not exist is not included in their job descriptions. Accepting both that God exists and that evolution is true, God must be using evolution to fulfill his plan for life and humanity.

Indeed, some late-nineteenth-century theologians saw evolution as rescuing the Judeo-Christian God from the deist God of the Enlightenment. Prior to Darwin, deism was difficult to refute. Why would a perfect God have to keep stepping in to adjust the cogs in the Newtonian world machine?

Scottish evangelical Protestant Henry Drummond (died 1897), whose 1874 book, *The Greatest Thing in the World*, sold 12 million copies, wrote in 1894, "The idea of an immanent God, which is the God of Evolution, is definitely grander than the occasional Wonder-worker, who is the God of an old theology."[65]

The noted twentieth-century theologian Arthur Peacocke described how a theist can view evolution: "God must now be seen as acting to create in the world, often *through* what we call 'chance' operating within a created order."[66]

Evangelical preacher Michael Dowd and his wife, Connie Barlow, have taken to the road to preach what they call the "'Great News' of a sacred view of cosmic, biological, and human evolution." In Dowd's book *Thank God for Evolution*, he argues that facts are "God's native tongue" and science is "our collective means for discerning God's ongoing public revelations."[67] But his God is still based on faith, not fact.

Until recent years, the common interpretation of evolution assumed that it has a direction toward higher purpose. If events appear random, that randomness is still somehow under God's control. He designed a system of law and chance and so he influences events without controlling them.[68] For liberal Christian theologians, God acts in human history through the person of Christ and allows nature to run itself. God acts in the personal lives of humans, not in the impersonal arena of nature. Religion is a way of life encompassing rituals and practices of a religious community. Bible stories are not to be taken literally, in the liberal view, but as guidance for a moral life.[69]

Of course, this is a major area of dispute between theists and atheists. If we are to learn our morality from the Bible, then we must own slaves, kill anyone who works on the Sabbath, and stone disobedient wives and children. The fact that we don't (anymore) demonstrates how irrelevant the Bible has become. Those who use the Bible as a reference for moral behavior are simply cherry-picking those teachings, such as the Golden Rule, that they have independently decided are moral for other reasons, while ignoring those teachings with which they disagree.

In any case, the evolution theologians are simply reviving natural theology that, as we have seen, takes everything we observe in nature as a window on God's plan for us. They miss the point that Darwinian evolution does not describe how God created life; it describes how complex life developed from simpler forms without God's help. If God set it up that way to begin with and then left the universe to take care of itself, then that is still a deist God, although a God who plays dice rather than the Enlightenment deist God who created the Newtonian world machine.[70]

SOCIAL EVOLUTION

Not very long after the publication of *Origin*, the notion arose that evolution by natural selection could also be applied to society. This became known as "social Darwinism," a term first introduced in 1877 and still used today— usually pejoratively. The idea is that the struggle for existence also applies to the competition between individuals and organizations in a capitalist society where only the fittest survive. In some minds, "survival of the fittest" justified eugenics, racism, and imperialism. Since the 1920s conservative Christians have blamed social Darwinism for most of the evils of the modern world, including communism, Nazism, and Fascism.[71]

Although Darwin used the term "survival of the fittest" in the fifth edition of *Origin* in 1869, he attributes it to Herbert Spencer.[72] Spencer is generally regarded as the founder of social Darwinism,[73] although his basic idea was that evolution could be applied to virtually all phenomena as a unifying principle.[74]

In his groundbreaking book *The Selfish Gene*, first published in 1976, biologist Richard Dawkins introduced the concept of the *meme*, which represents an idea, behavior, or other cultural entity that replicates and transmits itself analogous to the biological *gene*. Examples he gave include musical tunes, ideas, catch phases, clothing fashions, and ways of making pots or building arches.[75]

A *memeplex* is a set of memes that survive as a mutually supporting group. Social Darwinism is a memeplex, as is the similar and equally controversial notion of *sociobiology* advocated by the Pulitzer Prize–winning biologist Edward O. Wilson.[76]

Dawkins saw many religious beliefs as memes, especially blind faith that "secures its own perpetuation by the simple expedient of discouraging rational inquiry."[77] Dawkins lists eight different religious practices he identifies as memes. Besides blind faith these include belief in God and life after death, martyrdom, killing or ostracizing heretics and apostates, demanding that others respect their beliefs—even weird beliefs such as the Trinity. In addition, beautiful music, art, and literature serve as replicating tokens of religious ideas.[78]

The study of memes, called *memetics*, has not yet been accepted as a legitimate line of scientific inquiry, and its application to religion has been severely criticized by philosopher John Gray, among others. In a 2008 article in *The Guardian*, Gray calls the memetic theory of religion "a classical example of the nonsense that is spawned when Darwinian thinking is applied outside its proper sphere."[79]

Gray does not make clear exactly what is nonsense about the observation that ideas propagate in a manner very similar to genes, once one recognizes that in biology, the basic unit that is transmitted from generation to generation is information. As computer scientist Craig James points out, it is neither survival nor fitness of the individual that matters in evolution but reproduction so that the information in the DNA is carried on to the next generation. He gives the example of the black widow spider. The female bites off the male's head during mating and devours him. While the male doesn't survive, his DNA does.[80]

Information happens to be carried in the medium of the chemical structure of DNA; but it need not be. Your name contains information. You can speak it, write it on paper, type it on a computer, text it on a cell phone, or transmit it through the air with smoke signals. And thanks to a cultural meme, it is passed on to your children.

RELIGION AS A VIRUS

A number of authors have suggested an apt metaphor for religion that, while not a theory that can be tested empirically (at least not to my knowledge), helps us to organize our thoughts about how religion operates in society. In this metaphor, religion is a virus, or memeplex, of the mind that acts in the way a biological virus acts in living organisms.

Philosopher Daniel Dennett has suggested that religions exhibit behavioral control among people similar to the way parasites invade animals, providing "deleterious replicators that we would be better off without . . . but that are hard to eliminate, since they have evolved so well to counter our defenses and enhance their own propagation."[81]

Building on Dawkins's and Dennett's ideas, Craig James and psychologist Darryl W. Ray independently have proposed that the religion meme can be viewed as a virus. Ray lists five properties of viruses that also characterize religion. Viruses

1. Infect people
2. Create antibodies or defenses against other viruses
3. Take over certain mental and physical functions and hide within the individual in such a way that they are not detectable by the individual
4. Use specific methods for spreading the virus
5. Program the host to replicate the virus.[82]

According to Ray, a specific religion infects people by childhood indoctrination and proselytizing. It creates defenses by justifying its rightness and the wrongness of other religions. It takes over mental functions by establishing rules such as abstinence and dietary restrictions and inducing guilt. It has "vectors" such as priests and ministers that spread the virus. It programs its hosts with rituals such as baptism, first communion, Bar

Mitzvah, confession, confirmation, and Bible reading. It immunizes itself against other viruses by preventing followers from learning objectively about other religions.[83]

James outlines how monotheism developed as a series of eight meme "mutations":[84]

- Specialist god to general-purpose god
- Polytheism to monotheism
- Tolerance of other gods to intolerance
- Local gods to global gods
- Physical ("same stuff as us") to abstract
- Pragmatic/natural ethics to god-given rules
- Unlikable to kind
- Sexual to asexual

He explains:

> In primitive animism, each spirit has a very specific ecological niche—the bear spirit does bear stuff, and the cloud spirit does weather stuff, and the two don't compete for survival. They "live" in separate ecological niches. But . . . as time passed, the multitude of spirits narrowed down to fewer gods, and then to just a few gods. At the same time, these gods became more general, and therein lies the problem: At some point, their "ecological niches" started to overlap, and they started competing with one another for attention. Just as in nature, if two gods serve the same purpose, it's unlikely that both will survive."[85]

James provides a good answer to what he calls "the Atheist paradox." He quotes satirist Becky Garrison as asking, "If religion were a truly useless and destructive mechanism with no redemptive qualities whatsoever, then wouldn't it be extinct by now?"[86] No, because it's a virus.

Additionally, James discusses several facts about evolution, both genetic and memetic, that Garrison misunderstands: (1) evolution works on individuals, not groups, so what benefits an individual (or its DNA) can often be harmful to the group; (2) parasites can hijack mechanisms that evolved for a different purpose; (3) alien genes or memes can invade other species and take them over; (4) genes and memes can have both good and bad benefit so that the bad can survive as long as it is outweighed by the good.[87]

So we see that not only did Darwin give us an explanation of the development and variety of life on Earth, displacing the old religious myths, but Darwinism also helps us to understand the evolution of those myths. Furthermore, memetic evolution is far faster than genetic evolution, which must rely on random mutations. Memetic evolution is also Lamarckian, so that acquired characteristics are passed on, again a speedy process.

CHAPTER 5
TOWARD THE NEW PHYSICS

If the basic claims of religion are true, the scientific worldview is so blinkered and susceptible to supernatural modification as to be rendered nearly ridiculous; if the basic claims of religion are false, most people are profoundly confused about the nature of reality, confounded by irrational hopes and fears, and tending to waste precious time and energy—often with tragic results.

—Sam Harris[1]

PUTTING HEAT TO WORK

In this chapter and the next two, we will spend some time developing the worldview that is presented to us by modern physics and cosmology. We will see how, in the nineteenth and twentieth centuries, science gradually has filled in the gaps in natural knowledge that once seemed to provide a convincing scientific case for a creator. We will begin with thermodynamics.

In the nineteenth century, the industrial revolution produced machines that took over much of the physical burdens that were previously performed by humans and animals. These machines operated according to Newton's laws of mechanics, which could be used to predict their behavior, thus greatly aiding in their design. Machines required energy to operate. Although Newton had not used the concept, his equations could be used to derive the fact that a certain quantity called energy was conserved in a closed system. If you wanted a machine to do a given amount of work, such as lift a certain mass a given height, then you needed to provide at least as much energy equal to the work you wanted done. In physics, the definition of work is useful energy.

In practice, you need to put more energy in than the work you get out because some energy is always irretrievably lost as friction. Engineers could see that friction produces heat, and it was proposed that heat is a form of energy. A new science developed called *thermodynamics* to handle phenomena involving heat. Heat is also energy, but not all of it is useful. The *first law of thermodynamics* extended the principle of energy conservation to processes involving heat. It said that the change in the internal energy of a system must equal the work done by the system minus the heat produced, or, equivalently, plus the heat added to the system. These quantities can be positive or negative. For example, with a refrigerator or air conditioner, the work done by a system and the heat added to a system are negative; that is, work comes in and heat goes out.

It was observed that it seemed impossible to build a perfect heat engine, one that converted all the input heat into work. It also seemed impossible to build a perfect refrigerator that just lowered the temperature of a system without doing any work on the system. The latter also implied that heat always flows from a hotter body to a colder one in the absence of work. Although allowed by the first law of thermodynamics, since energy is still conserved, some heat processes such as this were irreversible. This suggested a second law of thermodynamics. There are several versions that are all equivalent: you cannot build a perfect heat engine; you cannot build a perfect refrigerator; heat always flows from high to low temperature.

For example, suppose two blocks of lead are inside a rigid, insulated box that does not allow any heat or work in or out. One block is at a higher temperature than the other. When you bring them in contact, heat will flow from the higher to the lower temperature body. Eventually the two will reach an equilibrium temperature in between the two initial temperatures.

Although allowed by energy conservation, you will never see the cooler body transferring heat to the hotter one, lowering its temperature further and raising the temperature of the hotter body even more without doing work. That would be a perfect refrigerator and we could use it to build a perfect heat engine. A perfect heat engine would be a perpetual motion machine. The second law of thermodynamics forbids these.

The second law was shown to be equivalent to the statement that a certain mathematical quantity called *entropy* must remain constant or increase in time in a closed system. That is, the entropy of a closed system can never decrease. This accounts for the fact that some processes are apparently irreversible. The reverse process would violate the second law. Entropy was seen to be a measure of the disorder of a system, which when left alone eventually decays and runs down. For example, when a living organism dies, it is no longer able to input energy and eventually dissolves back into dust.

Being a closed system, the second law implies that the universe will eventually run down in what is called "heat death." However, that will not happen for many trillions of years.

THERMOTHEOLOGY

As Danish historian Helge S. Kragh describes in her unique study, *Entropic Creation: Religious Contexts of Thermodynamics and Cosmology*, published in 2008, ancient thinkers debated whether the universe was eternal or had a finite lifetime.[2] Aristotle thought it was eternal while Stoic philosophers argued that the evidence for irreversible decay is all around us. Around 320 BCE, the philosopher Zeno of Citrium remarked that if Earth had always existed, erosion would have flattened out all the mountains.[3] Of course, we now know that Earth did not always exist; rather, it formed 4.5 billion years ago and, furthermore, mountains are continually regenerated by the collisions of tectonic plates.

Early Christian thinkers such as John Philoponus (died 570) systematically argued against the eternity of the world (read "universe") since it challenged the doctrine of the creation. During the scientific revolution it was widely believed that the universe was slowly deteriorating. This was consistent with theological notions. For example, Martin Luther said, "The world degenerates and grows worse and worse every day . . . [and] will perish shortly."[4]

The mechanical theory of heat and work has been part of the ideological debate between science and religion since the 1840s. The concept of energy was regarded as evidence for a spirit world in opposition to materialism. Both the first and second laws of thermodynamics provided good arguments for the existence of a creator until the twentieth century. The first law comes in when you ask where the energy that the universe currently contains came from in the first place. The second law comes in when you ask where the order of the universe came from in the first place.

If the universe came from nothing, it should have zero energy. Thus, the argument goes, the first law of thermodynamics must have been violated at the creation of the universe. The creation therefore appears to have been a miracle.

Likewise, if the entropy of the universe is continually increasing with time, then it had to be lower in the past. That is, the universe must have begun from a state of high order, implying it could not have grown out of chaos but must have been created with high initial order. So, the second law of thermodynamics was also apparently violated by a miraculous creation.

These two scientific arguments for the existence of a creator were good ones at one time because they were based on the best empirical knowledge of the day and could not be defeated by reason alone—only by further empirical knowledge. That was to come, as we will see in chapter 7.

THE WAVE NATURE OF LIGHT

Besides thermodynamics, another important development in nineteenth-century physics concerned the nature of light and the more general, related notions of waves and fields. This had implications for the common belief found among theologians and nontheistic spiritualists that the universe is one irreducible totality in which human consciousness plays a role. This profound notion will be discussed in chapter 6.

In 1803, English physicist and all-around genius Thomas Young (died 1829) demonstrated that light had the properties of a wave. At that time, the conventional view was that light was composed of "corpuscles," as proposed by Newton. While Newton's contemporary, Dutch mathematician and scientist Christiaan Huygens (died 1695), had proposed a wave theory of light back in 1678, Newton's corpuscular theory held sway until Young demonstrated otherwise.

Young had used a ripple tank to demonstrate the interference of water waves. He did several experiments with light that showed wave effects. Today, the classic experiment to demonstrate the wave nature of light is called *Young's double slit*, although Young did not perform the experiment in the same way. In the double slit experiment, an opaque plate with two thin slits is illuminated by a point source of light. The two beams that emerge produce an interference pattern on a screen behind the plate. Now that lasers are available to provide an intense beam of monochromatic light, this is a simple classroom demonstration.

FIELDS

The nature of light is related to the concept of the field. Fields play an important role in the interplay between spirituality and science, or holism and atomism. A particle is confined to a point, or at least to a small region, in space and is what we call "local." Atomism reduces all of matter to localized particles.[5] A field covers all of space and, although modern quantum field theory can be interpreted as "reducing" everything to fields (more on this later), the notion of the field is also taken by modern quantum spiritualists to exemplify that the universe is one undifferentiated whole. For example, the late Maharishi Mahesh Yogi, founder of Transcendental Meditation, used the Grand Unified Field proposed by physicists in the 1970s to claim a scientific basis for his notion of a cosmic consciousness into which the human mind can tune. The Grand Unified Field has still not been verified empirically.[6]

Atomic reductionism is anathema to both religion and modern spirituality. Whether they are promoting Christianity or the New Age, authors are tempted to indulge in pseudoscience to buttress their conviction that reality cannot be broken down into tiny, independent units but is one, united whole.

Although the field concept was not developed mathematically until the nineteenth century, it was suggested by Newton's law of gravity. Newton never provided an explanation for the nature of gravity—just a formula to use for calculating the force on a body some distance from another body. When pressed, he speculated that gravity might be the result of the flow of etheric particles between bodies, very much an atomic explanation. But he had no way to observe these particles and simply said, *"Hypotheses non fingo"* ("I frame no hypotheses").

In the nineteenth century, Michael Faraday (died 1867), James Clerk Maxwell (died 1879), and others introduced the idea of the field to describe electric and magnetic phenomena. The field concept was applied to gravity as well. Fields were viewed as continuous media that surround a body and produce forces by reaching out and interacting with another body some distance away. A body needs mass to have a gravitational field, it needs electric charge to have an electric field, and it needs an electrical current, that is, a moving electric charge, to have a magnetic field.

This was the concept. But, more important, Newton's law of gravity gave an equation allowing one to calculate the gravitational field surrounding any distribution of mass. From this one could proceed to calculate the force on a particle of mass m by calculating the field and multiplying it by m.

Faraday had shown empirically that the electric and magnetic fields were different aspects of the same phenomenon. This actually follows from Galilean relativity. An observer in the reference frame in which a charged particle is at rest will witness phenomena that can be described by an electric field. Another observer in a reference moving with respect to the charge will witness phenomena that can be described by a magnetic field, since the moving charge is an electric current and magnetic fields are produced by currents.

The two fields were united into a single *electromagnetic* field in equations derived by Maxwell and published in 1861 and 1862. These equations, along with the equation for calculating the force on a charged particle in an electromagnetic field derived by Hendrik Lorentz, provided everything that there was to know about classical electrodynamics. The theory was complete. Classical electrodynamics has been used up to the present day for computing the electromagnetic fields produced by macroscopic objects, such as a radar antenna.

When applied in a vacuum, Maxwell's equations had a solution that mathematically was a wave that travelled at a speed whose numerical value was the same as the speed of light in a vacuum, c. This implied that light can be modeled as an electromagnetic wave in which oscillating electric and magnetic fields travel through space. Further, the equations predicted that electromagnetic waves with frequencies outside the spectrum of visible light should also exist. In 1886, Heinrich Hertz (died 1894) verified this prediction by generating low-frequency radio waves that traveled at the speed of light.

Although treated by holists, paranormalists, and practitioners of complementary and alternative medicine as something mysterious,[7] the field is a very simple concept. It is a mathematical object that has a value at every point in space. The gravitational, electric, and magnetic fields are what are called *vector fields*, since they have both a magnitude and direction at each point. Another way to say this is that you need three numbers to represent a vector field at each point: one number to specify the magnitude and two numbers to specify the direction in three-dimensional space.

We can define other fields in physics. On the everyday scale, matter looks continuous. In that approximation, a material medium is described by measurable quantities such as density, pressure, and temperature that are each specified by a single number at every point within the medium. These are examples of *scalar fields* that are used in thermodynamics and fluid mechanics.

Notice, however, that these scalar fields describe matter as a continuous medium. We now know that familiar matter is composed of individual chemical atoms and molecules that move around according to the laws of particle mechanics. We can neglect the gravitational forces between these particles, and, if they are electrically neutral, they will also experience no electric or magnetic forces. They will then interact simply by colliding locally with one another, with no holistic processes involved. The whole is the sum of its parts and the scalar fields are reduced to particles the way a smooth, sandy beach can be reduced to tiny pebbles.

Even before the atomic nature of matter was empirically verified, nineteenth-century physicists were able to derive all the principles of thermodynamics and fluid mechanics from particle mechanics alone. For example, the pressure of a gas was derived as the statistical result of the impacts provided by the molecules colliding with the walls of the container. The internal energy of a gas is just the sum of the kinetic energies and other energies (vibrational, rotational, chemical) of the molecules within. A new science called *statistical mechanics* was developed that combined probability theory with particle mechanics.

The theological significance of statistical mechanics will be discussed in chapter 8.

In the twentieth century, it was found that all the fields of physics exist in one-to-one correspondence with particles called *quanta*. For example, the *photon* is identified as the quantum of the electromagnetic field. We still await a definitive identification of a quantum for gravity, although a quantum of the gravitational field has been postulated called the *graviton*. Once a quantum theory of gravity is established, the triumph of atomism will be complete.

THE ARROW OF TIME

These developments in nineteenth-century physics produced an important advance on the question of the nature of time, which also plays a major role in theology. Nowhere in classical mechanics can you find a fundamental mechanical principle that defines a direction for time. All physics equations work equally well in either time direction. They allow us to postdict the past as well as we can predict the future. For example, astronomers have postdicted that a full eclipse of the sun occurred over Asia Minor on May 28, 585 BCE (current calendar). This may have been the one that Herodotus said was predicted by Thales (see chapter 2). We will see later that time is also reversible in quantum mechanics.

Nevertheless, the fact that time has a direction is one that seems as commonsense obvious to us as the commonsense fact that Earth is flat. Near the end of the nineteenth century, Ludwig Boltzmann proposed that our conventional definition of the direction of time is a statistical one.

Boltzmann proved a theorem, the *H-theorem*, showing that if a closed system contains a large number of randomly moving particles, it will evolve with time toward a state of equilibrium that has maximum entropy. Once it reaches that state, it will remain there on average, although there will be statistical fluctuations away from equilibrium. These fluctuations will become smaller as the number of particles becomes larger. This was a proof of the second law of thermodynamics. The entropy of an isolated system tends to increase with time. There seems to be a natural trend from order to disorder, at least for a closed system.

In effect, Boltzmann showed that the direction, or arrow, of time is in fact simply *defined* as the direction of increase in the entropy of a closed system. Now, it would seem that the arrow of time could be different for different closed systems, but no realistic system of particles can be completely isolated from the rest of our universe; therefore these imperfectly closed systems of particles all have the same arrow of time. Different universes, independent and isolated from one another, could have opposite arrows of time.[8] In chapter 7, we will consider a scenario in which our universe appeared by *quantum tunneling* from an earlier universe. That earlier universe would have an arrow of time opposite to ours!

Make note that the definition of the direction of time is statistical. This implies that what are called "irreversible" processes in thermodynamics are actually still, in principle, reversible. For example, suppose you have a chamber in which most of the air has been pumped out. If you open an aperture in the chamber wall, air will rush through it from the outside. The reverse is never observed to happen. Once outside air has filled the chamber, in our experience it will not rush out the aperture at some later time, leaving behind a vacuum. This is an example of an "irreversible" process.

However, looking at it from the particulate viewpoint, it is not impossible for the air in the chamber to rush back out through the aperture. What has to happen is that all the molecules inside are moving in the direction of the opening the instant the aperture is opened. For example, suppose you are sitting in an auditorium with a thousand other people listening to me lecture. The doors are closed. Then a latecomer opens the door and all the air leaves the room, killing us all as our bodies explode from the lack of external pressure. This is highly unlikely, but technically not impossible.

Now, suppose we have two chambers with an opening between them. They contain just three or four molecules that bounce back and forth. Observing this we would be unable to specify a direction of time. Such is the case for the fundamental processes of physics and chemistry. For example, the chemical reactions $C + O_2 \rightarrow CO_2 + Q$, where Q is a certain amount of energy, and its reverse, $CO_2 + Q \rightarrow C + O_2$ are both observed. All fundamental processes are reversible, although not necessarily with equal probabilities. The probabilities in both directions just have to be high enough so the processes can occur either way in a reasonably short time. At the everyday level of experience, the probabilities are far from equal for those processes we classify as irreversible.

The theological implications of the fundamental reversibility of time are never discussed. Most theologians (and most scientists) assume that an intrinsic arrow of time exists "in reality." The central theological doctrines of creation, causation, and ultimate purpose are meaningless without a fundamental arrow of time. However, as we will see in chapter 7, there may be many different universes besides ours, some with the same arrow of time as ours and some with an opposite arrow. How can that be reconciled with the doctrine of creation?

SPECIAL RELATIVITY

As the twentieth century approached, most physicists pictured the electromagnetic field as a vibration of a continuous, invisible, frictionless medium called the *ether* (or *aether*) that pervaded all of space. Maxwell himself was cautious, suggesting that this might be just a mathematical model and not reality.

There was a problem. Maxwell's equations predicted that light traveled at the speed c in a vacuum independent of the motion of source or observer. This was true in all reference frames, seemingly violating the principle of Galilean relativity, discussed in chapter 3, which said that all speeds are relative. Certainly the speed of sound depends on relative motion. When you move toward a source of sound, your speed is added to the speed of the sound wave approaching you. When you move away from a source of sound, your speed is subtracted. This all follows from the theory of sound. It does not follow from the theory of light.

In 1887, physicists Albert Michelson and Edward Morley performed an experiment to measure Earth's motion through the hypothesized ether. They attempted to do this by measuring the speed of light along two perpendicular paths using an interferometer invented by Michelson. If light is a vibration of an all-pervading ether, the way sound is a vibration in air, then its speed should be different for different directions of the Earth's motion through the ether.

The expected change was a hundredth of the speed of light, well within the measuring accuracy of the instrument. But they found no difference in speeds between the two directions. This result is expected from Maxwell's equations but seems to violate Galilean relativity. Michelson and Morley failed to find any empirical evidence for the ether.

However, think about it another way. The principle of Galilean relativity says our physics models must be the same in all reference frames. If they were not, then they would be distinguishable from one another. Since Maxwell's model of electromagnetic waves has them traveling at the speed of light, then relativity *requires* that the speed of light be the same in all frames.

In 1905, Albert Einstein published a world-shaking paper titled "On the Electrodynamics of Moving Bodies."[9] He did not mention Michelson and Morley, although he must have known of the result. Einstein argued purely theoretically and perhaps saw no reason to sully his results with grubby data. He asked what the logical consequences would be if both the principle of relativity were correct and the speed of light were absolute. While Lorentz, Henri Poincaré, and others had pondered the problem and had made some headway, Einstein derived the daring conclusion that we must reconsider our notions of space, time, mass, energy, and other fundamental concepts of physics.

Einstein showed that time intervals measured on a clock and space intervals (distances) measured with a meter stick depend on your frame of reference. Observers moving at a constant velocity (constant speed and direction) relative to one another will measure different time intervals and different distances between the same two events. If you watch a moving clock go by, that clock will appear to you to run slower. This is called *time dilation.* An observer sitting on the clock will not notice anything different about her clock but will see yours moving more slowly.

The same is true for the measured distance between events. A meter stick is, by definition, a meter long as measured in a reference frame at rest with respect to the stick. If that meter stick is moving across your line-of-sight, you will observe it to shrink in the direction of its motion. This is called *Fitzgerald-Lorentz contraction.* An observer sitting on the meter stick will not notice anything different about his meter stick, but he will see one at rest in your reference frame shrink in the direction of its motion.

Unless you are using an atomic clock that can measure time with an accuracy of one nanosecond (a billionth of a second) or better, you will not notice the effect at the typical speeds of everyday experience. In his 1905 paper, Einstein only considered constant relative velocities, and the theory presented there is called *special relativity*. Many experiments have verified special relativity in the century following Einstein, including those done with atomic clocks at jetliner speeds.[10] In my own field of particle physics, special relativity is as much an everyday tool as a drill is to a dentist.

CAUSE AND EFFECT

In special relativity, two events in space and time are "local" if you can find a reference frame where they occur at the same position in space. Here's an example: You and your girlfriend go for a ride on a train. You kiss her when the train leaves the station and again when you arrive at your destination. Although these are two different spatial locations in Earth's reference frame, the kisses are still local because they occurred at the same location in space in the train's reference frame and on the same pairs of lips.

Two events are nonlocal if they cannot be put in the same reference frame without exceeding the speed of light. For example, suppose you are a Houston mission control scientist for a manned Mars mission and you discover a bug in a program that will cause the spaceship on Mars to explode in ten minutes. Any message you send to the astronauts on the planet will take eleven minutes. There is no way you can warn them since the event of which you are sending a message is nonlocal with the event of the explosion.

According to special relativity, two nonlocal events cannot be in causal contact. That is, one event cannot be the cause of the other. This has a profound implication for science and theology that is not always fully appreciated in either school. Most of science and virtually all of theology are deeply based on the notion of cause and effect. In Newtonian physics, a force is needed to cause a change in momentum. In medicine, smoking causes lung cancer. In theology, God causes everything.

But now we find that events can happen in the universe that cannot be connected by cause and effect. This is not a big problem for physics or the rest of science, because it just means that there could be multiple causes of events. But it presents a big problem for theology, which reduces everything to a single cause. Furthermore, as we will see in chapter 7, quantum events happen without an evident cause, that is, they happen spontaneously. So it is hard to see how the theological principle of a single being as the cause of all phenomena can be maintained in the light of relativity and quantum mechanics.

There is another profound implication of relativity that wreaks havoc with the traditional religious doctrine of creation. An absolute past and future that is the same for all observers cannot be defined. When two events are local, observers in every reference frame will agree on which occurred first. This will be true even though observers moving relative to one another will not agree on the spatial and temporal intervals. They will still agree on what is past and what is future.

However, different observers will not all agree on the time sequence of nonlocal events. Observer 1 might see event A before event B, while observer 2 might see event B before event A. Think of the implication. The two observers differ on which event is in the "past" and which is in the "future." We already saw in the previous section that no fundamental direction of time can be found in classical or quantum physics. Now we see that there is no universal past and future.

Of course, we and our fellow humans agree on a past and future in our everyday lives. But that is because our relative speeds are low compared to c and so, for the purposes of relativity, we are all in the same reference frame. However, consider an event occurring here on Earth and another occurring in the Andromeda galaxy two million light years away. There is a four-million-year interval—two million years before our present and two million years after—for which we cannot specify past and future for events in Andromeda. Similarly, if we go back 13.7 billion years in our past to the beginning of the universe, and another 13.7 billion years into our future, we cannot distinguish past from future for any event farther than 13.7 billion light years from us.

Now, in principle we can see out to 44 billion light years away, which is called our "light horizon." The universe is 13.7 billion years old, which means that the farthest object we can see, again in principle, was 13.7 billion light years away when the light left it. However, in that time the object has moved 44 billion light years away with the expansion of the universe. The upshot is that there are many more events in the universe with which we have no causal contact than events with which we do. And we cannot specify whether any of those events occurred in the absolute past or future since another observer might disagree.

GENERAL RELATIVITY

When two reference frames are accelerating with respect to each other we must use *general relativity*, published by Einstein in 1915. If you watch a clock go by that is accelerating with respect to you, it will appear to you to run more slowly.

In general relativity, Einstein also assumed the *principle of equivalence*, which says that if you are sitting on a tiny particle you can't tell whether it is accelerating or being acted on by a gravitational force. Thus general relativity became a *theory of gravity*. The equations of general relativity derived by Einstein made several predictions of phenomena, such as the bending of light by the sun and the precession of the perihelion of Mercury that did not follow from an application of Newtonian gravity. When these were shortly verified,[11] Einstein became the famous public figure we all know, who in 1999 was named *Time* magazine's "Person of the Century."

The general theory of relativity has replaced Newton's theory of gravity as our current working model of gravity. Newton's theory still applies for most practical applications; however, the global positioning system (GPS) in your car corrects for gravitational time dilation using the equations of general relativity. It would not get you where you want to go if it did not.

CHAPTER 6
PARTICLES AND WAVES

The reductionist attitude provides a useful filter that saves scientists in all fields from wasting their time on ideas not worth pursuing. In this sense, we are all reductionists now.

—Steven Weinberg[1]

Matter is an illusion. Only consciousness is real.

—Deepak Chopra[2]

THE DEMATERIALIZATION OF MATTER

Religious apologists and quantum spiritualists have grossly distorted the developments of twentieth-century physics, relativity, quantum mechanics, and relativistic quantum field theory. They want us to believe that these great scientific achievements have demolished the reductionist, materialistic views of the past when, in fact, they have done quite the opposite. Reductionism and materialism are stronger now than they ever were.

William Grassie tells us, "The concept of materialism deconstructed itself with the advent of quantum mechanics."[3] According to University of Notre Dame philosopher Ernan McMullin, twentieth-century physics resulted in the "dematerialization" of matter.[4] Theologian Philip Clayton concurs: "Physics in the twentieth century has produced weighty reasons to think that some of the tenets of materialism were mistaken."[5] He adds:

Somewhere near the beginning of the last century, the project of material reduction began to run into increasing difficulties. Special and general rela-

141

tivity, and especially the development of quantum mechanics, represented a series of setbacks to the dreams of reductionist materialism, and perhaps a permanent end to the materialist project in anything like its classical form.[6]

Before we get into the details of these philosophical claims, let us take a look at the actual science.[7]

THE HISTORY OF QUANTUM MECHANICS

We begin with a brief review of the early history of quantum mechanics that should be sufficient to enable us to see how spiritual implications are being wrongly inferred.

The quantum narrative begins in 1900 with Max Planck and his explanation of black-body radiation. A *black body* is an object that does not reflect light but is a perfect absorber and radiator of light.[8] What is observed for a black body at room temperature is a spectrum that peaks at infrared frequencies. The peak moves to the visible region as the object is heated and becomes red-hot. Past the peak, at higher frequencies, the intensity of the light gradually drops to zero. This fall-off is not accounted for in the wave theory of light.

The nineteenth-century wave theory of light had predicted that the electromagnetic radiation from black bodies should go to infinity at high frequencies. This is because, as frequency increases, wavelength decreases so more and more vibration modes can fit within the body. This "ultraviolet catastrophe" is not observed.

Planck solved the problem by proposing that light comes in discrete bundles called *quanta*, in which the energy of each bundle is proportional to the frequency of the light. This explained the shift of the spectrum to higher frequencies. A body can contain only so much energy in the motions of its constituent particles. These particles increase in kinetic energy as the temperature of the body increases. The jiggling of charged electrons in the body results in electromagnetic radiation. According to Planck, higher energy means higher frequency radiation. The dropoff at higher frequency occurs because conservation of energy limits how high the body's energy can go.

In 1905, the same year he published the special theory of relativity, Einstein proposed that Planck's quanta were composed of individual particles that were later dubbed *photons*. The energy of each photon is equal to the frequency of the light times Planck's constant, h. Planck had determined h from fitting observed black body spectra to his mathematical model. Einstein's proposal, using Planck's value of h, accounted quantitatively for the *photoelectric effect*, in which the electric current produced when light hits metal depends on the frequency (energy) of the light and not, as expected from wave theory, on the intensity of the light.

In the wave theory, light is electromagnetic radiation in the visible and nearby spectral bands, infrared and ultraviolet. The particle nature of light was further verified in 1923 when Arthur Compton showed that X-rays, electromagnetic radiation in the spectral band just above ultraviolet light, decreases in frequency when scattered from electrons. Wave theory predicts that X-rays should not change in frequency, but just reduce in intensity as they lose energy to the electron. In the photon theory, the frequency of the scattered X-rays is lower than the incident rays since that frequency is proportional to the energy of the corresponding photons. In all these examples, the photon theory agreed quantitatively with measurements, with the same value of Planck's constant fitting the data in each case.

The identification of light as being composed of localized particles was a big step in making physics even *more* reductionist, contrary to the claim of theists and quantum spiritualists that quantum theory has weakened reductionism. The photon theory (along with special relativity) eliminated the holistic ether.

In 1911, Ernest Rutherford proposed a model of the chemical atom patterned after the solar system in which electrons orbit around a nucleus much smaller than the orbits themselves. This model was able to explain why alpha rays from radioactive nuclei scattered at abnormally large angles from gold foil. Rutherford inferred that the alpha rays were bouncing off highly localized, heavy chunks of matter at the center of the gold atoms. The electrons in the atoms were insufficiently massive to produce the observed effect. Rutherford concluded that the atom was mostly empty space, with almost all of its mass concentrated in a tiny nucleus.

In 1913, Niels Bohr applied Rutherford's model to the hydrogen atom, where a single electron orbits a proton. He postulated that the orbital angular momentum of the electron must be a multiple of Planck's constant h divided by 2π, a quantity physicists label as $\hbar$ and call the *quantum of action*. This enabled Bohr to calculate the frequencies of the very sharp *spectral lines* that were observed when an electric spark was sent through hydrogen gas and heated the gas to high temperature. Again, the wave theory provided no explanation. Interestingly, while Bohr obtained the correct spectrum of hydrogen (to first approximation), his hypothesis was actually wrong and would be improved upon twenty years later by Heisenberg and Schrödinger (see below). This is not the only example in the history of science where theories that fit the data well turned out to be wrong.[9] We have no way of knowing whether our physics models provide us with a reliable picture of ultimate reality. They just fit the data.

While Einstein, following Planck, had shown that light is composed of particles traveling at the speed c, the wavelike behavior of light did not go away. In 1924, Louis-Victor-Pierre-Raymond, 7th duc de Broglie, proposed in his doctoral thesis that objects such as electrons that we are accustomed to calling particles also have wave properties. He noted that the wavelength of a photon—that is, that of the associated electromagnetic wave—equals Planck's constant h divided by the photon's momentum. De Broglie assumed that all particles have the same relationship between their momentum and the wavelength of their associated wave. Clinton Davisson and Lester Germer confirmed de Broglie's hypothesis in 1927 when they observed the wavelike diffraction of electrons in a crystal.

These developments led to what is called the *wave-particle duality*, in which particles such as electrons and neutrons also behave like waves while waves such as light also behave like particles. The wave-particle duality is behind most of the spiritual claims that are made for quantum mechanics.

Werner Heisenberg, with the help of Max Born and Pascual Jordan, put quantum mechanics on a mathematical basis in 1925 in a formulation using matrix algebra. However, most people who know something about quantum theory are more familiar with the alternative formulation using less advanced mathematics published by Erwin Schrödinger the following year. Schrödinger utilized the calculus of partial differential equations that is covered in the sophomore or junior years by physics, math, and chemistry majors. Thus students in these disciplines are able to apply Schrödinger's theory to the hydrogen atom and derive its energy levels, which remarkably turn out to be the same, to a first approximation, as those derived by Bohr with his far cruder and, as I pointed out, ultimately incompatible model.

Heisenberg and Schrödinger's theories were shown to be equivalent by Paul Dirac, who developed his own more elegant formulation of quantum mechanics employing linear vector algebra.[10] Most physicists today use Dirac's method. Once you can do the math (about junior level), it is amazing how simple quantum mechanics is and how few assumptions are needed to derive its full structure.[11] The so-called mysteries of quantum mechanics are in its philosophical interpretation, not in its mathematics.

Heisenberg is most famous for his *uncertainty principle*, published in 1927, which can be derived from standard quantum mechanics. Here is its simplest form: one cannot simultaneously measure both the momentum and position of a particle with unlimited precision. The product of the uncertainty (standard error) in momentum and the uncertainty in position must be greater than Planck's constant divided by 4π.

This result had the profound effect of demolishing the Newtonian world machine. Recall that Newton's laws imply that everything that happens is determined by prior events. In order to predict the motion of a particle you need to know its initial momentum and position and the forces acting on it. Ordinarily, on the macroscopic scale, this is no problem and is limited only by measuring precision. But on the submicroscopic scale it becomes serious.[12] For example, suppose you start out with an electron that has no forces on it but is confined to a volume of empty space equal to the volume of a hydrogen atom. It is a free electron, not bound in an atom. You cannot predict the position of that electron six seconds later within a volume equal to that of Earth.

To the delight of Christian theologians, this eliminates the Enlightenment deist god as a candidate for the creator. The deist god creates the universe and its laws but relies on determinism to make things come out the way he wants without intervening further. Some theologians think that quantum mechanics and the uncertainty principle open up a way for the Abrahamic God to act on the universe without performing miracles. But they have celebrated too soon. The uncertainty principle opens up the possibility of a different kind of deist god, one who "plays dice," using chance to let the universe develop its own way. More about this later.

Continuing with our historical review, in 1928 Dirac developed what we now call the Dirac equation, which describes the electron fully relativistically, that is, at all speeds up to near the speed of light. There were two remarkable outcomes of this work. First, Dirac proved that the electron has intrinsic angular momentum, or *spin*, equal to ½, a fact that had been previously inferred from experiment but did not follow from the nonrelativistic Schrödinger scheme. In quantum mechanics, a particle is either a *fermion* with half integer spin or a *boson* with integer or zero spin. The electron is a fermion. The photon is a boson with spin 1.[13] Today Dirac's equation is used to describe all spin ½ particles, such as muons, tauons, neutrinos, and quarks.

Second, Dirac showed that, in the scheme of things, the electron should be accompanied by an antielectron, a particle of the same mass and spin but with opposite charge. In 1932, Carl D. Anderson detected the antielectron in cosmic rays. He dubbed it the *positron.*

In 1948, Richard Feynman produced yet another formulation of quantum mechanics called *path integrals* that paved the way for the advances in fundamental physics that were to follow over the next thirty years. Feynman also showed that quantum physics was time-reversible, like classical physics. In Feynman's picture, the positron is indistinguishable from an electron going backward in time. This can be applied more generally for any particle and its corresponding antiparticle.

QUANTUM FIELDS

When in the early twentieth century light was recognized as being composed of photons, a theory of photons was needed that was consistent with Maxwell's equations of electromagnetism in the classical limit. Basically the electromagnetic field had to be "quantized." A quantization procedure had already been developed for going from classical equations to quantum mechanics. Essentially,

the same equations hold, with observables like momentum and energy represented by mathematical "operators" rather than by simple numbers.[14]

The free electromagnetic field—that is, a field in a region of space where
there are no charges and currents—is mathematically equivalent to an infinite
set of harmonic oscillators. The quantization of harmonic oscillators is standard stuff, so the electromagnetic field is easily quantized. That is, a mathematical function can be written down to describe the field and the photon can
be treated as an excitation, or "quantum," of the field.

In 1927, Dirac showed that the wave function for an electron in the Dirac
equation is also a quantum field with the electron being the quantum of that
field. The new quantum field theory was completely relativistic—that is, consistent with the special theory of relativity, which the older quantum theories of Schrödinger and Heisenberg were not. Einstein's theory of gravity, the
general theory of relativity, was not needed since gravity is negligible at the
subatomic scale, so quantum gravity was put off as a problem for future study.
It had no observable consequences with the technology of that time anyway.

Subsequent developments hit a snag, however, when attempts to produce
a theory of interacting electrons and photons using quantum field theory
led to calculations producing mathematical infinities in the results. As
World War II approached and scientists were needed for the war effort, the
problem was set aside. A solution to the infinities' problem was not needed
to design radar systems or build a nuclear bomb. Richard Feynman worked
at Los Alamos supervising a crew of human "computers" who used mechanical adding machines to solve numerical problems handed to them by the
scientists working on the bomb. Julian Schwinger worked on radar at MIT
(Massachusetts Institute of Technology).

After the war, the infinities' problem was solved by a procedure known as
renormalization, whereby one subtracts infinities from both sides of an equation. Dubious as this seems, it works. This led to the highly successful theory
of *quantum electrodynamics* (QED), independently developed by Feynman and
Schwinger in America and by Sin-Itiro Tomonaga in war-ravaged Japan. The
three methods were shown to be equivalent by Freeman Dyson, an English
physicist without a PhD working at Princeton. QED was able to quantitatively predict some tiny effects that could not be explained by previous the-

ories. The agreement was astounding. In one case, involving the electron's magnetism, the accuracy was one part in a trillion, agreeing with equally remarkable laboratory measurements.[15]

TOWARD THE STANDARD MODEL

I began my research career in physics in 1959 as a graduate research assistant at the University of California at Los Angeles (UCLA). At the time I entered a brand-new field called *high-energy physics*, also known as *elementary particle physics*, that had grown out of nuclear physics.

After Rutherford discovered the nucleus of the atom in 1911, and after the invention of the cyclotron by Ernest O. Lawrence in 1929 with particle accelerators, experiments with radioactivity probed nuclei with beams of protons or electrons of ever increasing energy. The experiments revealed that all nuclei were composed of protons and neutrons. The placement of each chemical element in the Periodic Table was determined by the number of protons in the nucleus of the corresponding atom. Protons are positively charged and so repel each other while neutrons carry no net electric charge. So they had to be held together by a previously unknown attractive force that was experienced by both protons and neutrons. Gravity is far too weak for such small masses. This new force was called the *strong nuclear force*.

Experiments also uncovered a fourth force called the *weak nuclear force* that accounted for the phenomenon of *beta-radiation*, in which a nucleus spontaneously emits an electron and a neutrino, thereby becoming a new nucleus with one more proton and one less neutron. The two nuclear forces are short range, operating only at distances smaller than the typical nucleus, in contrast to gravity and electromagnetism, which are of unlimited range. The energies radiated by our sun and other stars result from the weak force.

Intense efforts to apply quantum field theory to the forces between protons and neutrons using the same methods that proved so successful for quantum electrodynamics did not work. The infinities were intractable. People thought maybe quantum field theory was inadequate. For a few years around 1960, a radically new concept called *S-matrix theory* was explored that applied highly sophisticated mathematics to calculate the probabilities for particles to scatter from one another. Probability distributions were all that was being measured in accelerator experiments, so what else was there to explain? Even space and time were discarded, at least in the new model. The independent variables were momentum and energy, which are what is measured in particle collisions anyway. In this picture the particles themselves are not unique entities but are somehow composed of one another in what was called a "bootstrap." S-matrix theory had a holistic quality about it that, along with other unsupported claims about quantum mechanics, attracted comparisons with Eastern mysticism.

The most successful proponent of this philosophy in the popular mind was Fritjof Capra, who in 1975 wrote a bestseller titled *The Tao of Physics*.[16] Capra at the time was a young theoretical physicist from Austria who was working on S-matrix theory at the University of California at Berkeley, which was the center of that activity under the capable leadership of Geoffrey Chew. Although Capra's book can still be found in bookstores, by the time *The Tao of Physics* was published, S-matrix theory had been largely discarded for two reasons: (1) it never produced a useful prediction that could be tested empirically; and (2) it was supplanted by the newly developing standard model of particles and forces that restored the tried and true atomic scheme in which everything is reduced to elementary particles.

At first, however, S-matrix theory did not seem so far off the mark. In the 1960s, accelerators reached higher and higher energies and the thousand or so physicists around the world, including myself, who were working in this previously unexplored regime found that particle collisions produced a profusion of new particles that did not fit into any known model. These particles could not all be elementary, yet they were not made up of protons, neutrons, and electrons, which were thought to be the only elementary particles at the time. A breakthrough was made in 1964 when Cal Tech physicist Murray Gell-Mann suggested that most of the new particles were composed of more elementary constituents, which he called *quarks*.[17] Quarks were characterized by having less than unit electric charge, which had never been seen before in any particle.

The quark scheme proved to work beautifully, and by the early 1970s, experiments bombarding protons and neutrons with very high-energy electrons and neutrinos confirmed that the proton and neutron were also not elementary but exhibited a particulate substructure with fractional charge. These observations recalled Rutherford's discovery of the atomic nucleus in 1911.

The new picture led to a renewed attempt to use relativistic quantum field theory to understand the two nuclear forces in terms of the interactions of quarks and another set of particles that seemed to be elementary, the *leptons*. The electron, muon, and neutrino are examples of leptons. In the *standard model of particles and forces* that grew out of these developments, there are six types of quarks and their corresponding antiparticles along with six leptons and their antiparticles. They all are spin ½ fermions. The photon is also an elementary particle, joining a group of twelve spin 1 bosons called "force particles" because of their role in mediating the basic interactions.

The standard model combined the weak and electromagnetic forces into a single, unified force called the *electroweak* force. The strong force was described within the standard model by a model that was dubbed by Feynman as *quantum chromodynamics* (QCD).[18] These quantum field theories were all renormalizable, that is, their infinites could be subtracted out.

Further advances occurred swiftly, as they do in science when it is finally on the right track. When advances don't occur swiftly, science is probably on the wrong track. (String theory comes to mind). By the end of the 1970s, the standard model was found to be consistent with all observations. The only particle in the standard model that was not observed in experiments is the *Higgs boson*, named after the physicist Peter Higgs, who was one of several physicists who proposed its existence way back in 1964.[19] In the standard model the Higgs particle provides for one of the mechanisms by which elementary particles gain mass.

The standard model has remained unchallenged until the present day, when colliding beam experiments at the Large Hadron Collider (LHC) in Geneva are finally reaching the enormous energies that are needed to uncover the next layer of matter. As this manuscript was going to press, the preliminary results from the LHC presented at a conference in India were reported in online media. The Higgs boson has not been seen as expected, although there still are some hopeful signs. Far from being a disappointment, if the Higgs is not confirmed, physicists will be excited that at least, after a generation, we may be beginning to learn something new about the nature of matter.

For our purposes here, we need not go into further detail, especially on something that will be changing rapidly in the next few years. The point is this: Despite the claims of theologians and quantum spiritualists, the standard model provided yet another triumph for reductionism, and further developments are not likely to change this in the near future. All the visible matter of the universe, from everything we see around us to the most distant galaxies observed by our most powerful telescopes, to the tiny life forms we view with our most powerful microscopes, are composed of just three fundamental particles: two quarks whimsically called "up" and "down" and the electron. Add in the photon and we pretty much cover everything of interest to most people. Whether or not the Higgs particle exists is not going to affect many lives in the foreseeable future.

In the next chapter we will explore the cosmos and see that the matter in stars and galaxies that gives off light constitute a mere 0.5 percent of the total mass of our universe. Another 3.5 percent is nonluminous but is still made of the same ingredients—quarks and electrons. Of the remaining mass, 23 percent is *dark matter* and, by far the dominant component, 73 percent is *dark energy*. These components, which are clearly not composed of known quarks and leptons, are "dark" because they give off no light or anything else we have been able to directly observe with current technology, although many experiments are in progress seeking that end. While not observed directly, the existence of dark matter and dark energy is inferred from their gravitational effects. Since gravitation is one of the defining properties of matter, the dark matter and dark energy still fit into our materialist model. They do not exhibit any properties at this time that require us to revise our worldview to include nonmaterial ingredients.

QUANTUM SPIRITUALITY

Now let us bring in the claims of those who think they see in quantum mechanics evidence for a world beyond matter. Quantum mechanics has provided both theistic and nontheistic spiritualists with a way to imagine a holistic cosmic consciousness that includes the human mind. For example, theologian

Philip Clayton asserts that these two complementary ways of describing a phenomenon are incompatible and "depend on the interests of the observer and the experiment she designs."[20] That is, reality depends on human thought.

Seeking a scientific basis for a world that cannot be reduced to particles, quantum spiritualists argue that human consciousness can affect the outcome of events. This comes about from the wave-particle duality, discussed previously, where it is asserted that whether an entity is a wave or a particle depends on what you decide to measure. If you decide to measure a particle property such as position, then the entity is a particle. If instead you decide to measure a wave property such as wavelength, then the entity must be a wave. The decision about what to measure could occur long after the entity has left its source, in which case, it is claimed, your act of consciousness decided the nature of the entity. If the entity was a light from a galaxy 10 billion light-years away, your conscious act reached out 10 billion light-years in space and 10 billion years back in time, since the light took 10 billion years to reach Earth. Yes, that's what the quantum spiritualists really claim, although they seldom state it so starkly else their listeners call in the straightjacket squad.

Note that even if the act of measurement could control reality, human consciousness need not be involved. That measurement could be performed automatically, with a computer deciding randomly what to measure. The only way the quantum spiritualists can work their way out of that is to assume computers are conscious too.

The behavior of the quantum wave function has also encouraged the mystically inclined to see a holistic universe with the human mind (and, I guess, computers too) as part of this grand cosmic consciousness. When a measurement is made on a particle, its wave function is said to "collapse" instantaneously all over the universe to a new function that represents the particle's newly measured position. It is asserted that consciousness causes wave function collapse. Clayton tells us: "The phenomenon known as 'collapse of the wave function' suggests that the observer plays some constitutive role in *making the physical world become what we perceive it to be at the microphysical level*."[21]

This reasoning has led gurus of quantum spirituality, such as bestselling authors Deepak Chopra[22] and Rhonda Byrne[23] to conclude that we can change reality—past, present, and future—just by thinking about it. In chapter 9 we will discuss the recently published claim by psychologist Daryl Bem that he has empirical evidence for the human mind affecting the past. As with all such reports, and there have been many over the past century and a half, this contention has not stood up under critical scrutiny and was not replicated by an experiment that followed the same procedure.

DIVINE ACTIONS WITHOUT MIRACLES

Basic to all three Abrahamic religions is the belief that God is the controlling agent in the universe who acts as needed to make sure things work out the way he wants.[24] This was unnecessary for the Enlightenment deist god who created the Newtonian world machine. Why would a perfect God need to step in after he created a universe in which everything is already divinely predetermined?

To some theologians, the uncertainty principle, which eliminated the Newtonian world machine, suggests a way for God to act in the universe without the need to perform visible miracles.[25] No miracle has ever been confirmed by science. That is, no observation has been made that cannot at least be plausibly and more simply explained by known natural means. To be consistent with that fact, rather than simply denying it, theologians (to their credit) have attempted to show how God can act in such a way as to avoid his miraculous actions being detected.

Obviously God could have simply made his miracles observable to science. However, according to one argument, he needs to be devious, or mysterious, so that we accept him on faith. Why he wants to do that escapes me. As has been noted, a perfect God doesn't need to create a universe with troublesome human beings since he is already perfectly content. And even if he did, why would he want to spend eternity surrounded by clueless toadies? If he sought good conversation, you would think God would prefer the company of Bertrand Russell to Pat Robertson.

So, how does the uncertainty principle help? Macroscopic and even most microscopic objects have tiny uncertainties and behave deterministically to a high degree of precision even given the uncertainty principle. So to act without performing miracles, God must direct the motions of submicroscopic particles within the boundaries allowed by the uncertainty principle in such a way that his action is undetectable to humans. Now, changing the motion of one atom isn't likely to produce an important change on the everyday scale. God would have to really micromanage, simultaneously affecting the motions of a trillion trillion atoms at a time. Imagine how busy God would have to be, controlling the movements of 10^{79} atoms in the visible universe and the countless number of atoms beyond our light horizon and possibly an unlimited number of other universes. No wonder he doesn't have time to listen to prayers.

Physicist and Anglican priest John Polkinghorne and others have suggested a way to amplify the influence of a few atoms manyfold using the *butterfly effect of chaos theory*.[26] Under certain conditions, a tiny change in the initial conditions of an otherwise deterministic macroscopic system can produce widely different results. In the usual example that is given, it is as if a flap of a butterfly's wings can affect the weather next week. Such a system is called *chaotic* because of its unpredictability.[27]

Many authors misinterpret this result as implying that indeterminism extends to Newtonian physics. This is not quite right. The path of the chaotic system is still fully determined by the initial conditions and by the laws of classical mechanics. Chaos theory is properly called *deterministic chaos.* Unpredictability arises in practice because of the inability to measure the initial conditions with sufficient accuracy. If the accuracy exceeds the limit of the quantum uncertainty principle, then we have indeterminacy. But that indeterminacy is quantum in origin, not Newtonian.

THE END OF DUALITY

Now let us see why quantum spirituality fails. Along with Christian apologists such as Grassie, McMullin, and Clayton, quantum spiritualists Chopra and Byrne have misinterpreted the wave-particle duality. What quantum physicists discovered was that every physical entity has both particlelike and wavelike properties. In fact, *empirically they are all particles* while their so-called wavelike behavior does not exist for an individual particle but appears only as a property of a large ensemble of particles. Let me explain.

Recall from chapter 5 that in 1803 Thomas Young demonstrated the wave nature of light by observing the interference of beams of light. Today such experiments can be performed with far greater precision than was available to Young two centuries ago. In particular, we can use detectors that are sensitive to single photons. When the double slit experiment is performed with an array of such detectors one photon at a time, localized individual particle hits are registered. No wavelike interference pattern is seen until a large number of photons are accumulated. Then the pattern emerges as the statistical distribution of photon detections. But each individual photon itself does not behave like a wave. It behaves like a localized, nonholistic particle.

When we talk about the "wavelength of a photon," we are not referring to a property of an individual photon but to a characteristic of the mathematical function that describes a statistical ensemble of identical photons. The same experiment can be done with electrons or any other particle. The electron, photon, and all other submicroscopic objects are localizable particles and their wavelike effects refer only to the statistical behavior of a large number of them.

It does not matter whether you are trying to measure a particle property or a wave property. *You always measure particles.* Here is the point that most people fail to understand: Quantum mechanics is just a statistical theory like statistical mechanics, fundamentally reducible to particulate behavior. In statistical mechanics you use the average behavior of particles, following the laws of mechanics, to calculate collective quantities such as pressure and density fields. In quantum mechanics you use the average behavior of particles, following the laws of mechanics, to calculate collective quantities such as the wave function field. Neither theory specifies the motion of any given particle, only the statistical behavior of the ensemble. The wave function often (but not always) looks like a wave, hence its name. It is not the vibration of any medium, like a sound or water wave. The formulation of quantum theory developed by Schrödinger was originally called "wave mechanics," which, as mentioned, is the most familiar but also the least advanced.

However, there is an important difference between statistical and quantum mechanics. While we know the basic laws of particles behavior underlying statistical mechanics, we have no idea what those laws are for quantum mechanics, if any. We will talk more about this issue later.

In a few words, no incompatibility exists between the particle and wave picture. They are simply two different ways to describe the same phenomenon, namely, a beam of particles. A single particle is always a particle, never a wave. Physics teachers and authors use sloppy, incorrect language when they say, "An electron is either a particle or a wave." An electron is always a particle while an ensemble of electrons is treated as a wave.

Let me use another analogy that will be familiar to many. A communications engineer will sometimes describe a signal as a series of pulses localized in time, as if it were a beam of particles, or as a spectrum of frequencies, as if it were a combination of sine waves. The engineer's toolbox contains a mathematical device called the *Fourier transform*, invented in the nineteenth century by the brilliant French mathematician Jean Baptiste Joseph Fourier (died 1830), which enables the engineer to go back and forth between the two representations. The quantum uncertainty principle can be mathematically derived directly using the Fourier transform.[28]

Wave function collapse, described previously, seems to violate Einstein's rule that no signal can move faster than the speed of light. When the wave function collapses, it does so instantaneously throughout the universe. Einstein never believed it and called the whole idea a "spooky action at a distance."[29] But that's what quantum spiritualists are seeking—spooks.

QUANTUM REALITY

Philosophers of science have attempted to give names to the various points of view found in science about what science has to say about reality. Most physicists would probably agree with the following doctrines:

Commonsense realism

A reality exists that is independent of what people think about it.[30]

Scientific realism

The aim of science is to give us an accurate description of what reality is like, including aspects of reality that are unobservable.[31]

In addition, most physicists would hold to some form of:

Empiricism

Observation is the only source of knowledge about the world. This is not meant to include forms of abstract knowledge such as mathematics or linguistics, which do not exist (as far as I know) independently outside our heads.

Many experimental physicists also hold to a view opposed to scientific realism:

Instrumentalism

Scientific theories should be seen as instruments used to predict observations, rather than as an attempt to describe the real but hidden structures of the world that are responsible for the patterns found in observations.[32]

This is known in the field as "Shut up and calculate."

On the other hand, most theoretical physicists and mathematicians seem to follow:

Platonism

The mathematical functions and laws of the theories of physics are the "true reality" while our observations are shadows on the wall of Plato's cave.[33]

Except for the New Age gurus who claim we can "make our own reality" just by thinking we can, few rational people doubt that the objects we see with our naked eyes are real and independent of our thoughts. But what about those objects not visible to our naked eyes, such as electrons, atoms, molecules, and bacteria, which form much of the substance of physics, chemistry, and biology? These require the instruments of science to detect.

Now, few will argue that the bacteria a biologist sees with her microscope are not real, in the same way that few would deny the reality of a planet or star seen by an astronomer with his telescope. Many telescopes and microscopes today do not simply magnify images for the human eye but utilize more sensitive light detectors that send signals directly to a computer where the data can be stored and analyzed with greater precision and quantity.

And visible light is not the only means by which a scientist can observe an object. Astronomers utilize the whole electromagnetic spectrum, from radio waves to gamma rays. Furthermore, other particles besides photons are used. In 1987, a supernova in the Magellanic Cloud just outside our galaxy was detected in neutrinos. In the late 1990s, I worked on an experiment in Japan that "saw" the sun in neutrinos, right through the Earth in the middle of the night. Similarly, on a much smaller scale, microbiologists and chemists can use the electron microscope to see individual molecules.

What about atoms? Although they were first postulated to exist in ancient Greece, indirect evidence for their existence first began to accumulate in the nineteenth century. Nevertheless, the great physicist and philosopher of that period, Ernst Mach (died 1916), refused to believe in atoms because he couldn't "see" them. Even today I read in books that "atoms are invisible." This is wrong. If you look at the frontispiece in my 1990 book *Physics and Psychics*, you will see a picture of an array of chromium atoms taken with a *scanning tunneling microscope*.[34]

So, where do we draw the line? Surely what we see wearing glasses is real, and glasses are scientific instruments. How are fancier instruments, such as the scanning tunneling microscope, fundamentally different? As we get below the scale of atoms to elementary particles such as electrons, the detection process becomes more and more technical. During the early years of my research in elementary particle physics I worked with a now-obsolete device called the *bubble chamber*, in which charged particles left lovely trails of bubbles in a superheated liquid that we photographed and identified. I saw thousands of bubble tracks that tightly spiraled in the magnetic field inside the chamber and were easily recognized as produced by electrons. If they spiraled in the opposite direction, they were equally clearly identified as the tracks of antielectrons, or *positrons*. Were these electrons and positrons real? I don't see how you can say they were not real while insisting that the moon and that book in front of you are real.

The electron is one of the fundamental particles in the standard model of particles and forces. What about the other particles in the model? Few are observed as unambiguously as the electron and a number of similar particles such as the muon and tauon. Since they are uncharged, neutral particles such as photons and neutrinos do not leave tracks in our detectors, but they can be inferred fairly unambiguously from other tracks and the application of some well-established principles such as conservation of energy, momentum, and charge. In any case, the elementary particles were introduced into the standard model in order to explain empirical anomalies of one sort or another that could not be explained otherwise.

The more indirect the evidence, the more difficult it is to do decide whether or not a hypothesized particle is real. We can never know if some future theory will do away altogether with these currently favored objects, so we cannot prove they exist with complete certainty. However, in the case of the electrons, photons, neutrinos, and perhaps a few other key particles, I think we can reasonably presume that they will remain as ingredients in any future theory, and we can take them to be real "beyond a reasonable doubt."

If you are wondering about quarks, they have never been observed, as electrons and muons are, by the tracks they leave in particle detectors. In fact, the current standard model predicts that quarks never will be seen as free particles. Quarks are bound together in nuclei by a force that grows stronger as you try to separate them. It's like trying to separate two oppositely charged ions in water. They discharge as electrons flow from one to the other through the water.

So the reality of quarks is still somewhat open. But, the longer they remain part of any successful theory, including whatever eventually replaces the standard model, the greater the confidence we will have in treating quarks as real.

A more controversial issue is the status of the reality of some of the other ingredients of physics theories (only physics, and perhaps astronomy, has this problem) that many prominent theoretical physicists would like to consider "more real" than the particles of the standard model. These are the quantum fields that are the basic mathematical objects in the theory, while the particles are treated as the excitations (quanta) of these fields. These physicists, and I believe they are many, are "Platonic realists" who follow Plato and his doctrine of *forms* described in chapter 2. They consider the idealizations of our theories the true reality and the particles we actually observe the distorted shadows on the wall of Plato's cave.

Note, however, that quantum fields are not the "fields of energy" that you will often see referred to in popular literature as the reality with which quantum mechanics has replaced "material reality." Fields of energy are fields of matter because energy and matter are equivalent.

Applying Platonic metaphysics to quantum mechanics, the wave function, a type of quantum field, is "real" and so its simultaneous "collapse" throughout the universe grossly violates relativity. However, the much more rational and parsimonious position is that the wave function is simply a human-invented mathematical object that can do anything its inventors want it to do, so long as any calculations made using it agree with the data. We have never observed a wave function or quantum field, or even a classical electromagnetic field. All that our detectors ever register are localized hits that look very much like particles. Furthermore, the empirical fact remains that no information or object has ever been observed to travel faster than the speed of light.[35]

IS QUANTUM MECHANICS COMPLETE?

All his life Einstein argued that quantum mechanics had to be "incomplete." After a long series of debates with Bohr in which Einstein tried and failed to refute the uncertainty principle because of Bohr's brilliant responses, Einstein eventually accepted quantum mechanics as a "statistical theory." But he objected to the widely accepted philosophical interpretation of the theory, called the *Copenhagen interpretation*, championed by Bohr, in which properties have no meaning until they are measured.

In 1935, having just settled in at Princeton after leaving Europe, Einstein and two young colleagues, Boris Podolsky and Nathan Rosen, wrote a classic paper arguing that for a theory to be complete, every element in that theory must have a counterpart in physical reality.[36]

In conventional quantum mechanics, the momentum and position of a particle cannot be simultaneously measured. Yet it is generally assumed that a particle still possesses both properties. Note that because of the de Broglie relation, the momentum of a particle is inversely proportional to the wavelength of the corresponding wave. We have already seen how a beam of particles collectively has wave properties, which are associated with its wave function. Thus the beam intrinsically can have a well-defined momentum while each particle has a well-defined position.

In 1952, David Bohm proposed an alternate interpretation of quantum mechanics, based on an earlier idea of de Broglie's, in which the motions of particles are controlled by unobserved subquantum forces, or so-called *hidden variables*—or, as de Broglie called them, *pilot waves*.[37] The theory was deterministic in principle but probabilistic in practice, giving the same empirical results as the conventional theory. That appeared to make it untestable.

However, in 1964, physicist John Bell proved a remarkable theorem that showed how to empirically distinguish between conventional quantum mechanics and any theory of local hidden variables. Recall from chapter 5 that "local" means limited to information transfer at the speed of light or less. In 1982, the proposed experiment, usually referred to as the "EPR experiment," was performed in France by a team led by Alain Aspect.[38] The result agreed perfectly with conventional quantum mechanics and ruled out local hidden variables forever. Subsequent independent experiments have confirmed this result.

Bohm and his supporters were undaunted. They simply said that the Bohmian theory was nonlocal; that is, it allowed influences to move faster than the speed of light.[39] Basically, Bohm's subquantum force was described mathematically as being produced by a *quantum potential*, which was the combined effect of all the particles in the universe. The observed quantum behavior was the result of the simultaneous, superluminal (actually, infinite speed) interaction of the particle in question with the rest of the universe.

In the case of the EPR experiment, the results are usually interpreted to imply that a measurement performed on a particle at one point in space affects the results of a measurement at a distant point, even though any signal between the two would have to travel faster than the speed of light.

Here again we find the emergence of a holistic universe. Needless to say, the theists and spiritualists were again delighted, and Bohm himself took up spirituality and began writing on the "wholeness" of reality.[40] Quoting theologian Philip Clayton, "The [Aspect] experiments force upon us a view that lies well outside any commonsense conception of matter."[41]

Still, most physicists were not carried away with these results. After all, the Aspect experiment was just another one among thousands confirming the predictions of standard quantum mechanics. Furthermore, the experiment did not in fact demonstrate a superluminal connection between two particles. Nowhere has it ever been shown that a measurement of one particle specifically affected the measurement of its partner from the same source. Experiments always produce statistical distributions of the measurements of many particle pairs and compare them with statistical predictions of the theory of local hidden variables as opposed to quantum mechanics. We saw previously that in the double slit experiment, the wave picture does not apply to individual particles but to ensembles; the same is true here. Once again we find that quantum mechanics is a theory that calculates probabilities and not the behavior of individual particles.

While Bohm's model still has its supporters, the fact that it implies superluminal transfers of information violates relativity. This would be acceptable if superluminality were observed, but it never has been (reports of superluminal neutrinos are problematical). In fact, it can be proved that superluminal information transfer is forbidden in any theory consistent with the axioms of relativistic quantum field theory, which is the basis of all the achievements in theoretical particle physics since before World War II.[42] The "nonlocality" of quantum mechanics that is always bandied about by authors is in the minds of these authors. Until superluminality is observed, physicists should stick to the simplest model that is consistent with all existing knowledge.[43] The wave function is nonlocal, but it is also an abstract mathematical object like the price of a car. You can't drive a number.

And what about Einstein's objections to the conventional interpretations of quantum mechanics? Interpretations is pluralized here because there are more than one that still are conventional in the sense that they statistically predict what we observe, and as long as they do that successfully, it's the best we can do. Einstein was a scientific realist who believed that the ingredients of our models had to have "counterparts" in reality. But, we have no right and indeed no need to make that assumption. As has been mentioned several times already, and will be elaborated upon further in chapter 7, the models we use in physics are invented by physicists for the purpose of describing and predicting observations. They have to agree with the data, so they must have some connection to reality, but the true connection need not exist in one-to-one correspondence; it remains behind a veil that empirical science alone cannot penetrate. Models are artifacts. If we view quantum mechanics as an artifact, then as long as it successfully describes the data, it is complete. Why does every element in that theory have to have a counterpart in physical reality, as long as the theory does the job it is supposed to do?

As we saw in the previous section, we can more reasonably assign reality to localized particles than quantum fields or wave functions. Even when we are measuring a "wavelength" we are detecting particles, not waves. Furthermore, particles do not travel faster than light. And even if they did, this would not imply a holisitic reality. We can never determine with complete certainty whether particles or fields, or neither, are the ultimate reality, and so we should not be making grand metaphysical claims about a theory that is simply designed to describe observations and indeed does so very well.

CHAPTER 7

COSMOS AND CREATOR

Some foolish men declare that Creator made the world. The doctrine that the world was created is ill-advised, and should be rejected. If god created the world, where was he before creation? If you say he was transcendent then, and needed no support, where is he now? No single being had the skill to make the world—for how can an immaterial god create that which is material? How could god have made the world without any raw material? If you say he made this first, and then the world, you are face with an endless regression. If you declare that the raw material arose naturally you fall into another fallacy, for the whole universe might thus have been its own creator, and have risen equally naturally. If god created the world by an act of will, without any raw material, then it is just his will made nothing else and who will believe this silly stuff? If he is ever perfect, and complete, how could the will to create have arisen in him? If, on the other hand, he is not perfect, he could no more create the universe than a potter could. If he is formless, actionless, and all-embracing, how could he have created the world? Such a soul, devoid of all modality, would have no desire to create anything. If you say that he created to no purpose, because it was his nature to do so then god is pointless. If he created in some kind of sport, it was the sport of a foolish child, leading to trouble. If he created out of love for living things and need of them he made the world; why did he not make creation wholly blissful, free from misfortune? Thus the doctrine that the world was created by god makes no sense at all.

—Jinasena (ninth-century Jain master)[1]

CREATION MYTHS

Science and religion are not necessarily incompatible because the view of the cosmos we get from science differs so dramatically from that presented in sacred scriptures and cultural traditions. Only fundamentalists take those stories literally and that forces them to conclude that science is simply wrong. For them, science and religion are incompatible and no one has to write a book to prove it.

However, the majority of believers today recognize that these creation stories are myths. No doubt they were meant to convey some message, but whatever it is, it has nothing to do with science. Science does not conflict with religious myths any more than it does with *Harry Potter.* Although sometimes based on actual events and personages, myths are basically fictions.

We have already seen how theologians and lay believers have tried to come to grips with Darwinian evolution. Basically they say it is God-guided, but that is contrary in principle to the Darwinian model and is just an unacknowledged form of intelligent design. Furthermore, evolution implies humanity is an accident—in total disagreement with the universal religious belief that we are special. Many will disagree, but these are two places where science and religion are fundamentally incompatible.

In the last two chapters we discussed some of the physics issues that enter into the religion-science dialogue. We saw that considerable misunderstanding exists about the metaphysical implications, if any, of quantum mechanics. In later chapters we will evaluate the inferences people have made about a place for quantum consciousness, an end to reductionism, and the emergence of higher-level principles in complex systems that supposedly point toward a universe of purpose. We will see that the empirical facts simply do not support these pious hopes.

In this chapter let us continue the physics discussion by moving from the subatomic world to the cosmos.[2] Just as people have always found the argument from design convincing based on their observations of events around them, they have turned their attention to the heavens and asked, "How could all of this have happened by chance?"

In his 2006 bestseller, *The Language of God*, Francis Collins, who, as mentioned earlier, is director of the National Institutes of Health, former head of the Human Genome Project, and an evangelical Christian, gives the typical theist view: "I cannot see how nature could have created itself. Only a supernatural force that is outside of space and time could have done that."[3] Strangely he seems to be able to see how God created himself. The usual theological response to this question is that God always existed and so was not created. But then, why couldn't the universe itself have always existed and thus not have been created?

We saw that William Paley's natural theology was once a good scientific argument for a creator. Until Darwin came along, science had no explanation for the complexity of life. Now it does. Similarly, the argument for a cosmic creation was also quite reasonable at one time, a miracle seeming to be necessary to explain creation. However, it is still a God-of-the-gaps argument, and now science has filled all the known gaps with plausible scenarios for an uncreated universe. Let us run through the theological arguments first, and then we'll see why they fail.

BIG BANG THEOLOGY

As briefly mentioned in chapter 5, nineteenth-century physics and cosmology provided several good scientific reasons to argue that something miraculous had to happen for the universe to come into existence. Let us now discuss these in more detail, starting with the principle of *conservation of mass*.

Conservation of Mass

Until the twentieth century, all observations indicated that the total mass of a system could not change unless some mass was either inserted from outside the system or removed to the outside. Measurements with chemical reactions, for example, seemed to bear out this principle. The character of the individual bodies carrying mass could change, as in chemical reactions, but the total mass remained the same. The universe obviously has mass, so where did it come from? In the theological view, it came from nothing, *creation ex nihilo*, by the miraculous mass-creating act of God.

Conservation of Energy and the First Law of Thermodynamics

Similarly, the universe contains energy. The law of conservation of energy says that the total energy in an isolated system is conserved. Like mass, energy can change from one form to another as long as the total stays constant. For example, a falling body loses potential energy but it gains kinetic energy in an equal amount as it falls.

As we have seen, the first law of thermodynamics is a generalized form of the law of conservation of energy that distinguishes heat from other forms of energy and work. The origin of the universe seems to imply a miraculous violation of the first law.

The Second Law of Thermodynamics

The other great principle of thermodynamics, the second law, also once provided strong support for a created universe. Recall that the second law says the entropy, or disorder, of an isolated system must either stay the same or increase with time, that is, become more disordered. The universe is presumably an isolated system and now contains order, so it had to be at least as orderly or more orderly in the past. So it seems to follow that some intelligence must have done the ordering.

The Big Bang

Perhaps the most important cosmological claim of virtually every religion is the creation itself. For there to have been a creator, there had to have been a creation. And that means that the universe must have had a beginning in time. The big bang is claimed as evidence for such a beginning.

The idea behind the big bang was first proposed in 1927 by astronomer and Belgian Catholic priest Georges-Henri Lemaître, although he did not use the term *big bang*. He showed that an expanding universe was perfectly consistent with Einstein's general relativity.[4]

Einstein did not approve, however, reportedly telling Lemaître, "Your math is correct, but your physics is abominable."[5] Einstein still held the traditional belief that the universe is a static "firmament," as implied in the Bible and most other scriptures that present creation myths. But "static" here is not meant to imply that objects are all at rest. They are moving about, but their average distance apart stays the same.

Einstein had inserted into his gravitational equation in general relativity a factor called the *cosmological constant* that provided a repulsive force to counteract the gravitational attraction that otherwise should make the universe collapse. Although the cosmological constant is often referred to as a "fudge factor," this is a misnomer. Such a constant is required in Einstein's equation, although no value is given. If positive, it produces a gravitational repulsion. If negative, we have an additional attraction.

Lemaître's theory was not well recognized until 1930 when the eminent English astrophysicist Arthur Eddington wrote an article referring to Lemaître's expanding universe as a brilliant solution to outstanding problems in cosmology.[6]

In the early 1920s, astronomer Edwin Hubble, working at the Mount Wilson Observatory in California, discovered that many of the diffuse objects in the sky called *nebulae* were in fact distant galaxies. The universe extended well beyond our home galaxy, the Milky Way. Later in the decade Hubble and his assistant, Milton L. Humason,[7] estimated the distances to galaxies using a technique invented by Henrietta Swan. This they combined with measurements of the redshifts of the spectral lines from stars in the galaxies that had been measured by Vesto Sipher.

In chapter 6 we saw how the light emitted from a high-temperature gas is characterized by "spectral lines" of well-defined frequencies. Different gases have different spectra. These can appear two ways, either as bright *emission* lines in a darker background or as dark *absorption lines* in a light background. In the latter case the atoms in the gas selectively absorb light passing through it at well-defined frequencies. By observing the spectra of light from stars, astronomers are able to decipher the composition of the surface of the star. The element helium was observed this way in the sun before it was discovered on Earth, hence the name, derived from *helios*, Greek for "sun."

Hubble and Humason showed that compared to laboratory measurements, the pattern of spectral lines of galactic gases was most often shifted to lower frequencies. This is called a *redshift* since the lowest-frequency visible light is red in color. A few had "blueshifts" to higher frequency, notably the only galaxy visible as a diffuse object to the naked eye, Andromeda. Hubble and Humason found that the amount of redshift from a galaxy was roughly proportional to its distance from us, although there was a lot of scatter in the data points. The proportionality factor became known as the *Hubble constant.* However, two recent books have pointed out that Lemaître published an estimate of this factor two years earlier based on the same data.[8]

Lemaître provided an explanation of the observations, consistent with Einstein's equation: the universe is expanding, so as time goes by almost all galaxies are receding from us.[9] The observed redshift is the Doppler effect that results from these galaxies' recessional speeds.[10] Hubble's data, along with other published results, showed that the galaxies were moving away from one another as if from a giant explosion, where those galaxies with higher speeds have moved the farthest apart. This became known as the big bang, a derisive term introduced by astronomer Fred Hoyle who favored a steady-state universe.

When Einstein realized that the cosmological constant was not needed for agreement with observations, he called it his "biggest blunder." For many years the cosmological constant was assumed to be zero. However, no theoretical reason has yet been found for equating it to zero, and, as we will see, it may not be.[11]

Referring to the big bang, in 1951 Pope Pius XII told the Pontifical Academy, "Creation took place in time, therefore there is a Creator, therefore God exists."[12] Lemaître wisely advised the pope not make this statement "infallible." Theists make much of the fact that Lemaître was a priest and that his belief in a creation may have given him the idea of the big bang. Perhaps it did, but he was a good scientist and had excellent scientific reasons for proposing the big bang.

Believers were overjoyed when the big bang theory was confirmed in a whole series of astronomical observations, starting with the serendipitous discovery of the cosmic background radiation by radio astronomers Arno Penzias and Robert Wilson in 1964.[13]

Of course, the big bang theory looks nothing like the story of creation found in the Bible or in any other religious scripture. But, as promised, biblical errors and contradictions will not be used to argue the incompatibility of science so long as they are not taken literally. Still, theologians continue to use the big bang to argue that the universe had a beginning. Robert Jastrow, head of NASA's Goddard Institute for Space Studies, called this the most powerful evidence for the existence of God ever to come out of science.

> For the scientist who has lived by faith in the power of reason, the story ends like a bad dream. He has scaled the mountains of ignorance; he is about to conquer the highest peak; as he pulls himself over the final rock, he is greeted by a band of theologians who have been sitting there for centuries.[14]

However, it does not follow that a beginning requires a divine creator.

The Singular Universe

For many years, the Christian philosopher, apologist, and master debater William Lane Craig has argued that a theorem published in 1970 by famed Cambridge cosmologist Stephen Hawking and eminent Oxford mathematician Roger Penrose provides evidence that the universe had a beginning.[15] Using Einstein's general theory of relativity, Hawking and Penrose proved that the universe began with a *singularity*—an infinitesimal point in space of infinite density. Theologians argue that time itself had to begin at that point and so must have the universe.

More recently, Craig and US Navy warfare analyst James Sinclair have added two other arguments that the universe must have had a beginning:

1. The universe cannot be eternal because that implies a beginning an infinite time ago, in which case we would never reach the present.
2. New arguments from general relativity prove that the universe had an "absolute beginning."[16]

Craig quotes the great mathematician David Hilbert as saying: "The infinite is nowhere to be found in reality. It neither exists in nature nor provides a legitimate basis for rational thought. The role that remains for the infinite to play is solely that of an idea."[17] Here, on this specific point but not on his final conclusion, I find myself agreeing with Craig. While it is true that the word "infinity" appears often in scientific literature, it should be understood as meaning "very big" or "unlimited." When infinity appears in an equation, it is a mathematical abstraction, not reality.

Craig and Sinclair propose the syllogism:

1. An actual infinite cannot exist.
2. An infinite temporal regress of events is an actual infinite.
3. Therefore, an infinite temporal regress of events cannot exist.[18]

I hope you noticed the inconsistency here. On the one hand, Craig claims the universe started as a singularity of *infinite* density, and then he turns around and says that nothing *infinite* can occur in reality. (What about God?)

An eternal universe, one with no beginning or end, is anathema to Judeo-Christian-Islamic belief. If there was no beginning, there was no creator. In 1600, the monk Giordano Bruno was burned at the stake for proposing that the universe was infinite, among other heresies. As historian John Hedley Brooke points out, Bruno's teaching deprived humanity of a privileged place in the cosmos.[19]

The second argument for a beginning made by Craig and Sinclair is based on a theorem proved from general relativity and published in 2003 by mathematician Arvind Borde, physicist Alan Guth, and cosmologist Alexander Vilenkin.[20]

While Craig has been the primary theological source for these cosmological arguments for God, they can be found in virtually every popular Christian book that mentions cosmology. For example, Christian apologist and evangelist Ravi Zacharias stated in a 2008 book:

Big Bang cosmology, along with Einstein's theory of relativity, implies that there is indeed an "in the beginning." All the data indicates a universe that is exploding from a point of infinite density.[21]

In his 2007 book, *What's So Great About Christianity?* Dinesh D'Souza writes,[22]

In a stunning confirmation of the book of Genesis, modern scientists have discovered that the universe was created in a primordial explosion of energy and light. Not only did the universe have a beginning *in* space and time, but the origin of the universe was also a beginning *for* space and time.

The Argument from Fine-Tuning

The final argument that is widely presented as evidence for a creator is called the *argument from fine-tuning*. This is the claim that the constants of physics are such that any slight change in their values would cause life, as we know it, not to exist.

In 1987, theoretical physicist Tony Rothman wrote:

The medieval theologian who gazed at the night sky through the eyes of Aristotle and saw angels moving the spheres in harmony has become the modern cosmologist who gazes at the same sky through the eyes of Einstein and sees the hand of God not in angels but in the constants of nature. . . . When confronted with the order and beauty of the universe and the strange coincidences of nature, it's very tempting to take the leap of faith from science to religion. I am sure many physicists want to. I only wish they would admit it.[23]

Francis Collins writes:

The precise tuning of all the physical constants and physical laws to make intelligent life possible is not an accident, but reflects the action of the one who created the universe in the first place.[24]

The fine-tuning argument is an outgrowth of a long history of physicists puzzling over reasons for the values of fundamental physical constants.[25] This led physicist Brandon Carter to propose what he unfortunately dubbed the *anthropic principle*. At a 1973 symposium in Kraków, Poland, Carter made the observation, "Although our situation is not necessarily central, it is inevitably privileged to some extent."[26]

The Kraków symposium was in honor of the 500th birthday of Nicolaus

Copernicus. Carter was disputing the *Copernican principle*, which states that neither humanity nor anything else occupies a special place in the cosmos. The Copernican principle had grown out of Copernicus's insight that Earth revolved around the sun. As we have seen, the fact that Earth was not the center of the universe was a tremendous blow to the theology of the time—a grave threat to the notion that humanity is special. Now Carter was suggesting we are privileged in some way after all.

In 1986, astronomer John Barrow and physicist Frank Tipler published a highly technical tome called *The Anthropic Cosmological Principle*, which presents a detailed physics discussion, with equations, of what are called the *anthropic coincidences*.[27] Several other excellent popular books survey the subject without equations.[28] Although recognizing the theological, and teleological, implications, Barrow and Tipler were careful not to conclude anything about the source of the coincidences. Such a claim would be provided by theologians and Christian apologists.[29] A comprehensive list of the primary claims of divine fine-tuning can be found in the article by physicist and Christian apologist Hugh Ross.[30]

There can be no doubt that life, as we know it, would be different if the constants of physics were only slightly different. However, this does not rule out other forms of life that might have developed had the universe had a different set of constants. Referring to Ross, biologist Rich Deem lists five crucial parameters without the fine-tuning of which he claims no form of life of any kind would be possible.[31]

Five Crucial Parameters

1. The ratio of the electromagnetic force to gravity

This ratio is 10^{39} for the forces between an electron and a proton. If the strengths of these forces had differed by just a few orders of magnitude, the universe would have collapsed long before stars, galaxies, and life of any sort had a chance to form.

2. The ratio of the numbers of protons to electrons in the universe

If this number had been slightly larger, electromagnetism would dominate over gravity and galaxies would not form. If it had been smaller, gravity would dominate and chemical bonding would not occur.

3. The expansion rate of the universe

In his blockbuster 1988 bestseller *A Brief History of Time*, Stephen Hawking wrote: "If the rate of expansion one second after the big bang had been smaller by even one part in a hundred thousand million million, the universe would have recollapsed before it ever reached its present size."[32]

4. The mass density of the universe[33]

According to Ross, if the mass density of the universe were slightly larger, then overabundance of the production of deuterium (heavy hydrogen) from the big bang would cause stars to burn too rapidly for life to form. If smaller, insufficient helium from the big bang would result in a shortage of the heavy elements needed for life.[34]

5. The cosmological constant

In Einstein's gravitational equation, the cosmological constant is equivalent to an energy density in a vacuum, that is, in a space devoid of matter. By equating this density to the density of the zero point energy that is left in a volume after you remove all its particles, you obtain a number that is 120 orders of magnitude higher than what is observed. Such a high value would result in a universe that would so rapidly inflate that galaxies would have no time to form. Theists tout this as the best example of God's fine-tuning of physics so that we could exist. Physicist Leonard Susskind calls the cosmological constant problem "the mother of all physics problems" and "the worst prediction ever."[35]

This summarizes the cosmological case for God.

BIG BANG SCIENCE

Let us now look at what modern physics and cosmology have to say about these theological claims based on cosmology. Please keep in mind that nowhere is the argument being made that the big bang did not occur. The evidence that it did is overwhelming and from a number of independent observations that agree quantitatively with the model. While other models have been proposed that offer ad hoc explanations for this or that specific observation, none comes close to comprehensively fitting all the data the way the big bang model can. As we will see, the big bang simply did not require a miraculous creation.

Conservation of Mass

Early in the twentieth century, Einstein showed that mass is equivalent to energy. In his famous equation, a body of mass m has a *rest energy* $E = mc^2$, where c is the speed of light in a vacuum, a constant that merely changes the units of mass to units of energy and can be set equal to one. This implied that any form of energy can be used to create mass, or mass can be changed into any other form of energy. For example, the kinetic energy of two colliding particles can result in the creation of new particles. In chemical and nuclear reactions that produce energy, that energy comes from the rest energy of the initial ingredients. The nonconservation of mass is measurable in nuclear reactions, but also occurs in chemical reactions although there the mass differences are too tiny to measure directly.

Thus the law of conservation of mass is invalid and the mass of the universe could have come from energy. But then, where did the energy come from? The law of conservation of energy still would seem to require miraculous creation.

The First Law of Thermodynamics

As we have seen, the big bang is considered by theists as evidence for a creator. This is yet another invocation of the God-of-the-gaps argument—but we don't have to wait for science to fill the gap. Ironically, the big bang has filled its own gap by providing natural explanations for the apparent violation of the first and second laws of thermodynamics.

Our best current observational evidence strongly supports a picture first proposed in 1980 by physicist Alan Guth and others called *inflation.*[36] According to the inflationary cosmological model, the current big bang was preceded by a short period of rapid, exponential expansion lasting about 10^{-35} second. Inflation solved a number of problems with the prior big bang theory and furthermore implied that the total energy of the universe is zero.

Think of the three dimensions of space as analogous to the two-dimensional surface of a balloon. When that balloon is inflated by many orders of magnitude, its surface will be nearly flat. Similarly, inflationary cosmology results in space becoming very flat (Euclidean). A flat space is characterized by a balance in the kinetic, rest, and gravitational potential energies giving zero total energy.

The theoretical prediction from inflationary cosmology that the energy of the universe is zero was confirmed by astronomical observations late in the twentieth century. In short, zero energy was required to make our universe. Conservation of energy—the first law of thermodynamics—is not violated by a model in which the universe came from a state of zero energy, that is, came from nothing.[37]

Now, because of the quantum uncertainty principle the energy cannot be exactly zero; that is, there is always a tiny "zero-point energy." However, this does not affect our conclusion since no miracle is required, no law of physics was violated.

The Second Law of Thermodynamics

Likewise, big-bang cosmology, even without inflation, explains why the second law of thermodynamics also was not violated in producing order in the universe. The expansion of the universe does the trick. We can think of the visible universe as a sphere. For any spherical body, its maximum entropy is that of a black hole of the same size. Thus, as the universe expands, its maximum entropy increases, leaving room inside for order to form. In fact, observations are consistent with a model in which 13.7 billion years ago, the universe was confined to a region so small that it was equivalent to a black hole and possessed maximum entropy. That is, our universe was once in a state of complete chaos in which local order did not form until it started to expand. This did not violate the second law because while the entropy was maximal at that early time, it was also small and could increase as the maximum allowable entropy increased with the universe's expansion.

Note, by the way, that if the universe began in total chaos it possesses no memory of a creator, even if there were one. Thus modern cosmology not only does not require a creator, it shows that the only creator that is possible is one who simply tossed the dice and gave us a universe that was totally absent of any plan whatsoever. Such a creator might as well not exist since it has no effect on the world as we know it.

The Big Bang

Despite these results, theologians still look to the big bang as evidence that the universe had a beginning and thus a creation. However, modern cosmology does not require that everything that exists began with the big bang. Our visible universe, which began with the big bang, could be just one part of a vaster whole.

While it's true that *our* universe began with the big bang, we have no basis to rule out the possibility that it arose from an earlier universe. Here we are not referring to a particular model going back to 1948 called the *oscillating universe*, which is now ruled out by the data.[38] That model referred only to the universe we live in and not the modern cosmological picture that envisages our universe as just one of many universes in what is termed the *multiverse*.[39] And, as the ancient atomists first realized, there is no reason that any one universe should be special.

Prominent Christian theologian Richard Swinburne has vehemently objected to the notion of a multiverse: "To postulate a trillion-trillion other universes, rather than one God in order to explain the orderliness of our universe, seems the height of irrationality."[40] Irrationality is in the eye of the beholder. A trillion-trillion natural universes seems far more rational to me than one supernatural God of limitless power for which there isn't a shred of evidence. At least we can apply established physics and cosmology to specu-late knowledgeably about a multiverse. We have nothing but ancient supersti-tions to provide a basis for speculations about God.

On the other hand, cosmologist Don Page, an evangelical Christian, differs from most other theists in finding the multiverse proposal to be sup-portive of the God hypothesis: "God might prefer a multiverse as the most elegant way to create life and the other purposes He has for His Creation."[41] But then the Tooth Fairy might have preferred the multiverse, too.

Let us go through the theoretical arguments Craig and others have made for the universe necessarily having had a beginning. Two are based on theo-rems derived from general relativity, which is not a quantum theory and so is not applicable in the domain of distances on the order of 10^{-35} meter that existed at the beginning of the inflationary big bang and so should not even be used to discuss that realm. The third is mathematical.

Hawking and Penrose long ago admitted that their 1970 theorem proving that the universe began in a singularity, while not mathematically erroneous, did not apply at the origin of our universe. In *A Brief History of Time*, Hawking said,

> The final result was a joint paper by Penrose and myself in 1970, which at last proved that there must have been a big bang singularity provided only that general relativity is correct and the universe contains as much matter as we observe.[42]

Hawking continues:

> So in the end our work became generally accepted and nowadays nearly everyone assumes that the universe started with a big bang singularity. It is perhaps ironic that, having changed my mind, I am now trying to convince other physicists that there was in fact no singularity at the beginning of the universe—as we shall see later, it can disappear once quantum effects are taken into account.[43]

The more recent theological claim that Borde, Guth, and Vilenkin have proved that the universe had to have a beginning is also in error. Again, this theorem was derived from general relativity and so is inapplicable to the issue of origins. Furthermore, it is disputed by other authors.[44] I asked Vilenkin personally if his theorem required a beginning. His e-mail reply: "No. But it proves that the expansion of the universe must have had a beginning. You can evade the theorem by postulating that the universe was contracting prior to some time."[45] This is exactly what a number of existing models for the uncreated origin of our universe do.

The final argument for a beginning made by Craig is a mathematical one. He claims that the universe cannot be eternal because it would take an infinite time to reach the present. Implicit in Craig's argument is that the universe had a beginning and so an eternal universe would take an infinite time to reach the present from that beginning. However, recall that I agreed with Craig and his reference to mathematician Hilbert that actual infinities do no occur in the physical world. They simply appear in abstract equations that describe but do not constitute reality, at least as far as we can know. However, an eternal universe is not an infinite universe. In an eternal universe time is endless and beginningless. It did not have a beginning an infinite time ago. It had no beginning. The time from any moment in the past to the present is finite.

The Argument from Fine-Tuning

The cosmological constant problem described above is used by theists as the prime example of the fine-tuning of the universe that they claim as evidence for God. However, Don Page, the evangelical Christian cosmologist previously quoted, has pointed out that the apparent positive value of the cosmological constant is somewhat inimical to life, since its repulsion acts against the gravitational attraction needed to form galaxies. If God fine-tuned the universe for life, he would have made the cosmological constant slightly negative![46]

We can use established physics to put to rest the argument from fine-tuning. Again, fine-tuning is a God-of-the-gaps argument. Anyone using this argument must do more than just point to a gap in scientific knowledge. He or she must prove beyond a reasonable doubt that science can never fill the gap. This is not an impossible task. Let us consider a mundane (noncosmological) example of a God-of-the-gaps argument that could be proven beyond a reasonable doubt. If it could be shown by careful, repeated experiments that the pope can cure illnesses with his prayers and no one else can, science would be hard put to provide a plausible scientific explanation.

Barring such a case, we can defeat any proposed God-of-the-gaps argument by simply providing a *plausible* natural explanation consistent with our best existing knowledge to fill the gap. That argument need not be *proven*. In the case of fine-tuning, this can be done.

The simplest explanation for the anthropic coincidences is that our universe is part of a super-universe called the multiverse that was mentioned above. We just happen to live in that universe suitable for evolving our form of life. Such a multiverse is suggested by modern cosmology.[47]

Some theists and scientists have objected to the multiverse hypothesis as being nonscientific, since other universes are unobservable and nonparsimonious, because the hypothesis violates Ockham's razor. But these objections are not legitimate. Science talks about the unobservable all the time, such as quarks and black holes. They are components of models that agree with observations.

Wlliam of Ockham (ca. 1285–1349) was an English theologian and logician. He is frequently quoted as having said, "Entities should not be posited without necessity." However, this statement does not appear in his works, and the notion that we should always seek the simplest explanation probably goes back to an earlier time. In science, Ockham's razor is usually interpreted to mean that a theory should not contain any more premises than are required to fit the data. It does not mean that the number of objects in the theory must be minimum. The atomic theory of matter introduced a trillion-trillion more objects in a gram of matter than the theories that considered matter in bulk, yet it was more parsimonious with fewer premises and agreed better with the data.

Furthermore, proposals have been made for the possible verification of and even possible evidence for multiple universes.[48] The basic idea is that gravitational interaction with an outside universe might produce a detectable asymmetry in the cosmic background radiation in our universe.

Finally, there are some speculative theoretical arguments for multiple universes. As proposed by the eminent physicist Leonard Susskind in his 2005 book, *The Cosmic Landscape*, string theory has some 10^{500} possible solutions, each of which could correspond to a separate universe within the multiverse.[49] Nobel laureate physicist Steven Weinberg supports the notion: "Just as Darwin and Wallace explained how the wonderful adaption of living forms could arise without supernatural intervention, so the string landscape may explain how the constants of nature that we observe can take values suitable for life without being fine-tuned by a benevolent creator."[50]

While the notion of multiple universes should not be dismissed, no evidence for fine-tuning exists even when considering our universe alone. So we do not need multiple universes to refute the argument from fine-tuning.[51] Let us focus on the five anthropic coincidences that have been proposed as being so exquisitely fine-tuned that life of any sort would otherwise have been impossible.

1. The ratio of the electromagnetic force to gravity

The number 10^{39} is often quoted as the factor by which the strength of electromagnetism exceeds that of gravity. However, this is not a general statement. This number is just the ratio of the forces between an electron and a proton. It will be different for different pairs of particles since it depends on the charges and masses of the particles. There actually is no way to define a universal strength of gravity, either absolutely or relative to the strengths of the other forces of nature.

The smallest "natural" mass that can be formed from fundamental constants alone is the Planck mass, 2.18×10^{-8} kilograms. The equivalent rest energy is 1.22×10^{28} electron-volts. Compare this with the rest energy of the heaviest quark, the top quark, which is 1.72×10^{11} electron-volts. The gravitational force between two Planck-mass particles with the same charge as an electron would be 137 times *stronger* than the electromagnetic force.

So the puzzle is not why gravity is so weak but why the masses of elementary particles are so low compared to the Planck mass. This is explained in the standard model of elementary particles and forces. The masses of all particles are intrinsically zero and their observed masses are small (compared to the Planck mass) corrections that, in the model, arise from their interactions with other fields.

2. The ratio of the numbers of protons to electrons in the universe

If the universe came from nothing, then the law of conservation of charge would say that the universe currently has zero total electrical charge. It then follows that the number of protons and electrons will be exactly equal, to balance their positive and negative charges.

3. The expansion rate of the universe

Theist literature that attempts to use science to prove the existence of God almost uniformly refers to that statement by Stephen Hawking on pages 121–22 of *A Brief History of Time*, quoted earlier, that says the expansion rate is fined-tuned to "one part in a hundred thousand million million." However, the theist literature also uniformly ignores Hawking's explanation given a few pages later, on page 128:

> The rate of expansion of the universe would automatically become very close to the critical rate determined by the energy density of the universe. This could then explain why the rate of expansion is still so close to the critical rate, without having to assume that the initial rate of expansion of the universe was very carefully chosen.[52]

As we saw above, the inflationary model gives us a universe that has an exact balance between positive and negative energy. This is the critical density spoken of here.

4. The mass density of the universe

As shown previously, the mass density of the universe is related to the expansion rate. At any given time, it can be calculated. So both of these parameters are fixed.

5. *The cosmological constant*

The huge discrepancy of 120 orders of magnitude between the vacuum energy estimated from theory and its empirical value has to be the worst calculation in physics history. The multiverse is the currently favored solution to the problem among physicists.

However, the calculation is so obviously wrong that it should be ignored—certainly not taken as evidence for God. While physicists have not yet reached a consensus on the correct calculation, one possibility that agrees with observations is called the *holographic principle*.[53]

The calculation of the vacuum energy density of the universe involves a sum over all the zero-point energy states in the universe. The "worst calculation ever" assumed that the number of states is proportional to the volume. But now there is reason to believe that this is wrong. The universe can have no more states than that of a black hole of the same size. The number of states in a black hole is proportional to its surface area, not its volume. So the holographic principle says that the number of states in the universe is proportional to the surface area of our event horizon. The energy density calculated from this assumption is of the same order of magnitude as the vacuum energy density that is determined from observations.

At the minimum, we can still conclude that it is certainly wrong to sum over the zero-energy states in the volume of the universe rather than on the surface. The fine-tuning of the cosmological constant is another God-of-the-gaps argument in which the gap is being filled in by some purely natural mortar.

Other Parameters

A systematic study of the remaining parameters that are claimed to be fine-tuned shows that their values all have plausible explanations within the framework of existing knowledge. They fall into four classes:

1. Basic physics constants

Three basic physics constants that are often claimed to be fine-tuned cannot be because their values are arbitrary, depending only on the system of units being used. These include the speed of light in a vacuum c, Planck's constant h, and Newton's constant G.

2. The excited state of carbon nucleus

In 1952, astronomer Fred Hoyle used anthropic arguments to predict that an excited carbon nucleus, $_6C^{12}$, has an energy level at around 7.7 million electron-volts, which was later confirmed by experiment.[54] However, calculations show that a wide range of values for the energy level of that state produce sufficient carbon.[55]

3. Strengths of elementary forces

The strengths of the elementary forces are assumed in fine-tuning arguments to be constants that can take on a range of values that are then tuned for life. In fact, they vary with energy, and their relative values and energy dependences are close to being pinned down by theory in ranges that make some kind of life possible. When the standard model is supplemented by a concept called *supersymmetry*, the reciprocals of three strength parameters of the model vary linearly with the logarithm of energy and converge to a point at 3×10^{16} GeV.[56] Such an energy would have occurred during the early stages of the big bang.

4. Parameters of Earth and solar system

Many parameters of Earth and the solar system are claimed to be fine-tuned for life.[57] This fails to consider that with trillions of planets in the visible universe and the countless number beyond our horizon, a planet with the properties needed for life is likely to occur many times.

It is also to be noted that in almost all the literature that advocates fine-tuning, the authors make a serious analytical mistake by varying only one parameter at a time and holding all the others constant. This fails to account for the fact that a change in one parameter can be compensated by a change in another, opening up more parameter space for a viable universe. A proper analysis finds there is no evidence that the universe is fine-tuned for us or anything else.

NATURAL ORIGINS

So far in this chapter we have discussed the arguments for the existence of a creator based on cosmology and shown why they are invalid. Then where did the universe come from if it was not divinely created? The truth is that we do not know. But that does not mean that science has nothing to say about the question or that a purely natural origin to our universe is beyond the reach of science.

Most popular books by physicists give a hand-waving argument in which the universe arises as a "quantum fluctuation" from the vacuum. These books usually give the example in which a particle-antiparticle pair, such as an electron and a positron, are created in a vacuum—a process known as "vacuum polarization." The pair can have zero total energy, where the rest and kinetic energies of the particles are balanced by the electromagnetic and gravitational potential energies, in agreement with the observations mentioned previously in which the current energy density of the universe is balanced in just this way. However, theologians point out that the initial state of a vacuum is not the "nothing" that is associated with nonexistence. Rather, it is a perfectly definable physical state with particle pairs popping in and out of existence all the time. They reasonably ask, where did that state come from? The simplest answer is that it always existed, so it didn't have to come from anything. As we saw above, the argument that the universe can't be eternal is erroneous and, furthermore, an eternal multiverse is suggested by modern cosmology.

Let us go beyond hand waving and discuss some serious attempts that have been made to provide plausible scenarios for the natural origin of our universe. A number of models exist for an uncreated origin of the visible universe that have been published by reputable scientists in reputable, peer-reviewed journals.[58] They all are presented in mathematical detail and are consistent with established physics and cosmology. Rather than discuss all these models here, let me simply describe one straightforward scenario that utilizes a well-established mechanism called *quantum tunneling* that appears in physics in other contexts.

QUANTUM TUNNELING

Quantum tunneling is worked out mathematically in most lower division college physics textbooks. In classical physics, an object cannot surmount a barrier if its kinetic energy is less than the potential energy at the top of the barrier. However, in quantum mechanics, the mathematics of the Schrödinger equation allows a solution inside the barrier in which the wave function is not zero.

Normally, physical quantities such as a body's momentum and kinetic energy are defined "operationally" by how they are measured and the resulting measurements are represented by *real* numbers whose squares are positive. Inside the barrier, however, such quantities are represented mathematically by *imaginary* numbers, that is, numbers whose squares are negative. They are what we call *unphysical*. We can't measure them, but we can still talk about them mathematically. The wave function given by Schrödinger's equation, which I have emphasized is just an abstract mathematical object, simply has imaginary momentum and kinetic energy.

When we cross the opposite side of the barrier we return to the physical region where measurements can again be made. Schrödinger's equation gives us a solution in which quantities are once again real numbers and the probability for the object leaking through the barrier is nonzero. In short, quantum tunneling allows a physical body to pass through a wall. While that probability is very low on the everyday scale, which is why we don't witness people walking through walls, it happens on the subatomic scale and has been firmly established empirically. In an important example, the nuclear fusion process, in which two hydrogen nuclei combine to produce energy, results from the nuclei quantum-tunneling through the repulsive barrier that separates them due to their positive electric charges. Since fusion is the prime source of the sun's energy, we would not be here if it were not for quantum tunneling. The scanning tunneling microscope, mentioned earlier, with which we can see individual atoms, is an application of quantum tunneling.

TUNNELING FROM (OR THROUGH) NOTHING

Now let us apply quantum tunneling to the origin of our universe. To my knowledge, Vilenkin first worked this notion out mathematically in 1982.[59] An excellent account can be found in his 2006 popular book *Many Worlds in One: The Search for Other Universes*.[60]

Vilenkin's scenario calls for our universe tunneling out of "nothing," which he takes to be the region of chaos that existed before the big bang. However, trying to define "nothing," as we have seen, is a contentious matter. As with the quantum fluctuation model described above, cosmic creationists simply say that Vilenkin's "nothing" is still something that exists and so does not answer the question of how the universe came into existence from "nonexistence."

So, for my purposes here, I am going to avoid the issue by having our universe tunnel from "something," namely a prior universe that always existed. As we saw above, we have no reason to rule out an eternal multiverse.

We can envisage an earlier universe that is more or less a mirror of ours, on the opposite side of the time axis where the origin is at $t = 0$. Such a universe is not forbidden by any known principle and, furthermore, is accommodated by the same cosmological equations that we use to describe our universe. Just insert a negative t in the equations.

According to those equations, this earlier universe contracts down until it becomes as small as it can be according to quantum mechanics, a sphere of radius equal to the Planck length, 1.62×10^{-35} meter. At that point it has maximum entropy, as discussed earlier, and forms a region of total chaos.

However, like the region inside a barrier, this region of chaos is in an "unphysical" state, absent of any observable structure but which still can be described mathematically. The wave function of the universe (a well-defined quantity, believe it or not)[61] is able to tunnel through the chaos and reappear on the other side of the time axis and become our expanding universe.[62]

Now this scenario is not to be confused with older proposals about an "oscillating universe" in which our expansion is followed by a contraction and then by another expansion, ad infinitum. That proposal fails because of the second law of thermodynamics. The direction of increase in entropy of the universe does not reverse during the contracting phase but keeps increasing and eventually hits the limit of total chaos long before the universe has collapsed back to Planck dimensions.

In the new scenario, our mirror universe is only contracting from our point of view. Since, as we saw in chapter 5, the arrow of time is defined by the direction of increasing entropy, time's arrow in the mirror universe will point opposite to ours. Thus we can view the two universes as emerging from the central chaos and expanding in opposite directions. Theologians have not come to grips with the fact that time can flow in either direction. The whole notion of a creator assumes a unidirectional flow of time.

So, in this scenario, we do not have to explain how our universe came from "nothing." From our point of view in this universe, it tunneled from an earlier universe. Where did that earlier universe come from? From its point of view it tunneled from our universe.

Now, I do not claim that this is exactly how it all happened. But the fact that such a scenario can be fully formulated, with mathematical rigor based on existing knowledge, suffices to refute any claim that a miraculous creation necessarily occurred.

THE GRAND ACCIDENT

Stephen Hawking is no doubt the most celebrated scientist in the world. Deservedly so, since he has for decades made outstanding contributions to physics and cosmology and to their public understanding, while suffering from amyotropic lateral sclerosis that leaves him almost completely paralyzed. I have frequently referred to his 1988 bestseller *A Brief History of Time*.

More recently he has published another bestseller, *The Grand Design*, written with physicist Leonard Mlodinow.[63] Hawking's books, given their great exposure, usually lead to a media emphasis on what the books say about nonscientific issues, in particular, religion. In *Brief History*, Hawking stirred the pot by referring to God in many places. In his final paragraph of that book he talks about someday discovering a complete theory that tells us why it is that we and the universe exist. His concluding sentence: "If we ever find the answer to that, it would be the ultimate triumph of human reason—for then we would know the mind of God."[64]

Of course this was interpreted as meaning that Hawking believed in God. But he was misunderstood the way Einstein was frequently misunderstood when he mentioned "God" in his utterances. Both asserted their nonbelief in a personal God in no uncertain terms and were simply using the word "God" in the Spinozian sense as a name for the order of the universe.

In the introduction to *Brief History*, Carl Sagan explicitly summarized Hawking's main proposal: "A universe with no edge in space, no beginning or end in time, and nothing for a creator to do."[65] This refers to a model for the origin of the universe called the *no-boundary model*, which Hawking published in 1983 along with James Hartle.[66] The tunneling scenario presented earlier is a modified version of the Hartle-Hawking model. Their wave function is simply extended to the negative side of the time axis.

Hawking's new book is not just another gee-whiz recitation of recent developments but a profound proposal for the nature of reality that turns on its head the conventional view of how physics operates. Philosophers of science will give it a lot more attention than physicists, who generally tune out such discussions.

The authors have learned the lesson not to mention "God" too often, but the media has nevertheless jumped on their final conclusion: "Spontaneous creation is the reason there is something rather than nothing, why the universe exists, why we exist. It is not necessary to invoke God to light the blue touch paper and set the universe going."[67]

The Grand Design is an unfortunately misleading title, probably chosen by the publisher, as many titles are, with an eye toward selling the most books. People will read into this title the implication that there is still "something out there," some supernatural force that is behind everything. This is a common belief today among the many who are abandoning organized religion but are finding it difficult to accept the strict materialism and lack of design implicit in atheism. The book should have been called *The Grand Accident*, because that's what "spontaneous (creation)" (see preceding paragraph) refers to—an uncaused accident.

Hawking and Mlodinow begin by informing us, "philosophy is dead," that it has not kept up with modern developments in science, particularly physics, and that scientists "have become the bearers of the torch of discovery in our quest for knowledge."[68] In this I think they are wrong.

For centuries, the message of science has been that all knowledge of the world is obtained through direct observation. As we have seen, the revolutionary theory of quantum mechanics developed in the last century clashed with everyday experience but has proved enormously successful, passing with flying colors the many stringent laboratory tests to which it has been subjected for a hundred years.

However, no agreement has ever been reached on what the quantum models inform us about regarding the true nature of reality. This is by contrast to classical Newtonian physics, where no one doubts that an apple and the moon are real. Hawking and Mlodinow conclude: "There is no picture- or theory-independent concept of reality."[69] They call this view *model-dependent reality*.

This is not as world-shaking as it seems if you give philosophers some credit. As has been mentioned, most theoretical physicists still hold a naive realistic view of their theories, in which not only apples and the moon are real but so are electromagnetic fields and quantum wave functions. But, despite the disdain of Hawking and Mlodinow, philosophers have been telling physicists (to deaf ears) for years that all observations are "theory laden." Indeed, centuries ago the great philosopher Immanuel Kant pointed out that every human concept is based on observations that are operated on by the mind so that we have no access to a mind-independent reality. Hawking and Mlodinow say, "It is pointless to ask whether a model is real, only whether it agrees with observations."[70] This I do agree with, as has already been emphasized in several places.

The authors express their preference, as do many contemporary physicists, for the formulation of quantum mechanics proposed by physicist Richard Feynman while still a graduate student at Princeton in the 1940s, mentioned here and in chapter 6, that they refer to as "alternative histories." Actually, a better term is "sum over histories." In this model, one does not follow the traditional procedure of calculating the path of a particle, or system of particles, by applying some equation of motion. Rather, one assumes all possible paths exist and by summing a "probability amplitude" over all those paths, the probability for the particle or system ending in a specific state can be computed.

For example, in the double slit experiment, each photon from the source takes both paths available to it and goes simultaneously through both slits. The Feynman model enables you to calculate the observed interference pattern. If you place a detector along one of the paths, you then know which path was followed by the photon, and the model correctly predicts that the interference pattern will go away.

Hawking and Mlodinow apply Feynman's ideas to M-theory, a generalized extension of string theory. Although neither theory has come close to being testable in the laboratory, not only in the present but perhaps in the foreseeable future (if ever), Hawking and Mlodinow argue, disputably, that M-theory is the only candidate for an ultimate "theory of everything" that meets all the requirements for such a theory.

M-theory has eleven dimensions, ten of space and one of time. Since we only need three dimensions of space along with the one of time to describe observations, seven space dimensions are curled up at tiny distances far smaller than our most powerful instruments can probe. The shapes of these inner dimensions carry all the information needed to build the universe. The problem is, M-theory provides no unique shape but allows for 10^{100} possible shapes. (Some say 10^{500}, but what's a factor of 10^{400} between friends?)

This means that there are 10^{100} (or 10^{500}) possible universes. Applying Feynman's ideas, Hawking and Mlodinow argue that these universes constitute alternative histories that all exist and a sum over them should be taken to get the probability for the universe that we observe. As they put it, "The universe appeared spontaneously, starting off in every possible way."[71]

So it is no longer assumed that there is an initial state, presumably "nothing," from which a predetermined path is followed to the present. Rather, the universe is taken as it is at the present time and, somewhat like a detective reconstructing a course of events at a crime scene, the most probable path back to the origin is calculated. Note that the histories that contribute to the Feynman sum depend on our current observations. Thus they do not have an independent existence but depend on what we measure. Hawking and Mlodinow say, "We create history by our observations, rather than history creating us."

This unfortunate wording comes dangerously close to the claims of quantum spiritualists such as Deepak Chopra, who say that quantum mechanics tells us that we make our own reality (see chapter 6). But this is not quite the proposal made in this book, which is far more plausible. Human consciousness is not deciding what reality is. Reality is all the possible histories, and these can't be changed by thinking about them. All these histories contributed to our present state and our observations pin down that state, enabling us to compute the probability for each history reaching that state.

In the Feynman method, for classical macroscopic systems, the probability for one path far exceeds the probability for the others and so it is the one we witness the system taking. Only for quantum phenomena, such as photons passing through slits, are the probabilities for different paths comparable.

So, 10^{100} universes converged on the one that has the necessary structure to produce life as we know it. With such a huge number of possibilities, many would be expected to allow for some kind of life different from ours as well as for ours. Thus, the theological claim that our universe was exquisitely fine-tuned for us by a creator is once more refuted. Hawking and Mlodinow make it clear that the idea of multiple universes, the multiverse, was not invented to account for apparent fine-tuning, as often charged by apologists, but is "a consequence of the no-boundary condition as well as many other theories of modern cosmology."[72] And once you have one accidental universe, you will have many.

Some reviewers have criticized *The Grand Design* for basing its case on the highly speculative M-theory. Actually, even without M-theory, the alternate histories of Feynman, together with cosmological models such as the no-boundary model, are sufficient to provide a purely natural, noncausal explanation for the existence of our universe and our place in it. M-theory or no M-theory, the pieces of our universe fell into the places where they are, not because of a guiding hand and a grand design, but through mere accident.

THE UNIVERSE AS A QUANTUM COMPUTER

The metaphors people have used throughout human existence to describe reality depend on the models with which they describe their experiences. Seeing animals all around them, primitive people attributed natural phenomena to the action of animate spirits that were contained in all things, living and nonliving. When agriculture developed and people became settled in villages ruled by chiefdoms, the chief became the metaphor for the gods. That is, although supernatural, the gods were human with all our imperfections. When the villages combined into city-states ruled by kings, the gods gradually become more majestic, with a major god ruling over a royal court of lesser gods. With the rise of empires, the next step was monotheism, with a single God ruling over all. See *The Evolution of God* by Robert Wright.[73]

With Newtonian mechanics, the metaphor for God became the great craftsman of the Newtonian world machine. Today, with the dominance of computers throughout society, the universe is increasingly being described as a computing machine with the basic stuff of the universe being bits of information—or *qubits*, which are the information units in quantum computers. God is then the great quantum computer programmer.

In their 1992 book, *The Matter Myth*, physicist Paul Davies and science writer John Gribbin declared "the death of materialism." They write, "Matter as such has been demoted from its central role, to be replaced by concepts such as organization, complexity, and information."[74] In *Programming the Universe*, physicist Seth Lloyd, a pioneer in quantum computing, unsurprisingly agrees with the metaphor of a quantum computer to describe the universe.[75]

Theologians have found this new information ontology particularly congenial in their continuing attempts to come up with a model of God that is consistent with science. Niels Henrik Gregersen identifies the divine Logos in the first verse of the Gospel of John ("In the beginning was the Word") with information.[76] Wright calls Logos "the divine algorithm."[77] Gregersen asserts, "God is present in the midst of the world of nature as the informational principle (Logos) and as the energizing principle (Spirit). . . . It is only in the interplay between information (Logos) and energy (Spirit) that the world of creation produces evolutionary novelties rather than mere repetitions."[78]

Theologian John Haught has reconciled himself to the fact that the metaphor of God as the great craftsman in the sky has seen its day. He writes:

The image of God as a "designer" has become increasingly questionable, especially in view of evolutionary accounts of life. Is it possible, then, that the notion of "information" may be less misleading than that of design in theology's inevitable reflections on how divine purposive action could be operative in the natural world? . . . In place of design, I would suggest that natural theology may more appropriately understand divine influence along the lines of informational flow.[79]

Another prominent theologian, Keith Ward, comments:

In my opinion . . . one may hold a view that the universe is constructed on an informational pattern that is carried and transmitted by the mind of God. The God hypothesis is not contradicted by and is quite strongly supported by some, of the speculations of contemporary information theory. So my conclusion is that information is the ultimate ontological reality held in the mind of God.[80]

What is information anyway? Claude Shannon, the father of information theory, identified it with entropy.[81] When a signal is sent from one spot to another, the information transferred was defined by Shannon as being equal to the entropy change that takes place, when the latter is expressed in bits. The entropy lost is the information gained.

Entropy has been part of physics since the nineteenth century, so it is not clear what is new with the new information metaphors. The theologians quoted above suggest other forms of information, but they do not quantify them, so we should stick to Shannon's definition in order to keep things precise.

Now, it is true that entropy is a rather abstract mathematical quantity. I still remember as a sophomore in engineering school around 1952, sitting in a thermodynamics class with my classmates pressing the instructor, "What is entropy? What's it made of?" We wanted to see it and touch it. I don't recall getting a satisfactory answer, although his response was undoubtedly correct. I would have the same problem when I taught thermodynamics in later years.

Well, you can't see and touch entropy just as you can't see and touch the quantum wave function. And, you can't see and touch information. Entropy is just a measure of how many possible physical states exist. Information is its complement. It's how much the possibilities are reduced when we learn about the actual states. Information and entropy are out there in the Platonic world of ideal forms. They are just abstract objects we use in our mathematical descriptions of the world, and that world is still made of matter and nothing else—according to our best current knowledge.[82] If the universe is a digital computer, that computer still is made of elementary particles.

Theologian Gregersen makes a key observation:[83]

> The theological candidate of truth that the divine Logos is the informational resource of the universe would be *scientifically* falsified if the concept of information could be fully reduced to properties of mass and energy transactions.

Well, perhaps this does not constitute a falsification, but as far as we can tell, information does reduce to mass and energy transactions. Information can be embodied in many ways, but there is no disembodied information.

This is not to imply that the concept of information and the metaphor of the universe as a quantum computer are not useful. The great physicist John Wheeler, who inspired Lloyd, famously called the concept of casting physics in terms of information, "It from bit."[84] Lloyd says he has proved a conjecture originally proposed by Feynman, who was Wheeler's student, that "quantum computers could function as a universal quantum simulator whose dynamics could be the analog of any desired physical dynamics.[85] As far as I know, this is the case for classical computers as well.

Lloyd then calculates that the cosmic computer can store 10^{92} bits and has performed a maximum of 10^{122} operations on those bits in the age of the universe. These are large numbers, but they are finite. Furthermore, they grow as the universe expands, opening up more and more possible arrangements of matter as time goes on. This can happen because the internal energy of the universe increases as the universe expands. This does not violate the first law of thermodynamics because that energy comes from the work done *on* the universe by the negative pressure of the dark energy. No energy is needed from outside. In any case, the complexity of matter is growing all the time.[86]

As we have seen, the best we can tell from cosmological data is that our universe started out with maximum entropy, meaning minimum (really, zero) information and minimum complexity. At the first operationally definable moment called the Planck time, the cosmic computer had only two possible states: one bit on which one operation could be performed. But this was sufficient to start things off. As the universe expanded, more and more bits were generated, and more and more operations were performed. Thus, complex phenomena such as life were inevitable without any fine-tuning.

Far from providing aid and comfort to those who seek evidence for a creator, the computer universe makes the existence of a creator an even more unnecessary and baseless hypothesis.

THE COSMIC INSIGNIFICANCE OF HUMANITY

Since Copernicus, humanity's conception of its place in the universe has steadily diminished from the biblical teaching that we are the center of the universe to one in which we are but a miniscule speck in space and time. Once we had telescopes with which to peer into the sky, our view of the universe grew from originally that of a single star system and its planets to a galaxy of a hundred billion stars and on to a visible universe of 100 billion galaxies. And that was not the end of it. As we have seen, since the 1980s we have found good reasons to think that our visible universe is but a drop of water in an ocean of galaxies lying beyond our light horizon, perhaps a hundred orders of magnitude larger, that all resulted from the same big bang. Furthermore, this universe may be just one of countless others just as big.

While a god might still preside over all this, it becomes incredible to believe that he, she, or it put his, her, or its favorite creatures on this tiny planet and left the rest of this vast multiverse inaccessible to them.

CHAPTER 8
PURPOSE

[Nature] has no more regard to good above ill than to heat above cold, or to drought above moisture, or to light above heavy.

—David Hume[1]

CONTINGENCY OR CONVERGENCE?

As we saw in chapter 2, Aristotle introduced the principle of final cause in which objects move in the direction toward their ultimate goal. For example, of the four elements, fire and air move up, while water and earth move down, because each is heading toward its natural resting place.

Dinesh D'Souza is among many theists who claim that evidence exists for some teleological principle built into nature that causes matter to become increasingly complex with time. He quotes Freeman Dyson:

Before the intricate ordered patterns of life, with trees and butterflies and birds and humans, grew to cover our planet, the earth's surface was a boring unstructured landscape of rock and sand. And before the grand ordered structures of galaxies and stars existed, the universe was a rather uniform and disordered collection of atoms. What we see . . . is the universe growing visibly more ordered and more lively as it grows older.[2]

D'Souza observed that this is a "scientific description," not "theological speculation."[3]

Paleontologist Stephen Jay Gould had made the argument that evolution does not show signs of progress, that it was arrogant for humans to place

199

themselves at the top of the evolutionary ladder. However, D'Souza assures us that this is no longer "conventional wisdom" in biology and that some prominent biologists, notably Simon Conway Morris, leading expert on the fossils of the Burgess Shale, and Nobel laureate Christian de Duve have argued that "all this talk about randomness and contingency is overrated." They claim to see evidence for a plan in evolution.[4]

In his 2003 book, *Life's Solution: Inevitable Humans in a Lonely Universe*, Conway Morris assembled a collection of examples of what is called *convergent evolution*.[5] As he explains it, from very different starting points organisms "navigate" to the same biological solution. In the most commonly discussed example, the same one used by Paley, Conway Morris says the eye evolved on multiple occasions along separate evolutionary lines with the same goal in mind—sight. Conway Morris claims to see the same convergence everywhere, from molecules to social systems, and insists that it is very unlikely to have resulted from conventional Darwinian evolutionary processes.

However, in his review of *Life's Solution* in *The New York Times,* philosopher Elliott Sober pointed out, "You can't show that an event was inevitable or highly probable just by pointing out that it has happened many times. To estimate the probability of the camera eye's evolving, you need to know how many times it evolved and how many times it did not. Conway Morris never investigates how often convergences failed to occur."[6]

In fact, if humanity was so inevitable, why did only one of the ten billion or so species that ever lived on Earth develop anything close to human intelligence?

According to Conway Morris, evolution has reached its limit. In a 2009 letter to the *Manchester Guardian*, he betrays a religious predilection that may color his scientific conclusions:

> When physicists speak of not only a strange universe, but one even stranger than we can possibly imagine, they articulate a sense of unfinished business that most neo-Darwinians don't even want to think about. Of course our brains are a product of evolution, but does anybody seriously believe consciousness itself is material? Well, yes, some argue just as much, but their explanations seem to have made no headway. We are indeed dealing with unfinished business. God's funeral? I don't think so. Please join me beside the coffin marked Atheism. I fear, however, there will be very few mourners.[7]

While there is no doubt that some tendency toward convergence is present in evolution, its mere existence does not prove it cannot be a consequence of natural selection. Let us consider the evolution of the eye. The fact that different evolutionary paths lead to the same basic mechanism for sight can be trivially understood without the need for some guiding force. Just apply a little physics. Vision is obviously so important to an animal that strong selective pressures are naturally going to point it in that evolutionary direction. Light is made up of photons and there is only one way that photons can be detected: by knocking electrons in atoms either to a higher energy level or out of the atom altogether where they generate an electrical signal that can be transmitted to the brain. It follows that all evolutionary paths toward sight will have to converge at that point. However, beyond this basic physics there is little convergence evident in detail, with the vision problem solved by evolution at least ten different ways.

D'Souza is being grossly misleading when he says Conway Morris's view is somehow the new conventional wisdom in biology. The majority of biologists still adhere to basic Darwinism, in which life evolved by a process of chance and natural selection. In his 2009 book *Why Evolution Is True*, biologist Jerry Coyne notes that convergent evolution is a well-known process that is fully understood by conventional evolution. He says it "demonstrates three parts of evolutionary theory working together: common ancestry, speciation, and natural selection."[8]

On his blog, Coyne comments on the article by Conway Morris in the *Manchester Guardian* that was quoted above:

Conway Morris is way, way peeved at atheists. He mentions them several times in his piece. He thinks he has vanquished them with his "unanswerable" evolutionary arguments. But he has not. He is simply proposing a "God of the gaps" argument, and here the gap is our mind. It's Alfred Russel Wallace [who discovered natural selection simultaneously with Darwin, but became increasingly mystical] recycled. He is wrong: neither will atheism die, or even flinch a bit, and we will, I predict, some day understand, as Darwin believed, that the human mind is simply a product of the blind and materialistic product of natural selection.[9]

Purpose remains today a major area of conflict between science and religion. Religion still insists that the universe has a purpose, with humanity as its focus. However, no scientific evidence supports this yearning. The world and humanity look exactly as they should look if they are not a part of any divine plan.

TEILHARD'S OMEGA POINT

Perhaps the most original Christian evolutionary theologian was the early twentieth-century Jesuit priest and paleontologist Pierre Teilhard de Chardin.[10] Many of his ideas, though rather abstract and obscure, conflicted with church teaching and were banned from publication during his lifetime. Nevertheless, Teilhard became an important cultural figure, an archetype for the intellectual, dissenting priest in several novels and movies.[11] His best-known work is *The Phenomenon of Man*, published in 1959.[12]

Although Teilhard claimed to argue from science alone, Ian Barbour contends that Teilhard's view was not natural theology but a theology of nature. The Jesuit interpreted evolution metaphysically, writing that it applied to all beings including God.[13] The notion that God evolves is hardly a traditional one. Indeed it flies in the face of the universal theological concept of God as eternal and unchanging.

Teilhard was a scientist but primarily a mystic. He was not a dualist, again unconventional for a Catholic priest. But spirit, not matter, is the basic stuff for Teilhard. Reality is ultimately spiritual, with matter being composed of homogeneous units of "psychic energy." He asserted that the most significant result of evolution is the emergence of human consciousness. Although a recent event in cosmic history, consciousness is not to be regarded as a fleeting phenomenon destined to disappear when humans inevitably become extinct, but as being of primary, eternal importance.

Teilhard equated Christ/God with the evolving cosmos, unfolding and fulfilling himself toward a future *Omega Point*. He proposed a *law of complexity and consciousness* that gives meaning and purpose to everything. Consciousness develops as a property of complexity, which accelerates with time. Teilhard tried to reduce all phenomena to the information they contain (a recently rediscovered idea), which reveals the meaning, purpose, and goals of cosmic evolution and human destiny.[14]

According to Teilhard, the evolving universe emanated from God at the *Alpha Point* and humankind will eventually return to God as the final goal of cosmogenesis billions of years in the future at the Omega Point.

Not many theologians take the Omega Point notion seriously, but they have taken seriously the idea that evolution shows a movement toward progress, which suggests that some kind of divinely created law is behind a steady increase in complexity. They propose that God uses evolution to carry out his plans.

However, the modern consensus among evolutionary biologists is that evolution by chance and natural selection does not automatically imply an increase in complexity. A mutation might bring about a simpler system better able to survive, or at least not result in species extinction. Evolutionary biologist John Maynard Smith observed, "There is nothing in Darwinism which enables us to predict a long-term increase in complexity."[15] Paleontologist Gould agreed, saying, "Natural selection is a theory of *local* adaptations to changing environments. It proposes no perfecting principles, no guarantee of general improvement."[16] And biologist and historian William B. Provine summarizes his views on what modern evolutionary biology tells us loud and clear:

> There are no gods, no purposes, no goal-directed forces of any kind. There is no life after death. . . . There is no ultimate foundation of ethics, no ultimate meaning of life, and no free will for humans either.[17]

While it is true that evolution does not necessarily imply an increase in complexity, obviously such increases do occur and, as we saw in chapter 4, simplicity can beget complexity contrary to the claims of intelligent design theorists.

TIPLER'S OMEGA POINT

Physicist Frank Tipler has claimed a scientific basis for Teilhard's Omega Point. In a 1994 book titled *The Physics of Immortality*, Tipler proposed an incredible scenario, based on a rather loose interpretation of physics principles.[18]

In Tipler's scenario, human-created robots eventually fill the universe. These robots create others and the robot species becomes increasingly intelligent. Then, after about a billion-billion years, the universe will be uniformly populated with an extremely advanced form of life that will be capable of feats far beyond anyone's (except Tipler's) imagination.

At that point, Tipler assumes the universe will begin to contract toward what is called the big crunch, the reverse of the big bang. The advanced life form that evolved from our robot creations must collapse the universe in a highly controlled way. Assuming it can manage this, life then converges on the Omega Point. As this happens, time runs slower and slower. This is expected from general relativity, where Einstein showed that a clock in a gravitational field slows down. Clocks would run mighty slow if the entire mass of the universe were concentrated in a volume smaller than a proton. Note that this slowing down would not be observed in the clock's own reference frame; however, from the standpoint of the universe as a whole, the elapsed time would approach eternity.

Tipler associates the Omega Point, as did Teilhard, with God. Being the ultimate form of power and knowledge, the Omega Point would also be the ultimate in Love. Loving us, it would proceed to resurrect all humans who ever lived (along with their favorite pets and popular endangered species). This is accomplished by means of a perfect computer simulation, or "emulation." Since each of us is defined by our DNA, the Omega Point simply simulates all possible humans that could ever live, which of course includes you and me. Our individual memories have long dissolved into entropy, but Omega has us relive our lives in an instant, along with all the other possible lives we could have lived. Those deemed deserving by this Omega God will get to live even better lives, including lots of sex with the most desirable partners they can imagine. Those deemed undeserving by Omega will be put through purgatories, but if they perform satisfactorily they may gain heaven. So, we can all correct our mistakes. I will live a life where I learn to hit a curveball. Hitler will live a life in which he is Jewish. Barack Obama will be president over and over again until he finally gets it right.[19]

Now, it should be noted that most cosmologists currently do not expect that the big crunch will happen. The best guess based on current observation and theory is that the universe is open; that is, it will expand forever. In fact, since 1998, it has been established that the expansion of the universe is accelerating. Rather than a big crunch, we are heading toward an ultimately dead universe composed of only photons and neutrinos, or whatever are the remaining stable particles. So, attractive as it may seem for those who want to live forever, Tipler's scenario is not very promising.

REDUCTIONISM VERSUS HOLISM

A major contrast between religious or spiritual thinking and science concerns whether or not physical phenomena can simply be reduced to the sum of their parts.

As apologist William Grassie notes, the word *science* derives from the Latin *scire*, "to know," and perhaps also from the Latin *scindere*, "to split," and the Sanscrit *chyati*, "to cut off." Presumably these suggest the scientific method of breaking things up into parts. The word *religion* came from the Latin *religare*, "to bind together." Grassie suggests, "The concepts of reductionism and holism are embedded in the very etymology of the two words."[20]

In chapter 6 we saw that the current standard model of elementary particles and forces is fully reductionist. This notion disturbs those who see themselves as part of a great, integrated whole. While notions of a holistic science are bandied about, nothing much has come of them (recall S-matrix theory in chapter 6). The dominant methodology of science remains reductionist.

In the holistic view of life, every event is part of a grand scheme that applies, under divine guidance, to the whole system, from bacteria to humans and from billions of years ago to the present and indefinite future. In the view of quantum spiritualists, subatomic events are part of a grand scheme that applies holistically to every particle, from an electron in a french fry at

McDonald's to a photon in the cosmic background radiation billions of light-years away and billions of years in the past. In the reductionist view, physics as well as biology is broken down into a series of events local in space and time—collisions between subatomic particles such as electrons and photons.

Classical physics is reductionist. While direct proof of the existence of atoms was not found until the twentieth century, they were implicit in Newtonian mechanics, which was able to describe all of the behavior of macroscopic material systems—gases, liquids, and solids—in terms of the motions of their parts.

The great twentieth-century advances of relativity and quantum mechanics did not, as you often hear, "prove Newton wrong." Any scientific model is valid in a limited domain. While the advance of science since Newton has been remarkable, it is not as incredible as it is often blown up to be. Science evolves, so progress is to be expected. Nevertheless, the physics of Galileo and Newton, their predecessors such as Descartes, and their successors such as Laplace, Einstein, and Feynman, remains much the same in basic principles and methods. Today's physics students still study Descartes' analytic geometry and his Cartesian coordinates, Newton's and Leibniz's calculus, Galileo's relativity, and Newton's laws. They learn the same definitions of space, time, velocity, acceleration, mass, momentum, and energy that are found in Newton's mechanics.

It's true that some of the equations have been modified for high speeds, and some new equations have been added for small distances. Still, the equations of relativity derived by Einstein, both the special and general versions, reduce to the Newtonian equations when the speeds of the bodies involved are small compared to the speed of light in a vacuum, c. Similarly, the equations of quantum mechanics, derived by Max Planck, Niels Bohr, Erwin Schrödinger, Max Born, Pascual Jordan, Werner Heisenberg, Paul Dirac, Richard Feynman, and others, reduce to the Newtonian equations when Planck's constant h is set equal to zero.

The basic scientific method of observation and mathematical model-building remains unchanged. We physicists still model the world as material bodies and build mathematical models to describe what we see when we observe these bodies around us everyday or in the laboratory and cosmos.

New Age spiritualists and Christian apologists have appropriated quantum mechanics to claim a more holistic picture of nature. However, the standard model of particles and forces developed in the 1970s has agreed with all the data gathered at particle accelerators since then, and is only now being seriously tested at the Large Hadron Collider in operation in Geneva. Discoveries at the LHC are unlikely to change the general reductionist scheme.

In short, reductionism in science remains consistent with all the data. It isn't defeated just by the fact that it can't derive everything that happens. It still works. Holism has no evidentiary support. It doesn't work. Holism is nothing more than the wishful thinking of those who have the hubris to believe they are an important part of some cosmic plan.

WEAK EMERGENCE

Nevertheless, many observers think they sense a deviation from strict reductionism in the scientific observations made above the level of elementary particles. They argue that new principles "emerge" at those levels that do not simply arise from particle interactions. Grassie asserts, "The concept of emergence says simply that the whole is greater than the sum of its parts."[21]

Although comprising only 5 percent of the total mass and energy of the universe, up and down quarks, electrons, and photons are all that are needed as ingredients of conventional matter in a working model for those observable phenomena that are of direct concern to most humans. Whether you are a condensed matter physicist, a chemist, a biologist, a neuroscientist, a sociologist, a surgeon, or a carpenter, all the stuff you deal with in your work is made of up and down quarks, electrons, and photons. Only elementary particle physicists and cosmologists worry about the other 95 percent of matter.

The conventional reductionist picture envisages a series of levels of matter. From elementary particles (or strings, or whatever are the most elementary) we move to the nuclei of atoms, then to the atoms themselves and to the molecules that are composed of chemical atoms. While only on the order of a hundred different chemical atoms exist, the number of molecules is endless— especially the huge structures built around carbon that form the ingredients of life and our fossil fuels, as well as many synthetic materials, from plastics to polyesters.

The objects of our everyday experience are composed of molecules. Living organisms are an important component at this level, at least to us living organisms. How important they are on a cosmic scale is more dubious. Humans organize themselves into societies, so we can regard social systems, politics, and economics as a yet higher level of material existence. Beyond that we have Earth and its complex systems, the solar system, our galaxy, other galaxies, and whatever else is out there, such as black holes, the cosmic background radiation, dark matter, dark energy, and other universes.

In 1972, Philip W. Anderson, the eminent condensed matter physicist who won the Nobel Prize in physics in 1977, wrote an article in *Science* with the title "More Is Different." In the article, he complained about the implication of the reductionist notion that all the animate and inanimate matter of which we have detailed knowledge is controlled by the same set of fundamental laws. In that case, Anderson stated, "The only scientists who are studying anything really fundamental are those working on those laws. In practice this amounts to some astrophysicists, some elementary particle physicists, some logicians and other mathematicians, and a few others."[22] Anderson pointed out that many properties of complex systems cannot be derived from particle physics. Wow! No one had ever realized that before.

Obviously an elementary particle physicist cannot take her equations and produce a derivation of every physical property we observe at all levels. She cannot calculate the structure of DNA from "first principles" or predict the stock market. At every level of matter, from the smallest bodies to the largest, we have specialists developing the principles that apply at that level by applying the standard method of science—observation, model building, and hypothesis testing. The principles that apply at each level are said to "emerge" from the level below.

But the fact that we cannot derive these emergent principles from particle physics does not prove that everything cannot be just collections of particles. The practical irreducibility of emergent principles to particle physics is trivially an *epistemological irreducibility*. The key question is whether it is also an *ontological irreducibility*.

Many scientists who work at the higher levels of experience think emergence is ontologically irreducible. They argue that there is more to it than simply emergence from below. Since they never need to use any particle physics in their work, and are rarely exposed to it during their training, it seems to them that the regularities they uncover arise from an independent set of principles that apply at a higher level than quarks and electrons.[23]

For example, biologist Stuart Kauffman proposes that self-organization plays a role in biological evolution and the origin of life, which, he insists, cannot be reduced to chance and natural selection alone.[24]

Theologians and religious apologists are quick to agree. William Grassie writes, "From the surface tension of water in a glass to superfluidity and superconductivity in a physicist's lab, the behavior of huge numbers of particles cannot be deduced from the properties of a single atom or molecule."[25] This is misleading. Of course we don't deduce the behavior of a group of many particles from just the properties of a single particle, but we deduce that behavior by considering a system of particles with individual properties.

In chapter 5 we described two examples from physics that lend credence to the notion that higher-level principles emerge from those below with no further inputs required. In the nineteenth century, thermodynamics developed from the need to understand the engines of the Industrial Revolution. By the usual combination of creative thinking nourished by empirical data, principles such as the first and second laws of thermodynamics and the ideal gas law were discovered without any reference to the underlying nature of matter.

Then, James Clerk Maxwell, Ludwig Boltzmann, and Josiah Willard Gibbs showed that the principles of thermodynamics could be derived from the assumption that matter is composed of huge numbers of tiny particles that move around and collide with one another according to the laws of Newtonian mechanics. They knew it was impossible to predict the motion of each particle individually, so they used probability theory to predict the average behavior of many particle systems in what is called *statistical mechanics*.

Assuming the particles in a gas in equilibrium moved around randomly, Maxwell, Boltzmann, and Gibbs showed that the pressure in a container of a gas resulted from the momentum transferred when the particles collided with the walls of the container. The absolute (Kelvin) temperature of the gas was proportional to the average kinetic energy of the particles. The first law followed from conservation of energy. The second law followed from the tendency of a system of many particles to become more disorderly with time, as multiple collisions of particles with one another tend to wipe out any initial local regularity.

In short, thermodynamics emerged naturally from statistical particle mechanics. Every emergent thermodynamic principle that was originally introduced to explain macroscopic phenomena was derived from particle mechanics and statistics. The new picture was simpler, more elegant, and far more useful as physics entered the quantum era. And the whole was still equal to the sum of its parts.

A second example of "emergent" principles demonstrating how the whole is still equal to the sum of its parts comes from quantum mechanics. Two kinds of particles exist: fermions with half-integer spins, and bosons with integer spins. Both types of particles have diametrically opposed behavior when they are part of a collection of particles. No two identical fermions, such as electrons, can exist simultaneously in the same quantum state. This is called the *Pauli exclusion principle*.

Starting with hydrogen and adding, step-by-step, a proton to the nucleus of the corresponding atom and enough neutrons to keep the nucleus stable, electrons must be added to complete an electrically neutral atom. These electrons cannot all fit in the lowest energy level, and so some electrons must move to higher levels, or "shells." In this way the Periodic Table of the chemical elements is built up. Without the Pauli exclusion principle, the complexity of atoms would not exist, and neither would life as we know it.

On the other hand, identical bosons, such as helium atoms, tend to congregate in a single state, making possible phenomena such as lasers, superconductivity, superfluidity, and boson condensates.

These are all "emergent" phenomena that occur only for collections of particles. This might be regarded as the result of some holistic principle. However, their behavior can be derived from the basic physics of individual particles and their interactions with other individual particles.

As a third example, chaos theory, described in chapter 6, can be used to further demonstrate how the existence of emergent principles does not imply the whole is greater than the sum of its parts. Basic statistical mechanics applies to simple systems of many particles that are at or near equilibrium, that is, fairly uniform and homogeneous throughout with the same temperature and pressure everywhere within the system. However, most many-particle systems are not that simple. In particular, a special kind of system exists called a *non-linear dissipative system*, in which the motions of the bodies are so complicated that the mathematical equations describing those motions cannot be solved by standard methods and equilibrium statistical methods do not apply. Earth's atmosphere is a prime example, with its turbulence and large temperature and pressure variations from place to place and time to time.

Once computers were available it became possible to simulate the behavior of such systems, and in the process certain principles began to "emerge." The most important was what is termed the butterfly effect, whereby the tiniest change in the initial conditions of the system would result in a dramatic change in the behavior of the system. It was as if the flap of a butterfly's wings could change the weather two weeks from now. The phenomenon was somewhat misleadingly termed *chaos*.

As discussed in chapter 6, this feature of chaotic systems is often misunderstood as implying that some macroscopic systems are indeterministic, that is, their behavior cannot be predicted. In fact, chaotic systems obey classical Newtonian mechanics and so are fully deterministic and predictable *in principle*.

The fundamental predictability of chaotic systems can be demonstrated by computer simulations. As long as you start the system at *exactly* the same place in the simulation, it will end up at the same place. In fact, it was a loss of precision due to numerical rounding in a primitive computer simulation of the atmosphere developed by meteorologist Edward Lorenz in 1961 that led to the discovery of chaos.[26] Actual physical systems such as the atmosphere are unpredictable in practice because we can never measure the initial conditions precisely enough to make a prediction more than a short time into the future.

In short, while it is true that the principles at most of the levels of higher complexity above particle physics cannot be derived from the equations of particle physics alone, we have examples where we can determine so-called emergent principles, either mathematically or on the computer, from basic physics. The fact that we cannot derive the principles at every level does not imply that some new metaphysics must be behind those principles. One needs more than yet another God-of-the-gaps argument to justify such an extraordinary claim.

SELF-ORGANIZATION

No doubt self-organization occurs in nature. It is a common feature of complex systems. Many examples can be found in biology, but we need to emphasize that this feature is not limited to living organisms. When a wet foam composed of spherical bubbles dries out, the bubbles become polyhedral. When a liquid is heated from below, it spontaneously develops a pattern of hexagonal circulating cells. Turbulence in the atmosphere of the giant planet Jupiter has produced a coherent structure called the Great Red Spot that has persisted for at least three hundred years and probably much longer. From snowflakes to Saturn's rings, self-organization in material systems is ubiquitous. For many beautiful examples see *The Self-Made Tapestry: Pattern Formation in Nature* by Philip Ball.[27]

For a simple example, picture an expanse of sand on a beach near the waterline that has been smoothed by waves washing over it. Now, let the tide go out and let the sun dry the sand. Suppose the wind then picks up and blows across the sand. The wind obviously has no complex structure to it, but an intricate pattern of ripples in the sand will be produced. The spectacular sand dunes in a desert are examples of the same phenomenon.

In a second book, *Critical Mass: How One Thing Leads to Another*, Ball shows remarkable examples of self-organization in social systems.[28] One important point he makes concerns the role of randomness in the process. Many systems are *metastable*, that is, they stay the way they are until perturbed. For example, fast-moving traffic will flow smoothly in collective fashion until a random fluctuation causes the density of vehicles to increase above a certain critical value. The flow rate then drops drastically, the traffic becoming congested as the density increases until the traffic grinds to a halt.

This is an example of what is called *self-organized criticality*. Instead of change occurring smoothly, in many cases it occurs catastrophically. The snow on the side of a mountain piles up until it reaches a critical mass and produces an avalanche. No wonder that people, with their caveman brains, attribute agency to such events. The question of God's actions versus natural phenomena is a primary place where science and religion part company.

While it is true that we can't take the equations of the standard model of elementary particles and forces to derive the fact of the emergence of life, a clear line from physics to biology can be seen. Life still can be *reduced* to the same quarks and electrons.

An ancient belief that remains implicit in much theological thinking holds that some fundamental difference between life and nonlife exists, that some spark of life, an *élan vital*, or vital force—or soul—must be inserted into matter to make it alive. A remnant of this notion can be found in the way chemists still make a distinction between inorganic chemistry and organic chemistry, where the latter refers to all substances containing carbon. In 1828, German chemist Friedrich Wöhler (died 1882) synthesized urea in the laboratory, demonstrating that no vital force was necessary to generate organic compounds.[29]

None of the life sciences has ever found any difference in composition between living and nonliving matter. A living cell is made of the same quarks and electrons as a rock. A living cell and a rock are described by the same laws of physics and chemistry. No laboratory measurement has ever revealed a "living force" or "bioenergetic field" that makes life unique. Life seems to occur when a system of many particles reaches a certain level of complexity so that it is able to perform the various tasks we associate with a living organism.

Evolution itself, in its well-established formulation, demonstrates this progression from simple to complex—although, as we have noted, it need not always work in the direction of more complexity. Any given complex system would have to start as a simple system, if it were not divinely designed. If it were divinely designed, it would have to spring into existence with full complexity just as the Bible says. But the data, which D'Souza and Conway Morris would like to interpret as evidence for design, overwhelmingly support the opposite conclusion.

The natural progression from simple to complex can be found in all matter, not just the biological. In the absence of external heat (energy), the natural, unaided process of phase transitions in matter is from vapor to liquid to solid. This progression can be seen to occur from greater to lesser symmetry. Complex structure is basically broken symmetry.

These effects are easy to understand. Gases are highly symmetric and uniform because the energies involved in the interactions between molecules are weak compared to their kinetic energies. So the particles of the gas statistically fill all the position and momentum states available. That's why most gases behave much the same. But for systems with less kinetic energy (lower temperature), the molecular interactions are not just momentary collisions. The characteristics of the molecular structure and interaction come to dominate the macroscopic characteristics, and complexity is generated.

Life occurs in liquid water because there the molecules are close for a sufficient duration to interact but still have enough energy to change configurations. So, while the potential is there to produce life, that potential lies in the ability of carbon, hydrogen, oxygen, nitrogen, and so on, to form complex structures in liquid water. There's nothing mysterious about that—it is predictable from reductionist physics. No new principle beyond standard particle mechanics need be invoked.

I would like to correct a common misrepresentation that appears frequently in the apologetic literature concerning the positions that Dawkins and I have taken on the source of complexity.[30] In referring to my book *God: The Failed Hypothesis*,[31] chemist Edgar Andrews says, "Doesn't Dr. Stenger's idea that simplicity begets complexity totally contradict Richard Dawkins's argument that God, having created an exceedingly complex universe, must be even more complex and thus highly improbable?" Simply, no.

Here's exactly what Dawkins said:

> A designer God cannot be used to explain organized complexity because any God capable of designing anything would have to be complex enough to demand the same kind of explanation in his own right. God presents an infinite regress from which he cannot help us escape. This argument . . . demonstrates that God, though not technically disprovable, is very, very improbable indeed.[32]

The point Dawkins was making is that if Dembski,[33] Behe,[34] and their supporters in the intelligent design movement are correct that complexity can *only* arise from higher complexity, then God would be even more complex and an explanation would then have to be found for his complexity. But Dawkins does not believe for a moment that this is the case. I have personally checked with him to confirm that he agrees with my interpretation of his words. No one has been more eloquent than Richard Dawkins in describing how complexity arises from simplicity in biology, so it is ludicrous to say he supports the intelligent design view.

Incidentally, when Dawkins says the existence of God is "technically unprovable," he is also not disagreeing with me. I concur that we cannot disprove the existence of all conceivable gods by logic alone. The argument I presented in *God: The Failed Hypothesis*—that God does not exist beyond a reasonable doubt—is a scientific one. It is based on the fact that the theist (as opposed to the deist) God should be detectable by his actions in the world but has not been.

Now, as with the first cause argument of Aristotle and Aquinas, an infinite regress is avoided by saying that God had no prior cause or designer. But once you accept the logical possibility that an entity exists that had no prior cause or designer, then you have to prove why that entity can't be a godless universe.

In any case, Christian apologists are tilting at windmills. It is an incontrovertible fact about the natural world that simpler systems evolve into more complex ones without always needing the hand of a conscious craftsman.

STRONG EMERGENCE

Still, we find physicists, biologists, and theologians claiming some teleological principle in nature. In his 2004 book *The Cosmic Blueprint*, physicist, prolific author, and 1995 Templeton Prize winner Paul Davies says, "As more attention is devoted to the study of self-organization and complexity in nature, so it is becoming clear that there must be new general principles—organizing principles over and above the known laws of physics—which have yet to be discovered."[35]

South African cosmologist, Quaker, and 2004 Templeton Prize winner George Ellis has described two forms of causal action as we move from one level to another in the hierarchy of complex structure in matter. In *bottom-up action*, what happens at each level is caused by functioning from the next level below. In *top-down action,* the higher level of the hierarchy directs what happens at the lower levels.[36] Theologian Philip Clayton refers to the first as *weak emergence* and to the second as *strong emergence.*[37]

Weak emergence certainly occurs. The examples given previously, such as thermodynamics, exhibit bottom-up causality. The late prominent biochemist and theologian Arthur Peacocke defined *emergent monism* as the doctrine in which "naturally occurring, hierarchical, complex systems [are] constituted of parts that themselves are, at the lowest level, made up of the basic units of the physical world."[38] This sounds like weak emergence, but later in the same article he amplifies his definition to encompass strong emergence: "[Emergent monism views the world as] a hierarchy of interlocking complex systems and it has come to be recognized that these complex systems have a determinative effect, an exercising therefore of causal powers, on their components—a whole-part influence."[39]

Peacocke recognized that weak emergence is not what theologians are seeking. They want "a more substantial ground for attributing reality to higher level properties and the organized entities associated with them" and for this to possess a distinctive "causal (I [Peacocke] would say rather 'determinative') efficacy of the complex wholes that has the effect of making the separated, constituent parts behave in ways they would not do if they were not part of that particular complex system."[40] Theologians want strong emergence.

Ellis gives as an example of strong emergence the gas in a cylinder with a piston, in which the position of the piston determines the gas pressure and temperature. This is easily dismissed since the temperature and pressure can be calculated using the equations of thermodynamics derived in statistical particle mechanics.

What Ellis, Clayton, Peacocke, and other supernaturalists are desperately seeking is an example from science where a high-level system produces an effect at a lower level that cannot occur without some new principle coming in at that level. Then they can extend that notion to the very top level—to heaven with God himself acting down to control everything that happens below. Let's face it, downward causation is just Aristotelian "final cause," renamed because of the disrepute that the term has suffered at the hands of the scientific revolution.

Ellis's other examples include nucleosynthesis in the early universe, the collapse of the wave function, evolution, biological development, influence of mind on the body, and influence of mind on the world. Ellis doesn't mention one that I have heard before: how spinning a bicycle wheel causes the particles in the tire to move in a circle. All but one of Ellis's examples can be accounted for reductionistically within existing materialistic models.

The one example that requires some thought is the arrow of time. Ellis and authors before him note, correctly, that no direction of time is singled out in elementary particle physics. Yet an arrow of time is deeply embedded in our macroscopic experience. According to Ellis, who refers also to Paul Davies[41] and Heinz-Dieter Zeh,[42] the macrosystem acts down to the micro level, forbidding the time-reversed solutions to occur at that level.

However, they are trying to prove a principle that doesn't exist. Time-reversal is not forbidden at any level—submicro, micro, or macro. Air from outside can rush through a puncture in a flat tire to reinflate the tire. A broken glass can reassemble. These can happen. They are just very unlikely. As we saw in chapter 5, chemical reactions happen in both time directions. As Richard Feynman showed, antiparticles are indistinguishable from particles going "backward" (by our everyday standard) in time.

Peacocke cites several other examples that he claims show "the parts of these systems would not be behaving as observed if they were not part of that particular system ('the whole')."[43] Right. Like the electron in a bicycle tire. People who think this way simply have to get into the habit of picturing what is actually happening—particles are colliding with particles. Holistic thinking is not just wrong—it keeps you from grasping what is really very simple.

Peacocke lists the Belousov–Zhabotinsky reaction, which is part of a class of chemical reactions occurring far from equilibrium that exhibit spatial and temporal oscillations. These reactions are very interesting, but they are well understood in terms of the underlying chemistry.[44] They have been simulated on computers in terms of the interaction of their parts.[45] Nothing is holistic about any of them. None prove top-down causation.

Again, we have one of those questions that does not divide the burden of proof equally on both sides. Currently science has found no need for a special principle to provide for the movement from simple to complex. Computer simulations can reproduce many of the patterns using only existing principles plus randomness. No experiment has ever isolated a teleological principle in action, the way experiments can isolate the actions of gravity and electromagnetism. We can safely say, beyond a reasonable doubt, no such principle exists.

So, is some unacknowledged principle in operation at higher levels that is not the simple result of the interactions of quarks and electrons? Everyone but arrogant particle physicists would like to think so, but none has come close to providing a viable theory, much less evidence to necessitate discarding, modifying, or adding to the reductionist paradigm.

THE DEFINITION OF LIFE

One of the basic issues that divides science and religion and has great importance for the future of life on this planet, because of its huge impact on population growth, is the definition of life. The conflicts over reproductive rights and stem-cell research arise from different understandings of the definition of life. Most religions insist on the notion that humans have immaterial souls independent of their material bodies. If a belief system does not include a soul, it is not a religion. Buddhism is technically an exception, but most Buddhists hold to some sort of spiritual belief.

Belief in a soul leads to the conclusion that human life begins precisely when God inserts the soul into the body. Current Catholic dogma holds that this is at the moment of conception, so even a single fertilized egg is a living human being. Note that this is not the traditional position of the Church's leading theologians. Both Augustine and Aquinas held that the soul doesn't enter until "quickening."[46]

We can see that the notion of the soul harks back to the ancient belief that there is some animating force that gives life to an organism. As we found above, science has not found any evidence for such a force.

So, how is life defined in science? Biologists generally define an object as being alive if it is capable of metabolism and reproduction and has the ability to adapt to its environment. My colleague in the philosophy department at the University of Colorado, Carol Cleland, has cautioned scientists not to be too parochial in defining life when considering the possibilities of life on other planets—that it may be dramatically different from life on Earth. Basically, living things tend to be complex and to take energy from the environment, which they use for growth and reproduction.[47] They do not necessarily have to be composed of carbon molecules and DNA.

In their 2009 book, *Beyond Cosmic Dice: Moral Life in a Random World*, marine biologists Jeff Schweitzer and Giuseppe Notarbartolo-di-Sciara go carefully through all the various characteristics usually associated with living things and conclude:

> *There is no single unambiguous definition of life.* Most examples of life are complex; most metabolize, grow, reproduce, and evolve over time. But not all do, and not all have all of these functions present. Some physical systems also share the same characteristics. That fact is not troubling: it reflects the reality of nature. "Life" is an *arbitrary label* we apply to distinguish extremes of complexity along a continuum. We know that a block of pure quartz is not alive and that a screeching kid in the restaurant is; whatever label we paste on all those cases in between is a convenient convention but in no way reflects any fundamental break or division between the living and nonliving.[48]

So there is no scientific basis for religious arguments made against stem-cell research, early abortion, and other forms of birth control. These are areas where religious belief comes in to work against the best interests of individuals and society. Another place is in the religious objection to same-sex marriage. That objection is not based on any thought-out philosophical principle but on some obscure passages from the Bible. We have already seen that the Bible is so full of errors, contradictions, and immoral acts by its characters, including those of an ancient tribal sky-god named Yahweh, that it is virtually useless as a guide for human behavior. Abortion is still a moral issue, and most people would support rational alternatives. Just think of how better off everyone would be if couples of the same sex had all the legal rights of those in conventional opposite-sex marriages and so could adopt children who might otherwise have been aborted.

The dangers this planet faces from overpopulation hardly need detailing. All one has to do is look at the figures on population growth. In 10,000 BCE there were just 1 million human beings on Earth. In 1810 there were 1 billion. Today, we are almost at 7 billion. The estimate for 2050 is 9 billion. There simply is no way that we can sustain this rate of growth. Something has to give, and give soon. And a major cause of this problem can be laid directly at the feet of religion and its unsupportable positions on when life begins, reproductive rights, environmental controls, as well as its general distrust of science.

THE ORIGIN OF LIFE

The origin of life is certainly one of the most important outstanding questions in science today; it is not part of the current theory of evolution, which relies on the genetic system of life already being in place. In 1953, graduate student Stanley Miller, working under the guidance of the eminent chemist Harold Urey, made an amazing discovery in a very simple laboratory experiment. He found that the primary building blocks of life, the amino acids as well as other organic compounds, can be synthesized spontaneously by sending an electrical spark through a gaseous mixture of methane, hydrogen, ammonia, and water.[49]

At the time these were thought to be the ingredients of primitive Earth's atmosphere. Theists since have been quick to point out that the exact ingredients in the early Earth may have been different. They have concluded, therefore, that Miller's results were no evidence for a natural origin of life. Indeed, life did not spontaneously form from the ingredients in Miller's experiment, and no one has yet managed that feat. However, many experiments have confirmed Miller's findings with a wide range of gases.

The Miller-Urey result provided excellent evidence for a more general principle in nature that many people still find difficult to accept. They demonstrated, in the laboratory, that simple molecules can assemble naturally into more complex molecules. As described previously, this feature of simplicity evolving into complexity, which can be found in physics and other sciences besides biology, disproves the argument made by proponents of intelligent design creationism that the existence of complexity in the universe cannot be explained naturally. In the case of life, since purely natural processes can generate the basic elements of life, no obstacle exists to the ultimate assembly of these basic elements into more complex forms.

Certainly the current biological cell itself is far too complex to have arisen by purely random processes, and no biologist claims it did. However, it is now well established in the laboratory that chemicals themselves evolve independently of the specific mechanisms of biological evolution.[50] If the spontaneous formation of *some* complex system that had at least *some* of the properties of life were observed, then it would provide evidence that life had a natural origin.

The exhaustive review of the literature on life's origins provided by Albrecht Moritz in the *TalkOrigins Archive* contains many examples of proposed mechanisms and provides extensive links to references. Moritz is conservative: "After critical study of the scientific literature I conclude that advances in our knowledge, with particularly exciting findings in the last decade, have now made the spontaneous origin of life a plausible assumption."[51] Recall that it only takes a plausible assumption to defeat the God-of-the-gaps argument.

See also the beautifully illustrated article "Life on Earth" by Alonzo Ricardo and Jack W. Szostak in the September 2009 issue of *Scientific American.*[52] Here are the key concepts summarized by the editors:

> Researchers have found that the genetic molecule RNA could have formed from chemicals present on the early Earth.

> Other studies have supported the hypothesis that primitive cells containing molecules similar to RNA could assemble spontaneously, reproduce and evolve, giving rise to all life.

Scientists are now aiming at creating fully self-replicating artificial organisms in the laboratory—essentially giving life a second start in order to understand how it could have started the first time.

CHAPTER 9

TRANSCENDENCE

Epicurus has set us free from superstitious terrors and delivered us out of captivity . . . while we worship with reverence the transcendent majesty of nature.

—Cicero, *De Natura Deorum*

THE AFTERLIFE[1]

Life after death can be identified with the ancient notion that the human mind is not purely a manifestation of material forces in the brain but has a separate, immaterial component called the soul that survives the death of the brain and the rest of the body. To read the case for an afterlife made by two popular contemporary authors, see the books by Deepak Chopra[2] and Dinesh D'Souza.[3] For the philosophical case against an afterlife, see *The Illusion of Immortality* by Corliss Lamont.[4] For a history of belief in life after death, see *Life After Death* by Alan F. Segal.[5]

While the Torah, the Jewish scriptures that correspond to the first five books of the Old Testament, contains no mention of an afterlife, immortality was adopted into Judaism sometime before the first century BCE. While Plato held that the soul escapes the body after death, the Persians introduced the notion that the whole person, body and soul, survives death, which view the Jews then adopted.[6] This idea was adopted in turn by Christianity and Islam, in which it was given a far more central role than it has in Judaism.

The enormous Greek influence on Christianity that was initiated by Paul (the New Testament was written entirely in Greek) led many Christians to adopt the Greek view that only disembodied souls survive death. With the Copernican revolution in the Middle Ages, heaven was no longer a place beyond the stars, and hell was no longer inside Earth but, rather, these were viewed as immaterial places inhabited by immaterial souls. Nevertheless, bodily resurrection is still anticipated by both the Catholic Church and many Protestant sects.

As we saw in chapter 2, Augustine's view was that God created time along with the universe and is himself outside of time. Later Christian theologians formulated life after death to take place in an eternal realm disconnected from space and time. Actually, this realm should not even be characterized as "eternal" since that is a temporal term. It's kind of a constant "now." Of course, this is the Christianity of theologians, not the faithful in the pews, who are kept in the dark about theology by their preachers.

Life after death is an Eastern as well as Western idea, though the forms are quite different. The traditional belief in the East is some form of reincarnation in which souls have many lives on Earth.[7] A revised understanding of Hinduism was instituted in the *Upanishads*, a philosophical work from 2,500 years ago referred to as Vedanta, or "post-Vedic" Hinduism. According to this text, humans break out of the endless cycle of reincarnation while all our souls merge into a single reality. Buddhism adopted the Hindu notion of reincarnation (yet another version of an "afterlife") for souls that hadn't yet "achieved enlightenment," whereby the cycle ends in Nirvana and the soul, as in Vedanta, merges with reality while the individual ego is lost.

Although belief in the afterlife is a widespread notion, it is not unanimous. After all, perhaps 1 to 2 billion living people don't believe in it. We can identify four different perspectives: (1) the Eastern view of the disembodied soul undergoing reincarnations in new bodies and then eventually merging into a single ultimate reality; (2) the Western view in which the soul, even without a body, remains individually differentiated; (3) the Western view of survival of the whole person, body and soul together; and (4) denial of an afterlife. Christians who don't believe in bodily resurrection but in a heavenly realm beyond space and time still expect to meet their departed loved ones and favorite pets there as individual souls. Interestingly, this difference between East and West is a characteristic of the respective cultures, with individualism a prime trait of Americans and Eurasians, while East Asians place more emphasis on everyone harmonizing with their culture. Even East Asian beliefs in heaven or hell correspond to their cultural expectations and thus differ substantially from Western notions. You would think that if the notion of an afterlife were based on any kind of divine revelation there would be one universal version. But then there would be one religion rather than thousands.

BEYOND MATTER

If we were to try to think of the most basic disagreement between science and religion it would be on what is termed *transcendence*. The evidence gleaned from the earliest prehistoric burial sites, the historical record, and the global society today make it clear that what we commonly label as religion is characterized by the notion that something exists beyond the world that addresses our senses.

Notice that we do not require belief in God or gods in this definition. Buddhism, Taoism, and Confucianism did not have gods originally, although many of their adherents today worship gods of one sort or another. Nevertheless, the ancient sages of the East all had some concept of the transcendent.[8]

Science, on the other hand, deals only with the world of our senses. Of course, science has its theories and utilizes mathematics, which are just as much in our heads as the mantra in the head of a Tibetan monk or the prayer in the head of a Catholic nun. But to be science, thoughts must be tested against observations, while religious thoughts need not be. Indeed, when they are, as with the creation story in Genesis or intelligent design, they invariably fail the test.

Now, this does not mean that science is in anyway proscribed from applying its methods to seek empirical evidence for something transcendent. However, there exists a widespread notion, promulgated at the highest levels of the scientific community itself, that science has nothing to say about God or the supernatural. In an otherwise excellent and comprehensive report on the science of evolution and the issues involved around teaching it in schools, the National Academy of Sciences takes a strong NOMA position (non-overlapping magisteria; see chapter 1). Let me quote from the section titled "Religious Issues."

Aren't scientific beliefs based on faith as well?

> Usually "faith" refers to beliefs that are accepted without empirical evidence. Most religions have tenets of faith. Science differs from religion because it is the nature of science to test and retest explanations against the natural world. Thus, scientific explanations are likely to be built on and modified with new information and new ways of looking at old information. This is quite different from most religious beliefs.
>
> Therefore, "belief" is not really an appropriate term to use in science, because testing is such an important part of this way of knowing. If there is a component of faith to science, it is the assumption that the universe operates according to regularities— for example, that the speed of light will not change tomorrow.[9] Even the assumption of that regularity is often tested— and thus far has held up well. This "faith" is very different from religious faith.
>
> Science is a way of knowing about the natural world. It is limited to explaining the natural world through natural causes. Science can say nothing about the supernatural. Whether God exists or not is a question about which science is neutral.[10]

Now let us see why the above statement from the top science organization in America can be proven scientifically false on the grounds that it disagrees with the empirical facts.

RELIGION AND HEALTH

The National Academy of Sciences' assertion that science has nothing to say about God or the supernatural is refuted by the fact that highly respected institutions such as the Mayo Clinic, Duke University, and Harvard University have engaged in studies on the efficacy of remote intercessory prayer. If it could be shown scientifically that prayer really works, and no natural explanation can be found, then we would have an empirical case for transcendence.

The largest prayer experiment, called the *Study of the Therapeutic Effects of Intercessory Prayer* (STEP), was performed by a collaboration of researchers from six academic medical centers.[11] The study focused on 1,802 patients recovering from coronary-artery bypass surgery in six hospitals. They were divided into three groups: (1) 604 patients who received intercessory prayer after being informed that they may or may not receive prayers; (2) 597 patients who did not receive intercessory prayers after being informed that they may or may not receive prayers; (3) 601 patients who received intercessory prayers after being informed that they would receive prayers.

The investigators found no effect of intercessory prayer on the recovery of the patients examined. In fact, those patients who knew they were being prayed for fared worse![12]

Earlier Mayo Clinic[13] and Duke University[14] studies also failed to find any benefits of intercessory prayer. While several other published papers had previously reported positive results, these are the only three studies that were conducted with the highest scientific standards.

The three well-conducted studies found no significant effects *but they might have*, which would have provided evidence for the supernatural if not for God himself. The fact that patients in the STEP study fared worse after being prayed for cannot be taken as evidence for a vengeful God. They probably thought, "I must be in lot worse shape than they are telling me, if people are praying for me." And it is well established that patients' attitudes can have a strong effect on their health.

Another area of study on the health effects of religion and spirituality concerns the possible benefits of religious practice. The American public widely believes that religion and spirituality can reduce the risk of disease for adherents. A huge literature exists claiming scientific evidence that this is indeed the case. A number of reviews of the literature have strongly supported this conclusion.[15]

On the other hand, psychiatrist Richard Sloan and epidemiologist Emila Bagiella have examined a large number of such reports and concluded that there is little empirical support for these claims.[16]

A considerable effort to study the connections between religion and health is centered at the Center for Spirituality, Theology, and Health at Duke

University. Its founder, psychiatrist Harold G. Koenig, is a respectable scholar who has authored or coauthored numerous books and articles on religion, health, and ethical issues in medicine that promote the benefits of religious practice.[17] I have communicated with him personally and have found him to be a competent researcher.

Koenig coauthored one of the prayer studies mentioned above that found no evidence that prayer improves health. In an interview on the website Beliefnet, Koenig says about the Duke study, "The results are very consistent with good science and good theology." The experiment was well designed and well conducted, so it was good science. The negative results were also "good theology," according to Koenig, "because God is not predictable, he's not part of the material universe." Koenig insists that the studies "tell us nothing about the effectiveness of prayer."[18] However, if the results had been positive he would be singing another tune. Can't you see the headline? "Science Proves God Exists."

The one area where a fairly convincing case for the beneficial health effects of religious practice can be made is in churchgoing. A review of published papers on religion and health conducted by epidemiologist Lynda H. Powell and collaborators has supported this conclusion.[19]

The authors carefully chose from a large number of publications, tossing out many that failed to meet stringent controls or otherwise lacked proper scientific protocols. They labeled as "persuasive" the evidence that church attendance improves mortality. They did not say whether or not it guaranteed immortality.

However, churchgoing was pretty much the only religious practice that improved health outcomes. Here's the summary of Powell's conclusions:

> In healthy participants, there is a strong, consistent, prospective, and often graded reduction in risk of mortality in church/service attenders. This reduction is approximately 25% after adjustment for confounders. Religion or spirituality protects against cardiovascular disease, largely mediated by the healthy lifestyle it encourages. Evidence fails to support a link between depth of religiousness and physical health. In patients, there are consistent failures to support the hypotheses that religion or spirituality slows the progression of cancer or improves recovery from acute illness, but some evidence that religion or spirituality impedes recovery from acute illness.[20]

In sum, the claim that religious practices are beneficial to the health of practitioners is greatly overblown. I have a serious criticism of all these studies. From the reports I have seen, none include a control group of nonbelievers. Thus the researchers have no way of knowing what role religious belief and practice really plays in whatever effects they observe, positive or negative. Perhaps church attendance is of some value, but that can be attributed to the healthy lifestyle that is implied. There is no smoking in church, and no drinking except for the little sip of watered-down wine during Communion.

SPIRITUAL ENERGY

Energy is another physics term that is misused by those who seek a rational basis for their belief in the duality of matter and spirit. Ancient Chinese medicine is based on the concept of ch'i, which is thought of as a vital energy flow in the body. An acupuncturist regulates the flow with her needles. Modern complementary and alternative medicine also bases many of its claims on the imagined existence of human "bioenergetic fields." In 1998, nursing professor Elissa Patterson wrote in a peer-reviewed nursing journal, "We are all part of the natural harmonious energy of the universe." According to Patterson, the human energy field is part of this universal energy field and "is intimately involved with human life, often called the 'aura.'"[21]

One form of energy healing that was widely promoted in the nursing community late in the twentieth century was called Therapeutic Touch, in which the practitioner claims to regulate a patient's bioenergetic field by moving her hands over the patient's body, without actually touching the body. Therapeutic Touch has been thoroughly debunked by a wide array of researchers.[22]

Of course, claims of spiritual energy are all said to be based on quantum physics. Veterinarian Joanne Stefanatos, for example, informs us, "The principles of energy medicine originate in quantum physics."[23] In response to such claims, veterinarian David Ramey has edited a book titled *Consumer's Guide to Alternative Therapies in the Horse* that challenges the use of energy medicine in treating horses.[24]

In all these applications, energy is imagined to be some kind of immaterial, spiritual phenomenon. However, as we have already seen, energy is purely material in nature, a property of physical matter. No special form of biological energy associated with living things has ever been observed. Our bodies radiate tiny amounts of normal, purely physical electromagnetic waves such as those observed in an electroencephalogram (EEG). We also have an "aura" of infrared radiation that can't be seen with the human eye but is easily seen with infrared sensitive devices such as those carried by soldiers on the battlefield. This is simply the black-body radiation that results from us being warm bodies. We discussed black-body radiation in chapter 6. There is nothing spiritual about it.[25] A rock at body temperature has exactly the same spectrum as a live human. A rock at ambient temperature has exactly the same spectrum as a dead human.

RELIGIOUS EXPERIENCES

Another area where science has something to say about the question of a world beyond matter is in the claims of religious or spiritual experiences. These are very real to the people who have them. They often result in an epiphany that leads to a major change in the life of the one having such an experience.

If these experiences are truly profound, more than just inside the head but reaching out in some way to a separate reality, then they can be tested scientifically. And they have been. Let us review some of the evidentiary claims that have been subjected to scientific testing[26]

NEAR-DEATH EXPERIENCES: HISTORICAL DATA

One of the major examples of religious experiences for which people claim empirical support is the *near-death experience* (NDE). NDE events have attracted a large number of investigators who have a peer-reviewed journal of their own, the *Journal of Near-Death Studies*. Researchers Janice Miner Holden, Bruce Greyson, and Debbie James have assembled a comprehensive handbook on NDE research.[27] They begin with a review of thirty years of research on the subject, which can be briefly summarized as follows.

By the early 1970s, resuscitation technology had advanced to the point where many more people were being brought back from the brink of death than ever before in history. Perhaps 20 percent reported experiences of what they were convinced was another reality, a glimpse of "heaven." These reports began to get the attention of nurses and physicians. In 1976, medical student Raymond Moody published a book about these phenomena called *Life after Life*, in which he coined the term "near-death experience." Moody's book became a sensational bestseller, with 13 million copies sold by 2001.[28]

Holden and her colleagues list a number of earlier references to NDEs in popular, medical, and psychical research and many publications dating back to 1975. Almost all of these reports are anecdotal (a designation the authors avoid in favor of the term "retrospective") and are hardly likely to convince skeptics and mainstream scientists that they provide evidence for an afterlife. However, it can be safely concluded from these anecdotes that the near-death experience itself is a real phenomenon, somewhat like a dream or hallucination, but perhaps not exactly the same. The issue is whether such experiences provide any real evidence for a world beyond.

Psychologist (and reformed parapsychologist) Susan Blackmore, in her 1993 book on near-death experiences, *Dying to Live*, proposed that the phenomenon was the result of loss of oxygen in the dying brain.[29] Professional anesthesiologist Gerald Woerlee thoroughly confirms Blackmore's findings in his 2003 book, *Mortal Minds*.[30]

Many features of the NDE, especially the tunnel vision, can be simulated with drugs, electrical impulses, or acceleration such as during a ride in a centrifuge used for training fighter pilots. In his 2011 book, *The Spiritual Doorway to the Brain: A Neurologist's Search for the God Experience*, Kevin Nelson associates NDEs with a blending of two states of consciousness—wakefulness and sleep—that normally remain separate. In the case of pilots, the tunnel vision that is experienced goes away when goggles are used to apply pressure to the eyes, indicating that the phenomenon is related to the lowering of blood pressure in the eye.[31]

Despite the evidence that NDEs arise from natural brain processes,

Holden and coeditors are not quite ready to give up their quest of the afterlife. In the summary of their handbook they say:

> If it appears that the mental functions can persist in the absence of active brain function, this phenomenon opens up the possibility that some part of humans that performs mental functions might survive death of the brain.[32]

Nevertheless, they have to admit, "No single clear pattern of NDE features has yet emerged."[33] Furthermore, they do not make clear how they know that the "mental functions" they write about occurred simultaneously with the absence of brain function.

VERIDICAL NDEs

From my viewpoint as a retired research scientist, only veridical NDEs are worth studying. These are NDE experiences where the subject reports a unique perception that is later corroborated.[34] Researchers also define *apparently non-physical veridical NDE perception* (AVP) as veridical perceptions that apparently could not have been the result of inference from normal sensory processes.[35] If demonstrated by the data, these could provide the kind of evidence for consciousness independent of the body that we might begin to take seriously.

In chapter 9 of *The Handbook of Near-Death Experiences*, editor Holden reviews the attempts to verify AVP under controlled conditions. You would think the setup should be simple. Place a target such as a card with some random numbers on it facing the ceiling of the operating room so that it is unreadable not only to the patient on the table but to the hospital staff in the room. Then if a patient has an NDE that involves the commonly reported sensation of moving outside the body and floating above the operating table, he or she should be able to read that number. These out-of-body experiences (OBEs) are not always associated with NDEs, and they are treated as independent phenomena that also imply the existence of a soul independent of the body.

Holden has noted that this ideal situation is difficult to achieve with the operating room staff often glimpsing the target information, thus compromising the protocol. What's more, they have more important matters to attend to without the interference of experimenters who certainly can't control what goes on in the operating room. Holden can only report on five studies that were conducted with proper controls. She concludes, "The bottom line of findings from these five studies is quite disappointing: No researcher has succeeded in capturing even one case of AVP."[36] Note that Holden reveals her personal desires in this quotation. If she were a skeptic she might have called the result "gratifying." In either case it's best to keep an open mind.

Holden tells of receiving an e-mail from prominent NDE researcher Kenneth Ring:

> There is so much anecdotal evidence that suggests [experiencers] can, at least sometimes, perceive veridically during NDEs . . . but isn't it true that in all this time there hasn't been a single case of a veridical perception reported by an NDEr under controlled conditions? I mean, thirty years later, it's still a null class (as far as we know). Yes, excuses, excuses—I know. But, really, wouldn't you have suspected more than a few such cases *at least* by now?[37]

MARIA AND THE SHOE

Dinesh D'Souza is deeply impressed by NDEs, saying, "On the face of it, they provide strong support for life after death."[38] Few researchers in the field have gone so far.

D'Souza tells us of the case of a Seattle woman named Maria who experienced an NDE after a heart attack. Maria told social worker Kimberly Clark that she had separated from her body and floated outside the hospital. There, on the third floor ledge near the emergency room, she saw a tennis shoe with a worn patch. Clark checked the ledge and retrieved the shoe.[39]

However, there is no independent corroboration of this event. We only have Clark's report. No one could ever trace down Maria to corroborate her story. We have to take Clark's word for it. Later investigators found that Clark had embellished the difficulty of observing the shoe on the ledge. By placing their own shoe in the same position, they found it was clearly visible as soon as you stepped into Maria's room.[40]

THE BLIND SHALL SEE

Probably the most sensational claims in NDE research involve blind people reporting out of body experiences in which they were able to see. Physician Larry Dossey is the author of several popular books that promote spiritual healing such as prayer; I have clashed with him on occasion.[41] In his book *Recovering the Soul*, he claimed that a woman named Sarah had an NDE in which she saw

> a clear, detailed memory of the frantic conversation of the surgeons and nurses during her cardiac arrest; the OR [Operating Room] layout; the scribbles on the surgery scheduling board on the hall outside; the color of the sheets covering the operating table; the hairstyle of the head scrub nurse; the names of the surgeons on the doctors' lounge down the corridor who were waiting for her case to be concluded; and even the trivial fact that the anesthesiologist that day was wearing unmatched socks. All this she knew even though she had been fully anesthetized and unconscious during the surgery and the cardiac arrest.[42]

And, on top of that, Sarah had been blind since birth!

Kenneth Ring and Sharon Cooper report that, when asked by other investigators to give more details, Dossey admitted this was a complete fiction.[43] Susan Blackmore also uncovered Dossey's fabrication.[44]

Ring and Cooper note that Blackmore "reviewed all the NDE evidence and concluded that none of it holds up to scrutiny." According to Blackmore, "There is no convincing evidence of visual perception in the blind during NDEs, much less documented support for veridical perception."[45] Ring and Cooper's later investigations also provided no veridical evidence.

NEAR-DEATH EXPERIENCES: RECENT DATA

In 2010 a new book on NDEs appeared, *Evidence of the Afterlife: The Science of Near-Death-Experiences*, by Jeffrey Long, MD, "with" journalist Paul Perry.[46] Thanks to considerable media hype, this book moved quickly to the bestseller lists. Long is a radiation oncologist who, with his wife, Jody, gathered thousands of accounts of near-death experiences. They did this by setting up a website asking for personal narratives of experiences. Besides providing their personal stories, respondents filled out a 100-item questionnaire "designed to isolate specific elements of the experience and to flag counterfeit accounts." The result is the largest database of NDEs in the world, with more than sixteen hundred accounts.

Long claims that medical evidence fails to explain these reports and that "there is only one plausible explanation—that people have survived death and traveled to another dimension." After studying thousands of cases, Long concludes: "NDEs provide such powerful scientific evidence that it is reasonable to accept the existence of an afterlife."[47]

In fact, there is little or no science in Long's book. It is based totally on anecdotes collected over the Internet where you can find limitless unsupported testimonials for every kind of preposterous claim. Not all anecdotes are useless. They can point the way to more serious research. But when they are the *only* source of evidence they cannot be used to reach extraordinary (or even ordinary) conclusions. To scientifically prove life after death is going to require carefully controlled experiments, not just a lot of stories. The plural of anecdote is not "data."

The question raised by near-death experiences is whether they provide evidence that mind and consciousness are more than just the product of a purely material brain. Such a conclusion contradicts the mass of evidence gathered so far in the neurosciences (see chapter 10) and will be accepted only when the data are totally convincing.

PROBLEMS WITH NDEs

There is no objective evidence that brain function stops entirely during a reported NDE. That an NDE actually occurred during a flat EEG (rather than before or after) is impossible to prove anyway. But even a flat EEG does not signal brain death, as many people mistakenly believe, since it just responds to the outer portions of the brain and does not catch activity deep inside the brain. If the properties traditionally attributed to the soul reside solely in the material brain and nervous system, then this is sufficient to rule out life after the death of the brain.

There are several excellent books and papers presenting strong, detailed arguments showing why the data from NDEs does not provide any evidence for an afterlife. Besides Blackmore's *Dying to Live* and Woerlee's *Mortal Minds*, there is *Religion, Spirituality, and the Near-Death Experience* by Mark Fox.[48] In 2007, Keith Augustine, then executive director of the Internet Infidels,[49] published an exhaustive three-part series of articles in the *Journal of Near-Death Studies*.[50] Each of these articles is accompanied in the same volume with several criticisms from researchers in the field followed by a response to those criticisms from Augustine. An updated, unified version of all three of Augustine's papers is available on the Secular Web.[51]

Let me mention just a few of Augustine's observations, along with those of other researchers that are particularly compelling. Refer to Augustine's paper to get the details and references to the original work on which he relies.

- 80 percent of those who come as close to death as possible without dying do not [recall having] an NDE. So it is not a common experience.
- Existing research presents no challenge to the current scientific understanding of NDEs as hallucinations.
- NDE studies, taken as a whole, strongly imply that whatever these experiences are, they are characterized by features that one would expect of internally generated fantasies but not of any putative "disembodied existence."
- As encounters with living persons repeatedly crop up in NDEs (one out of ten times), the less NDEs look like visions of another world and the more they appear to be brain-generated hallucinations triggered by a real or perceived threat to the experiencer's well-being.
- The only NDE experiences that are common among all cultures are encountering other beings and other realms. Otherwise the details depend on culture.
- Electroencephalograms and imaging techniques indicate that epileptic activity in the temporal lobe of the brain, specifically the TPJ,

or temporo-parietal junction, consistently results in out-of-body experiences (OBEs). Furthermore, many of the experiences reported by epileptics and those who have had their temporal lobe electrically stimulated match those of OBEs. Since the TPJ is a major center of multisensory integration of body-related information, it is not surprising that interfering with neural processing or cerebral blood flow in this area, or providing conflicting somatosensory inputs, results in dysfunctional representation. This provides strong evidence that OBEs are brain-induced and are localized in the temporal lobe. As mentioned, OBEs are often, but not always, associated with NDEs.

- Despite repeated assertions of quite frequent paranormal abilities (healing powers, prophetic visions of the future) manifesting after NDEs, often endorsed by NDE researchers, no experiencer has had their alleged psychic powers tested in a controlled experiment. The prophecies have been either vague or dramatically wrong. For example, in *Saved by the Light*, Dannion Brinkley reports his NDE and makes many predictions about the future.[52] The book was adapted in 1995 for a Fox Television movie starring Eric Roberts and was one of the highest rated television movies in that network's history.[53] But not one of Brinkley's predictions came to pass.

Many NDE researchers still hope to find evidence for an afterlife despite their own sincere admission that the data, so far, are simply not there. Augustine is careful to note that NDE researchers' beliefs are not to be confused with their actual findings. While the great majority of NDE researchers are no doubt honest and do not hide data that fail to confirm their beliefs, they are hardly disinterested in the question of survival after death. Who wouldn't be motivated by the possibility of discovering an afterlife?

Several authors have suggested that those who have NDEs cannot distinguish whether a private experience is a brain-based hallucination or a peek into the afterlife; therefore, the afterlife hypothesis is not falsifiable. This is wrong. These authors are like those who say science can never prove God exists. The existence of a realm beyond matter could be easily demonstrated by one returning from an NDE, OBE, or other religious experience if that individual has important information about the world that she or no one else could possibly have known; such knowledge could be verified scientifically. With millions of such experiences yearly you would expect a few to result in verifiable knowledge, if such experiences had anything at all to do with an immaterial reality. So far none have. Once again we have the absence of evidence *that should be there* if a certain phenomenon exists, which can be taken as evidence that the phenomenon does not exist.

REINCARNATION

The scientific study of reincarnation is a minefield of unsupported claims and lucrative hoaxes, such as the infamous fifty-year-old case that resulted in the bestselling book *The Search for Bridey Murphy* by Morey Bernstein.[54] Although thoroughly debunked,[55] Bernstein's book has gone through four editions, the most recent appearing as late as 1991.

The reincarnation debate was taken to a more serious level by the work of Ian Stevenson, a psychiatrist and University of Virginia professor. Deepak Chopra, in his 2006 book about immortality titled *Life after Death: the Burden of Proof*, cites Stevenson as providing strong empirical evidence for reincarnation.[56]

Over the years Stevenson collected thousands of cases of children in India and elsewhere who talked about their "previous lives." Many seemed quite accurate and sometimes the child had marks or birth defects that corresponded closely to those of the deceased person the child claimed to remember.[57]

Leonard Angel has written a review of Stevenson's monumental two-volume tome *Reincarnation in Biology*.[58] Angel says, "Close inspection of Stevenson's work shows that time after time Stevenson presents tabular summaries that claim evidence was obtained when, in fact, it was not. . . . Stevenson's case, irreparably, falls apart both in the presentation of evidence and in his analysis of evidence supposedly obtained."[59]

INVESTIGATING THE PARANORMAL

For more than 150 years, attempts have been made to find scientific evidence for special powers of the mind that violate the principles known to science. These purported powers came to be classified into three types of *psychic* or *paranormal* phenomena: (1) *extrasensory perception* (ESP), in which minds communicate outside the normal physical channels; (2) *telekinesis*, in which the mind can move physical objects; and (3) *precognition*, in which minds predict the future.

While the observation of any such violation might still be explicable in terms of some new natural principle, as is usually done whenever a conventional scientific experiment or observation cannot be explained with existing knowledge, the motivation of paranormal researchers, now called *parapsychologists*, has always been something even more profound and world-shaking. They want to prove the existence of the soul. This is clear from their personal musings, which are strongly spiritual in nature.

In my 1995 book *Physics and Psychics*, I reviewed the history of psychic phenomena, or simply *psi*, from a physicist's viewpoint.[60] For a philosopher's perspective, read *The Transcendental Temptation* by Paul Kurtz, published in 1986.[61] While much has been written on the subject since these books, there is little that is new.

The scientific search for the soul began in the late nineteenth century with experiments on so-called spiritualist mediums conducted by prominent physicists William Crookes and Oliver Lodge. Since then the history of paranormal studies has been a series of extraordinary claims of convincing evidence for psychic phenomena, enthusiastically reported in the news media, followed by the collapse of those claims under the scrutiny of skeptics, and, more important, the failure of such claims to be independently replicated. To the present day, paranormal studies have been plagued by those flaws that arise in any investigation whenever investigators who have an emotional interest in a particular outcome fail to take sufficient care to rule out other, usually more mundane, possibilities.

Rather than going over old ground, allow me to discuss a recent claim that, like most paranormal claims, attracted considerable media attention before fizzling out, which is typical of the whole history of the subject.

In 2011, Daryl J. Bem, a psychologist at Cornell University, published a paper titled, "Feeling the Future: Experimental Evidence for Anomalous Retroactive Influences on Cognition and Affect" in the peer-reviewed *Journal of Personality and Social Psychology*.[62] Quoting from the abstract:

This article reports 9 experiments, involving more than 1,000 participants, that test for retroactive influence by "time-reversing" well-established psychological effects so that the individual's responses are obtained before the putatively causal stimulus events occur. Data are presented for 4 time-reversed effects: precognitive approach to erotic stimuli and precognitive avoidance of negative stimuli; retroactive priming; retroactive habituation; and retroactive facilitation of recall. All but one of the experiments yielded statistically significant results.

In chapter 5, we saw how a direction of time cannot be found in the fundamental principles of physics, classical or quantum. Our everyday, familiar "arrow of time" is a statistical effect that results from the fact that we and our surroundings are composed of trillions of particles that move around largely randomly. When we are dealing with a few particles at the quantum level, time reversibility can occur, and, in fact, this helps explain a lot of "spooky" quantum behavior.[63] But Bem's subjects and their brains are macroscopic objects in which quantum mechanics plays no significant role (see chapter 11). So the observation of reverse causality at the macroscopic level would constitute a dramatic violation of our best understanding of the natural world.

Other psychologists immediately jumped on Bem, disputing his claim that his results were empirically significant. Eric-Jan Wagenmakers and three collaborators at the University of Amsterdam reanalyzed Bem's data using a statistical technique called the "Bayesian t-test" and concluded that "the evidence for psi is weak to nonexistent."[64] The authors urged the whole psychology community to use better analytical methods.

Jeffrey N. Rouder, of the University of Missouri, and Richard D. Morey, of the University of Groningen, also performed a Bayesian statistical evaluation of Bem's data and found that the data "yield no substantial support for psi effects of erotic or neutral stimuli." Some of Bem's other data tested positive, but they were insufficient for the authors to be convinced of the viability of psi.[65]

Researchers Stuart J. Ritchie, Chris French, and Richard Wiseman have rerun Bem's experiment and failed to replicate his results. The *Journal of Personality and Social Psychology*, which published Bem's paper, rejected this failure to replicate. According to the British science journalist Ben Goldacre, the reason given was that the editors of that journal "never publish studies that replicate other work."[66] This is very strange, since the result was not replicated. In all my years of research I have never previously come across a case where a failure to replicate an extraordinary published result has been refused publication in the same journal where the original result appeared.

In a detailed article in *Skeptical Inquirer*, University of Toronto psychologist James E. Alcock has carefully examined Bem's claims. After discussing many methodological and analytical problems in the nine experiments, Alcock concluded,

> Overall, then, this is a very unsatisfactory set of experiments that does not provide us with reason to believe that Bem has demonstrated the operation of psi. All that he has produced are claims of some significant departures from chance, and these claims are flimsy.[67]

Alcock's article also has a nice summary of five other well-publicized claims of psychic effects, beginning with the work of the famous Duke University researcher Joseph Banks Rhine, who coined the term "ESP." In a 1934 book titled *Extra-Sensory Perception*, Rhine wrote: "It is independently established that on the basis of [Rhine's card-guessing experiments] alone that Extra-Sensory Perception is an actual and demonstrable occurrence."[68] Alcock notes, "Methodological problems with [Rhine's] experiments eventually came to light, and as a result, parapsychologists no longer run card-guessing studies and rarely even refer to Rhine's work."[69]

Despite a century and a half of failure to demonstrate the existence of psi in the laboratory, a huge literature exists claiming scientific evidence for special powers of the mind that at least hint at something beyond matter that is accessible to humans. This literature suffers from many problems. Much of it is anecdotal, and thus virtually useless scientifically, since we have no way of checking the veracity of such testimony. Only carefully controlled experiments that provide *risky* empirical tests will convince the scientific skeptics, and until the skeptics are convinced, the hypothesis will remain unproven. Despite the common charge, skeptics in science are not dogmatic. As mentioned several times, they will readily follow where the evidence leads. They always have.

While paranormal studies often involve controlled experiments, few meet the stringent standards found in the basic sciences. For example, positive effects are often claimed at such a low level of statistical significance that a simple statistical fluctuation would reproduce the observation as often as once every 20 times the experiment is repeated (p-$value$ = 0.05).[70] In this case, one must accept the more parsimonious explanation that the effect was a statistical artifact rather than the occurrence of a miracle. While this low standard is often used in biomedical research, such a weak criterion is unacceptable in those sciences that deal with extraordinary phenomena. In physics, a claimed new effect is not considered too seriously until it is shown that it would not be reproduced as a statistical artifact once in 10,000 cases, (p = 0.0001). While we can sympathize with the pressure on medical researchers to publish any promising therapy that might save lives, these researchers would do better and avoid useless effort and the numerous false reports that appear in medical literature by setting their limit to, say, 1 in 100.[71] Like physicists, researchers of the paranormal are not in the business of saving lives but of seeking out new phenomena. Thus parapsychologists should be subject to the same standard demanded of physicists.

Attempts have been made to use a technique called *meta-analysis* to try to glean statistically significant results from individually insignificant data.[72] This is like the old joke told by Ronald Reagan (blessed be his name) about the kid on Christmas morning digging through a pile of horse manure because "there has to be a pony in there somewhere." Meta-analysis is totally unreliable and a waste of time in searching for a phenomenon not evident in individual experiments.[73] While the procedure can be useful for discerning trends, it must be applied with great caution. I cannot think of a single major discovery in science that has been made with meta-analysis alone.

No positive claim of psi has ever stood up to the same critical scrutiny applied in the mainstream sciences whenever an extraordinary event is observed. We can safely conclude that psi does not exist beyond a reasonable doubt.

THE PHYSICS OF CHRISTIANITY

If you thought that Frank Tipler's *The Physics of Immortality* described in chapter 8 was off the wall, you should read his followup, *The Physics of Christianity*.[74] There he claims to provide Christians with a rational basis for their faith based solely on physics.

Tipler rejects the definition of a miracle as an event that violates natural law. He asserts that miracles are perfectly natural phenomena, just highly improbable or unexpected. He then provides physical mechanisms for all the important miracles associated with the life of Jesus, from his virgin birth to his resurrection. How plausible they are is another matter.

Tipler says that virginal conceptions occur in nature and result in perhaps 1 in 30 human births. He notes that it is easy to induce egg cell division in the laboratory, so why not in a female body? It is thought that virgin births should always result in a female child, since all the genes are from the mother and females have two X chromosomes while males generally have the combination XY. However, Tipler notes that 1 in 20,000 males have two X chromosomes wherein their maleness comes from a key gene that has been inserted into an X chromosome. He conjectures that this gene may have been present in one of Mary's X chromosomes but inactivated in her case, becoming activated with Jesus. Tipler writes that a DNA analysis of the blood on the Shroud of Turin is XX, thus providing empirical verification for the virgin birth of the body that was covered with that cloth.

However, this claim is dubious, to say the least. There could have been contamination from anyone who ever touched the shroud, as the scientist who discovered it in the first place explained. Besides, the so-called blood on the shroud has been identified as paint.[75]

In Tipler's scenario, the miracles performed by Jesus during his lifetime, such as walking on water and raising the dead, were produced by a single mechanism—the annihilation of protons and electrons into neutrinos and nothing else.[76] In the case of Jesus walking on water, protons and electrons in the normal matter in a layer of water under his feet are annihilated. The neutrinos produced go off invisibly downward with high momentum, the upward recoil enabling Jesus to keep from sinking. (Or, he walked on rocks.)

As for the resurrection, the whole body of Jesus is converted to invisible neutrinos, with some heat also released to burn the burial cloth. To reappear before his disciples, Jesus reversed the process, as he did to raise the dead.

Ridiculous? Absolutely. Then why mention it? Tipler is actually a very knowledgeable and accomplished theoretical physicist. I would like to think he is pulling our legs, but an e-mail he once sent me leads me to think that he is serious. Either way, what Tipler has done is to try to provide natural explanations for the claimed miracles that are so central to Christian faith. You can see how hard, if not impossible, that is.

THE REMATERIALIZATION OF MATTER

As we have seen, religious apologists and quantum spiritualists claim that relativity and quantum mechanics have "dematerialized" matter, replacing the classic form of reductionist materialism with a new holistic paradigm in which body and spirit form a united whole. However, the facts are that twentieth-century physics has replaced classical physics with something even more materialist and reductionist. The classical physics of the nineteenth century still had some continuous fields such as the electromagnetic field. Quantum physics has only discrete particles like photons all the way down. Quantum, after all, means "discrete."

Twentieth-century physics did not dematerialize the universe—it rematerialized it. As we saw in chapter 5, at the end of the nineteenth century, electromagnetic waves were thought to be vibrations of an invisible, continuous medium called the ether that supposedly pervaded holistically throughout all of space. In 1905, Einstein did away with the need for the ether with his theory of special relativity. In the same year he showed that light was composed of particles that later were named photons.

Apologist Ernan McMullin interprets relativity to imply that energy can be transformed into what he calls "massless" energy and the reverse.[77] This is a gross misunderstanding of the meaning of Einstein's famous equation $E = mc^2$, which equates mass with the energy a particle has in the reference frame at which it is at rest. The massless energy he refers to is the kinetic energy carried by photons. While photons are indeed massless, they are still *matter* since they carry momentum and energy and can impart a force on an object by colliding with it, the same as an electron or quark does. Photons are particles. The processes McMullin refers to, the creation of electron-positron (antielectron) pairs by the collision of two photons and the reverse annihilation of pairs into photons, are totally particulate—two particles in, two particles out.

Furthermore, photons are affected by gravity, another defining property of matter. For example, in famous experiments that verified Einstein's general theory of relativity, photons were observed to be deflected gravitationally by the sun.[78]

A PERSONAL REMEMBRANCE

I was raised as a Catholic in Bayonne, New Jersey, and was surrounded by Catholics in my family and in the neighborhood. Although I attended public schools, I had catechism training and was fully confirmed as a Roman Catholic. Childhood memories are weak, but I don't remember ever being taught that I had to await the Second Coming to reach heaven. It certainly was the belief of every Catholic I knew that when we die, our disembodied souls go *immediately* to heaven. Few Christians then, and none now, figured on going to hell—even temporarily in what is called purgatory.

I always imagined my dead grandparents looking down on me when I did something bad. My parents, aunts, and uncles all believed that their departed loved ones, especially deceased children, were already with Jesus. And their priests certainly never dissuaded them from that view.

Yet the official doctrine of the Catholic Church remains one in which only Jesus, Mary, and a small number of already spiritually perfected souls such as the apostles and saints go straight to heaven upon death. Those with mortal sins on their souls go immediately to hell to be tortured forever with no hope of mercy.[79] The rest, even the most pious nuns who spend a lifetime reciting the Hail Mary and whose thoughts have little time for anything else, but who do not quite achieve spiritual perfection, must have their souls purified in the hell-like fires of purgatory.[80] Their time in purgatory can be mitigated by friends and family offering masses (at a price), lighting candles (at a price), and praying. When their souls are sufficiently purified, after perhaps a thousand years of torture, God releases them into heaven. But some may have to await the final judgment associated with the Second Coming, which has not yet occurred two thousand years after Jesus promised it would come in a generation.[81] That could still be a billion years away.

During my life I have gone to countless Catholic funerals and to funerals in many Protestant denominations. I have never heard a Catholic priest or Protestant minister tell the bereaved that their departed loved one was suffering in purgatory awaiting the last judgment. The grieving were uniformly assured that their loved ones were right at that moment in paradise, though few that I recall had achieved "spiritual perfection."

I also have the memory that most of these believers did not really think in terms of a full body resurrection, complete with hangnails and sexual organs. They imagined some kind of spirit, like a ghost, who might still come down from heaven and walk among us, along with angels and demons, who were also invisible and not made of matter. Anyway, the point is that there is a huge gap beween what the average Christian believes and what Christian theologians teach.

SCIENCE AND EASTERN MYSTICISM

We have seen how physicist Fritjof Capra and others have tried to draw parallels between modern physics and Eastern mysticism. Mysticism is associated with all religions, but Eastern religions, in particular Hinduism and Buddhism, are more deeply based on the mystical experiences obtained during meditation. While we have been primarily concerned with the conflict between science and the Abrahamic religions, in order to be complete, at least a brief discussion of science and Eastern mysticism is called for. Do science and mysticism conflict? Are science and Eastern mysticism compatible? Here I arrive at perhaps an unexpected conclusion. Science and Eastern mysticism do not conflict, but they are still incompatible.

I need to make clear that the mysticism discussed in this section is the real thing, not the phony variety I have referred to as "quantum spirituality." While some genuine mystics have been bamboozled into thinking quantum physics somehow confirms their beliefs, traditional Eastern mysticism predates science and has only an indirect connection with it today. That connection essentially just involves imaging studies of the brains of meditating mystics by neuroscientists (see chapter 11) and other observations, and not any sharing of insights about reality.

Empirical evidence seems to exist that meditation has some physical and psychological benefits, but this does not imply any supernatural elements. Meditation is not necessarily a transcendent spiritual experience.

Let us first see why genuine mysticism does not conflict with science. In his 2010 book *Piercing the Veil: Comparing Science and Mysticism as Ways of Knowing Reality*,[82] religious historian and philosopher Richard H. Jones notes that science uses sensory experience to divide the world into distinguishable structures in order to see what things are and how they work. That is, science reduces everything to its parts. But more than that, as we have emphasized, those parts are described in terms of mental concepts that need not have any direct one-to-one correspondence with whatever reality is out there.

By contrast, mystics attempt to use meditation either to eliminate the structures and concepts that our minds have created in order to see things as they "really are" (*mindfulness*) or to empty the mind altogether of conceptual, emotional, sensory, and any other internal distinguishable content (*depth-mystical experience*). To the extent that mysticism is strictly a mental experience absent of any formation of concepts about the nature of the physical world, then it does not conflict with science.

Jones points out that although many physicists have studied mysticism, especially those associated with the development of quantum mechanics in the early twentieth century, notably Schrödinger, they were not themselves mystics. Furthermore, because of the diametrically opposed approaches of science and mysticism, no contribution to the conceptual development of science has ever been made by mysticism.[83]

Then how can science and mysticism be incompatible? Once out of the meditative state, mystics feel the need to interpret their experiences, and they generally do so within the frameworks of their individual religious and cultural traditions. That is, they differentiate just as scientists do. As a result, a wide range of disagreements occurs, even within the major religions, as well as disagreements with the scientific view of nature.

Jones tells us, "Mystics claim their experiences give a knowledge of fundamental reality."[84] They believe that the reality they experience is transcendent and the source of all existence. Supernatural beliefs that have no basis other than impressions gained during meditation are widespread. Even Buddha accepted rebirth and karma.

A general agreement among mystics does seem to exist that "reality consists of a hierarchy of emanations from a transcendental source in the opposite direction of materialism: spirit appears first, then consciousness, then life, and lastly matter."[85] If not in conflict with scientific facts, this differentiated set of concepts strikes me as inconsistent with the mystical claim of the "unity of all things" supposedly gained under meditation. In any case, the mystical hierarchy is incompatible with the hierarchy science infers from its data: matter $\rightarrow$ life $\rightarrow$ consciousness, with no spirit and perhaps even no separate entity we can associate with consciousness. Without any evidence or other rational basis, science cannot accept the notion of transcendence. Indeed, as we will see, all the evidence points to the mystical experience being all in the head (see chapter 11).

CHAPTER 10
BEYOND EVOLUTION

We do not act rightly because we have virtue or excellence, but we rather have those because we have acted rightly.

—Aristotle

JUSTICE AND MORALITY

If we possess only a material mind, where do our moral ideas come from? How can particles moving around in the brain produce notions of right and wrong, good and evil? Theists say they can't.

An almost universal conviction among theists is that, without God, the world would be a place of utter depravity in which "anything goes." Many refer to a quotation attributed to the great Russian author Fyodor Dostoevsky (died 1881), purportedly from *The Brothers Karamazov*: "If God does not exist, everything is permitted." This line expresses a popular sentiment.

According to Dinesh D'Souza, one of the major reasons so many people seek an afterlife is they want to believe that the universe is just.[1] Life in this world is obviously unjust, with many rewards for the wicked and few for the virtuous. An afterlife based on justice makes it all come out even after your soul transfers from the material world to the spiritual world.

D'Souza argues that this is "why humans continue to espouse goodness and justice even when the world is evil and unjust. We seek to repudiate the laws of evolution and escape control of the laws of nature." Why do we do this? Because we have made "the presupposition of an afterlife and the realization of the ideal of cosmic justice makes sense of our moral nature better than any competing hypothesis."[2]

According to D'Souza, a presupposition is a hypothesis that says, "This is the way things have to be to make sense of the world." It is tested by asking, "How well does it explain the world?" His specific hypothesis is that "there has to be cosmic justice in the world in order to make sense of the observed facts about human morality."[3]

So he is cleverly turning the morality issue into a scientific argument, which is fine by me because that puts it on my home ground. Forget what theologians say. Forget what moral philosophers say. According to D'Souza, scientific observations of human behavior provide evidence for the existence of cosmic justice. And, since justice is obviously unavailable in this life, it follows that there must be an afterlife to provide it.

It seems to me that D'Souza has the argument turned around. If people believed that cosmic justice will be meted out in the afterlife, they wouldn't have any need to worry about justice in this life. On the other hand, people who don't believe in justice in the afterlife have a strong reason to see that justice is done in this life. Thus, belief in the afterlife has a negative impact on society. The hypothesis of no afterlife makes much more sense of observations than does D'Souza's hypothesis.

DID MORALITY EVOLVE?

Starting with Darwin[4] and his "bulldog" Thomas Huxley,[5] many authors have speculated about how morality may have evolved by evolutionary processes.[6] In 1976, Richard Dawkins ("Darwin's Rottweiler") introduced the term *selfish gene* to the lexicon of evolution.[7] Built on earlier work by George C. Williams,[8] the idea is that the basic unit of evolution is not the individual organism but the *gene*, the collection of molecules that carry on the individual's genetic information to the next generation. This is what really "wants" to survive, if we can use that metaphor.

The selfish gene explains why most parents would readily trade their own lives for their children's. And the selfish gene model of morality doesn't just predict that you will only act altruistically toward your family; it says you will tend to act altruistically toward those who most resemble you. It predicts racism as well as charity. This is not morality. This is not spirituality. This is pure, reductionist, materialist, natural selection.

Evolutionary biologist Robert Trivers observes that humans and other animals behave generously toward others when they expect something in return. Natural selection provides survival instincts to those who engage in mutually beneficial exchanges.[9]

Francis Collins is skeptical, however, asserting: "The hardwired behavior of the worker ant is fundamentally different from the inner voice that causes me to feel compelled to jump into the river to save a drowning stranger, even if I'm not a good swimmer and may myself die in the effort."[10] He doesn't report an actual instance here, just a pious opinion of his own righteousness. In any case, Collins's point is that the stranger will not have Collins's genes, so there is no evolutionary advantage to saving him.

Similarly, D'Souza insists that evolution cannot explain "the good things we do that offer no return." He provides as examples people giving up their seats on a bus to the elderly, donating to charities, or agitating for animal rights or against religious persecution in Tibet.

Group selection has long been postulated, controversially, as a way to reconcile evolution with moral behavior. Patriots frequently sacrifice their lives for their friends and countries. But D'Souza claims the argument has a fatal flaw. He asks, how would a tribe of individuals become self-sacrificing in the first place? Cheaters would be more likely to survive than their more altruistic fellow tribesmen.

D'Souza recognizes that there can be an ulterior, selfish motive to being recognized as a moral person. However, he says we still must confront the Machiavellian argument that "the man who wants to act virtuously in every way necessarily comes to grief among the many that are not virtuous." He claims that true morality, true virtue, rises above all this, acting without regard to self-interest.[11]

Evolution, according to D'Souza, cannot explain how humans became moral primates. He tells us, "Humans recognize that there is no ultimate goodness and justice in this world, but they continue to hold up these ideals." Why? Because they expect to be rewarded in the afterlife. Thus, according to D'Souza, the existence of the afterlife is "proved" by the observation of altruistic behavior in humans despite the nonexistence of earthly reward.[12] It's not clear to me why people could not be just fooling themselves.

Note that D'Souza's hypothesis implies that the motivation for altruistic behavior is self-interest after all! Is it not the extremity of self-interest to want to live forever in the first place and to expect a special reward for your righteousness when you get to the afterlife?

What are the observations that D'Souza takes as evidence for cosmic justice? He admits that morality is almost universally violated. However, universal criteria and standards that everyone refers to nevertheless characterize human behavior. Why should these criteria exist at all? D'Souza claims that they defy the laws of evolution, so they can't be natural. He asserts, "Evolution implies that we are selfish creatures who seek to survive and reproduce in the world."[13] This is contrary to moral behavior. Moral behavior frequently operates against self-interest.

D'Souza's hypothesis also predicts that only those who believe in an afterlife will exhibit altruistic behavior. That hypothesis can be easily tested. We just need to gather a sample of those who don't believe in an afterlife and see whether they are significantly less virtuous than those who believe.

WHAT DO THE DATA SAY?

There are at least a billion people in the world today who do not believe in God or immortality. Are they exceptionally immoral compared to believers? Do they not experience love? Are they all cannibals? Hardly. There is no evidence that nonbelievers are less moral than others. In fact, there are some indications that they are more moral, at least as measured by the types of societies they live in.

Skeptic publisher and *Scientific American* columnist Michael Shermer addressed this question in his important book *The Science of Good and Evil*. He reports, "Not only is there no evidence that a lack of religiosity leads to less moral behavior, a number of studies actually support the opposite view."[14]

Here is a list of examples Shermer gives from the literature:

- A 1935 survey found that honesty *decreased* with religiosity.[15]
- A 1950 survey of 2,000 YMCA associates discovered that atheists and agnostics are more willing to help the poor than those who rated themselves as being deeply religious.[16]
- A 1969 survey reported no difference in the self-reported likelihood to commit crimes between children who attended church regularly and those who did not.[17]
- A 1975 study showed that college-age students in religious schools were no less likely to cheat than their atheist and agnostic counterparts in nonreligious schools.[18]
- A comprehensive survey of studies in the psychology of religion revealed that there is a consistent positive correlation between "religious affiliation, church attendance, doctrinal orthodoxy, rated importance of religion, and so on" with "ethnocentrism, authoritarianism, dogmatism, social distance, rigidity, intolerance of ambiguity, and specific forms of prejudice, especially against Jews and blacks."[19]

Bringing these reports up to date, in 2005, freelance researcher Gregory S. Paul published the results of surveys that correlate religiosity in various first-world nations with social health. He concluded that "in almost all regards the highly secular democracies consistently enjoy low rates of social dysfunction, while pro-religious and antievolution America performs poorly."[20] Similar results are reported by sociologists Pippa Norris and Ronald Ingelhart in their 2011 book *Sacred and Secular*,[21] and by sociologist Phil Zuckerman in his 2009 book *Society without God.*[22] The facts are clear and simple: many people behave well, even better, without belief in God.

For facts about the negative social impact of religion see "The Social Implications of Armageddon" by Kimberly Blaker.[23] We will uncover more evidence in chapter 14.

Those who claim God as the source of morality cite two sources: scriptures and inner feelings. The Bible and Qur'an contain many moral teachings that, without getting too pedantic about defining good and evil, most of us would regard as good. Both the Old and New Testaments and the Qur'an teach the Golden Rule. Leviticus 19:34 says, "But the stranger that dwelleth with you shall be unto you as one born among you, and thou shalt love him as thyself; for ye were strangers in the land of Egypt." Mathew 7:12 says, "Therefore all things whatsoever ye would that men should do to you, do ye even so to them: for this is the law and the prophets." (See also Luke 6:31 and Luke 10:25–28). Muhammad said, "That which you want for yourself, seek for mankind."[24]

But the Golden Rule can hardly be attributed to the Abrahamic God. Confucius said, "Here certainly is the golden maxim: Do not do to others that which we do not want them to do to us."[25] Going back even earlier, Thales of Miletus said, "Refrain from doing what we blame in others for doing."[26] And how about this from an Egyptian papyrus from the Late Period (640–323 BCE): "That which you hate to be done to you, do not do to another."[27]

Okay, so the Abrahamic God could have planted these ideas in people's minds from the beginning. However, the truth is that we cannot point to the scriptures of Jews, Christians, and Muslims as the unique origin of the Golden Rule. More likely, thoughtful human beings reasoned it out in their own minds as a useful principle that humans could live by to make a better society for all.

The Old Testament is full of what any thinking person today would regard as evil acts committed by God or in the name of God: slavery, murder, genocide, floods, famines, plagues, earthquakes, and every conceivable cruelty and injustice.[28]

Furthermore, the New Testament is hardly the handbook for righteous behavior that Christians think it is. Jesus was not exactly a paragon of morality—at least not as most people today would interpret his behavior. According to the Gospels, Jesus told us to abandon our families and follow him, and he threatened us all with eternal damnation if we do not. From Matthew 10:34: "Think not that I come to send peace on earth: I come not to send peace, but the sword. For I am to set a man at variance against his father, and the daughter against her mother, and the daughter-in-law against her mother-in-law." And in Matthew 12:30: "He who is not with me is against me." Of course, apologists can cook up far-fetched explanations for these moral inconsistencies, but we should read them for what they say.

If God defines what is good and what is evil, then those who follow God's commands are morally justified to commit similar atrocities.[29] History shows the result: holy wars, burning of heretics, the Crusades, the Inquisition, the Thirty Years' War, the English Civil War, witch hunts, cultural genocide, brutal conquests of the Aztecs, Incas, and Mayans, ethnic cleansing, slavery, colonialist tyranny, and pogroms against the Jews eventually leading to the Holocaust.[30]

ATHEIST TERROR

Theists try to counter all this by pointing to the mass-murdering atheists of the twentieth century: Stalin, Mao, Pol Pot, Nicolae Ceausescu, Enver Hoxa, and Kim Jong-Il, as if this somehow justifies the religious mass murders that they can hardly deny.[31] Hitler is usually included in the litany, but he was a Catholic. Indeed, the Catholic Church never excommunicated a single Nazi, but in 2010 it excommunicated nun Margaret McBride for allowing an abortion that was necessary to save the life of a pregnant woman suffering from pulmonary hypertension.[32]

Religion scholar Hector Avalos has studied documents from the Stalin era that only recently became available. He points out that there is no documented statement in which Stalin justified his actions by saying something such as, "I don't believe in God, therefore I am committing violent act X." On the other hand, in all of the examples we saw above of terrorists associated with some religion, you can find direct statements of the form, "God wants X, therefore I am committing violent act Y." Avalos says, "We cannot find any direct evidence that Stalin's own personal agenda killed because of atheism."[33]

Now you might argue that while Stalin did not kill in the name of atheism, his godlessness failed to provide any restraint on his behavior. But then, neither has godliness provided much restraint to the murderers of history.

Avalos does not deny that Stalin committed many antireligious acts. But the predominant acts of violence committed during the period 1932–39, called the *Great Purge*, or the *Great Terror*, were clearly political in nature. Religion played a minor role. If a church went along, it was left alone. If it objected, it was persecuted along with everybody else who refused to cooperate.

By 1943, in the middle of the war with Hitler, Stalin found it useful to normalize relations with the Russian Orthodox Church (ROC) and even provided it with state funding. The church had a long history of cooperation with the highly oppressive czarist regimes and so it easily fit back into that mold. There never was any church-state separation in the old Russia, and there wasn't in World War II. The Church has continued its close cooperation with the Russian government today, but the Russian people have not exactly been hastening back to religion in droves.

As for the numbers killed by the twentieth-century mass murderers, Jared Diamond, the Pulitzer Prize–winning scientist and author, has observed:

> It's true, of course, that twentieth-century state societies, having developed technologies of mass killing, have broken all records for violent deaths. But this is because they enjoy the advantage of having by far the largest populations of potential victims in human history.[34]

Of course not all mass murderers commit their horrific acts in the name of God. They are all psychopaths anyway. That is, they have a brain disease. But I don't know of a single mass-murdering atheist who did so in the name of godlessness.

SLAVERY

In any case, the notion that religion is the source of morality is dubious. Consider the example of slavery. That both the New and Old Testaments condone slavery and even regulate its practice is incontrovertible.[35] It is often claimed that Christianity provided the moral force in abolishing slavery.[36] William Wilberforce (died 1833), a zealous evangelical Christian, is credited as leading the movement to abolish slavery in the United Kingdom. Evangelical Christians also participated in the abolitionist movement in America. On the other hand, during the American Civil War, the rebels led by Confederate president Jefferson Davis (died 1889) used the Bible to justify slavery.

In his 2011 book, *Slavery, Abolitionism, and the Ethics of Biblical Scholarship*,[37] Avalos shows how the Bible has been used throughout history to maintain and extend slavery and how abolitionists in the nineteenth century found they could not rely on the Bible as their guide because the proslavery biblical-based arguments were at least as effective. After an exhaustive study of the historical evidence, Avalos concludes:

> While some pre-Christian groups had outlawed slavery, Christianity continued it and expanded it worldwide. There were probably far more people enslaved (tens of millions) under Christian empires than in all pre-Christian empires combined. . . .
>
> Aside from unprecedented geopolitical and demographic developments, the major difference between previous eras of Christianity and the period between 1775 and 1900 was the marginalization of the Bible as a sociopolitical authority. That period witnessed the rise of biblical criticism, which undermined the authority and perceived reliability of the Bible in Europe and America. It was in that period that Americans invested their textual authority in a Constitution made by 'We, the People' instead of by a deity. Influential abolitionists such as Granville Sharp, William Wilberforce, Thomas Clarkson, and Frederick Douglass were part of that shift away from the Bible. Even if they did not all accept the new biblical criticism, they certainly realized the problems that using the Bible posed to abolitionism.[38]

Clearly, those who make reference to scriptures when they express moral judgments are cherry picking those teachings that they agree with and ignoring those with which they disagree. But if they didn't get those ideas from the Bible, where did they get them?

DOES MORALITY PROVE GOD EXISTS?

Modern religious thinkers have twisted the notion that morality comes from God into one in which God comes from morality. Francis Collins writes, "Moral Law still stands out for me as the strongest signpost for God.[39] He tells of reading C. S. Lewis as a youth and being "stunned by the logic" of one of Lewis's arguments for the existence of God: "If there was a controlling power outside the universe, it could not show itself as one of the facts inside the universe. . . . The only way in which we could expect it to show itself would be inside ourselves as an influence or a command trying to get us to behave in a certain way. And that is just what we do find inside ourselves."[40]

Collins adds, "Here, hiding in my heart as familiar as anything in daily experience, but now for the first time emerging as a clarifying principle, this Moral Law shone its bright white light into the recesses of my childish atheism."[41]

It is hard to see how a highly trained scientist such as Collins could be "stunned" by Lewis's argument. Why couldn't the controlling power of the universe make itself known by "facts inside the universe"? Such an entity is all-powerful, after all. Indeed, the Christianity that Collins embraces asserts many such "facts," and he mentions them throughout his book.

Perhaps moral law is hiding in the hearts of most people—psychopaths excepted—for natural reasons. However, evolution is not the whole story. As Sam Harris notes:

> While the possibilities of human experience must be realized in the brains that evolution has built for us, our brains were not designed with a view to our ultimate fulfillment. Evolution could never have foreseen the wisdom and necessity of creating stable democracies, mitigating climate change, saving other species from extinction, containing the spread of nuclear weapons, or of doing much else that is now crucial to our happiness in this century. [42]

Harris quotes psychologist Steven Pinker as observing, "If conforming to the dictates of evolution were the foundation of subjective well-being, most men would discover no higher calling than to make daily contributions to their local sperm bank."[43]

Christian apologist William Grassie makes a good point when he says:

> We humans increasingly transcend our biological origins. We are not slaves to our genes, nor need our morality be a slave to mere survival and repro-duction. Humans are a transcendent species. On one level, we are simply another mammal; on another, we are more like a whole new phylum in the epic of evolution.[44]

While he is probably thinking of "transcendent" as meaning spiritual or supernatural, the same sentiment would hold if we only viewed humans as perfectly natural organisms who have gone far beyond—transcended—any of the billions of other species with which we share the planet.

If evolution is not the sole source of our moral judgments, then where else did our inner sense of good and evil come from, if not from God? Well, while the behavior of an ant is governed by its genes, mammals developed brains that enabled them to alter their behavior based on experience. The better the brains, the more some kind of moral behavior is evident. Apes, monkeys, dolphins, elephants, and whales all exhibit protomoral behavior: attachment and bonding, cooperation and mutual aid, sympathy and empathy, altruism and reciprocal altruism, awareness and response to the social rules of the group.[45]

But scientist Collins is apparently unaware of these scientific facts. He notes, "In many instances, other species' behavior seems to be in dramatic contrast to any sense of universal rightness."[46] And, as Harris points out, "No other species can match us for . . . sadistic cruelty."[47]

Where, then, did our inner sense of good and evil come from? We learned it. Unfortunately, living among us are psychopathic individuals who did not learn it. They seem, to a greater or lesser extent, to lack any sense of good and evil. They commit the most heinous acts without a pang of conscience. In fact, they get great pleasure from these acts. One estimate is that 1 percent of the population, 3 million people in the United States, are psychopaths.[48]

However, a psychopath most likely has a brain disease. According to neuroscientist Harris, neuroimaging on both psychopathic and nonpsychopathic subjects has found that the psychopaths exhibit significantly less activity in regions of the brain that generally respond to emotional stimuli.[49]

THE POSTEVOLUTIONARY PHASE

Let us assume for the sake of argument that humanity has developed a moral instinct that cannot be attributed either to strictly biological evolution or to belief in an afterlife where one's virtue will be rewarded. Consider the following possible natural source. Suppose humanity has entered into a *postevo-*

lutionary phase in its development that is far from complete. The human body and brain have undergone only minor evolutionary changes in the last hundred thousand years. However, by that time our brains had evolved to the point where we could use them to overcome the negative consequences of our biological evolution.

In the last few pages of *The Selfish Gene*, Richard Dawkins says, "We have the power to turn against our creators. . . . Let us understand what our own selfish genes are up to because we may then at least have the chance to upset their designs."[50]

D'Souza mocks this notion, calling it "absurd." He asks how can the "robot vehicles of our selfish genes," namely us, rebel against our masters. "Can a mechanical car turn against the man with the remote control? Can software revolt against its programmer?"[51]

Maybe not, at least not until they become sufficiently complex to become intelligent life. Didn't D'Souza see the movie *2001*? As we have noted, no special "spark of life" is needed to inject life into a complex material system. It just has to grow highly complex. This is not widely understood, but we now know enough about what characterizes a living thing, indeed, an intelligent living thing, that we have no reason to believe that a machine cannot be intelligent. And as history shows, modern humans have always exhibited their ability to overthrow tyrants. So, why can't a machine?[52]

CHAPTER 11
MATTER AND MIND

The mind is inherently embodied.

Thought is mostly unconscious.

Abstract concepts are largely metaphorical.

These are the major findings of cognitive science. More than two millennia of *a priori* philosophical speculation about these aspects of reason are over. Because of these discoveries, philosophy can never be the same again.

—George Lakoff and Mark Johnson[1]

BODY AND SOUL

As I have by now often noted, science finds no need to include any nonmaterial entities in describing the world that presents itself to our senses. Still, every religion of which I am aware teaches that the human mind has access to a world beyond matter. Furthermore, many believers claim that since science still does not have a confirmed theory of consciousness, this leaves open the possibility that it never will develop such a theory based on matter alone. In yet another example of the argument from ignorance, they "cannot see" how matter can think.

Apologist Dinesh D'Souza writes, "The progress of evolution on earth shows an unmistakable trajectory from matter to mind." He asserts that while the mind may have arisen out of matter, it is "manifestly immaterial." In D'Souza's view, minds have qualities such as thoughts and ideas that are "different from those of material things." He argues that since humans are made up of perishable matter and yet have the capacity to generate imperishable ideas, "It is possible that our individual destiny might follow nature's destiny in moving from one type of existence to another."[2]

Now, rocks are material, yet they have imperishable attributes such as hardness and color. Will they join us in the afterlife along with our favorite pets?

D'Souza tackles "reductive materialism at its core" by asking whether "the mind can be reduced to the operations of neurons in the brain."[3] He tells us that as humans we experience two kinds of things: the physical and the mental. This suggests that we occupy two separate realms, as in the philosophical doctrine of *dualism* that goes back to René Descartes. If we could prove that dualism is true, then life after death becomes plausible. On the other hand, if matter is all there is, then life after death is highly improbable.

D'Souza admits that philosophers today have largely rejected dualism because they cannot figure out how an immaterial mind could interact with a material brain. He gives the example of a ghost trying to move a wall when it is in the nature of a ghost to go through walls.[4]

If I might make a silly aside that nevertheless illustrates the point, the behavior of ghosts such as the cartoon Casper can be reconciled with physics. Since Casper is pure spirit, he has no matter. So he can walk through a wall. We can't because the material particles of our bodies collide with the material particles in the wall. Casper has no particles to collide. Now you might ask, if he walks through walls, why doesn't he fall through the floor? You and I don't do so because the force of the floor "pushes up" at us to balance the force of gravity. Again, if Casper has no matter he has no gravity acting on him to make him fall through the floor. But then, how can Casper do anything to a material object? He can't kick a football because he has no particles to collides with those of the football.

There is a real problem for something of pure spirit such as the hypothesized immaterial mind to move material objects. Sure, you can make a model in which spirit has physical powers, but then we need some way to test that model. A model that is untestable is useless. In chapter 9 we saw how for almost two centuries now researchers have performed laboratory experiments to test for special powers of the mind. Not a single claimed positive result in all that time has stood up to the same scrutiny applied to all extraordinary scientific claims. None have been independently replicated, at least to the satisfaction of the scientific community. This includes instances of so-called mind over matter, where objects are supposed to be moved by thought alone.

Of course we still do not have a scientific consensus on the nature of mind and consciousness. Today, the leading approach in neuroscience is *functionalism*, in which the mind is what the brain does.[5] D'Souza associates the mind, in this view, with the software operating on the brain's hardware. In this mind-as-software model, if the hardware of the brain breaks down, "there is no reason that the software should not be able to find new instantiations."[6] To achieve immortality, we just have to download our minds onto a new computer, an idea that has been discussed by other authors.[7]

But even if we could download the full contents of our minds to another platform, that platform would have to be made of matter, and it would still not be either perfect or eternal as souls in the Christian heaven.

THE MATERIAL BRAIN

Considerable evidence exists that the phenomena we call mind and consciousness result from natural mechanisms in a purely material brain.[8] If we have disembodied souls that are responsible for our thoughts, decisions, dreams, personalities, and emotions, then these should not be affected by drugs. But they are. They should not be affected by disease. But they are. They should not be affected by brain injuries. But they are. As brain function decreases we lose consciousness, as when under full anesthesia. Why would that happen if consciousness arises from an immaterial soul?

Despite what D'Souza says, brain scans today can locate the portions of the brain where different types of thoughts arise, including emotions and religious thoughts. When that particular part of the brain is destroyed by surgery or injury, those thoughts disappear. In fact, different aspects of the mental religious life can be associated with particular functional networks localized within the brain. An analysis by neuroscientists at the National Institutes of Health identified the following patterns of brain activation that relate to specific components of religious thought:[9]

1. Notions of God's involvement and God's anger engage the prefrontal and posterior regions of the brain.
2. Doctrinal religious knowledge engages the networks processing abstract semantics.
3. Experiential religious knowledge engages networks involving memory and retrieval.
4. Adoption of religious belief engages networks providing cognitive-emotional interface.

While this does not prove that the soul does not exist, it shows that a disembodied entity is not what is doing the thinking.

Twenty-five-hundred years ago the Greek physician Hippocrates called epilepsy the "sacred disease." Today it is well known that temporal-lobe epilepsy results in visions that resemble those experienced by religious sages who claimed to talk to God. The study of this phenomenon was pioneered in the 1950s by Canadian neurosurgeon William Penfield.[10] In a remarkable book, *Did Man Create God?* physician David Comings discusses a number of Penfield's case reports.[11] Since then this has been a subject of considerable study.[12] Here's how prominent neuroscientist V. S. Ramachandran describes his observations in his 1998 book *Phantoms in the Brain*, written with science writer Sandra Blakeslee:

> But most remarkable of all are those patients who have deeply moving spiritual experiences, including a feeling of divine presence and the sense that they are in direct communication with God. Everything around them is imbued with cosmic significance. They may say, "I finally understand what it's all about. This is the moment I have been waiting for all my life. Suddenly it all makes sense." Or, "Finally I have insight into the true nature of the cosmos."[13]

Ramachandran comments further,

> I find it ironic that this sense of enlightenment, this absolute conclusion that Truth is revealed at last, should derive from limbic structures concerned with emotions rather than from the thinking, rational parts of the brain that take so much pride in their ability to discern truth and falsehood.

It really would be ironic if the great spiritual sages of history who have attracted billions of followers—Buddha, Ezekiel, Saint Paul, Muhammad, Joseph Smith, Ellen White—may have all suffered from temporal-lobe epilepsy or another neurological abnormality.[14] Not only is religion in the brain, it is a brain dysfunction.

However, apologist William Grassie speaks for many theists when he says, "So what if Mohammad or Saint Paul had temporal-lobe epilepsy? If God wanted to use that mechanism to transmit His revelation, then so be it."[15] But why would he? And why wouldn't he be able to keep his story straight so both would preach the same one true religion?

Physician Andrew Newberg has become famous and controversial for his claims about discovering neurophysical mechanisms for religious and spiritual experiences by using brain imaging on meditating Buddhist monks, praying Franciscan nuns, and other subjects.[16] He has been cagey about not interpreting his data as evidence that God is simply a brain phenomenon. In fact, with coauthors Eugene D'Aquili and Vince Rause, he contends that neurological processes have evolved "to allow humans to transcend material existence and acknowledge and connect with a deeper, more spiritual part of ourselves perceived of as an absolute, universal reality that connects to all others."[17]

However, Newberg and his colleagues can provide no empirical basis for this assertion. As I have already emphasized, if our minds had access to some deeper reality, then that claim could be easily demonstrated by reporting some fact about reality that can be tested scientifically and shown to not have arisen from inside the head alone.

In the 1980s, neuroscientist Michael Persinger claimed that he could stimulate spiritual experiences in a subject by applying a weak magnetic field to the temporal lobes.[18] A research group in Sweden failed to confirm this effect.[19] Persinger has disputed their findings.[20] However, despite considerable media coverage, his results have not been independently confirmed in any published study.

A significant line of research has developed in which much higher magnetic fields than those used by Persinger are seen to have significant mental effects. A technique called *transcranial magnetic stimulation* (TMS) is now being explored as a possible treatment of psychological disorders.[21] TMS operates on the principle of electromagnetic induction discovered in the mid-nineteenth century by Michael Faraday (died 1867). Faraday demonstrated that a time-varying magnetic field will generate an electric field. Induction is the principle behind electric motors and generators.

By placing an electric coil at a specified location on a subject's head and rapidly changing the current in that coil, the magnetic field thus produced induces an electric field in the brain that affects localized neural activity. The subject experiences little or no discomfort.

Brain imaging techniques such as functional magnetic resonance imaging (fMRI) provide amazingly detailed images of the brain and can locate sources of high brain activity. However, these techniques do not affect any changes in such activity. TMS, on the other hand, can actually interfere with cortical activity, interrupting the task being performed with the introduction of "neural noise."

TMS so far has been used mainly as an investigative tool in neuroscience for studies of perception, attention, learning, plasticity, language, and awareness. TMS has become one of the major tools of brain research and a promising possible treatment of brain ailments. It is beginning to find applications in the treatment of movement disorders, epilepsy, depression, anxiety disorders, stuttering, and schizophrenia.[22] A search with Google Scholar on January 3, 2011, came up with 20,300 papers published on the subject of TMS since 2000.

Again, this can be taken as evidence that the types of human behavior religion attributes to the action of an immaterial soul can be controlled by physical intervention.

Sam Harris provides a nice summary of the situation:

With respect to our current scientific understanding of the mind, the major religions remain wedded to doctrines that are growing less plausible by the day. While the ultimate relationship between consciousness and matter has not been sealed, any naive conception of a soul can now be jettisoned on account of the mind's obvious dependency upon the brain. The idea that there might be an immortal soul capable of reasoning, feeling, love, remembering life's events, etc., all the while being metaphysically independent of the brain, seems untenable given that damage to the relevant neural circuits obliterates these capacities in a living person. Does the soul of a person suffering from total *aphasia* (loss of language ability) still speak and think fluently? This is rather like asking whether the soul of a diabetic produces abundant insulin.[23]

CONSCIOUSNESS

Most people associate consciousness or self-awareness with something spiritual. They think that when we perform a deliberate action, such as lifting a spoon to our lips at the dinner table, some "ghost in the machine" is telling our arm and hand what to do. However, laboratory experiments pioneered in the 1980s by physiologist Benjamin Libet have shown that before we are aware of making a decision, our brains have already unconsciously made that decision for us.[24] While the interpretation of Libet's original results was controversial, continuing research has strongly confirmed the phenomenon.[25] We will see below the impact these observations have on the questions of free will.

Only recently have new data appeared that make the material option a far more plausible alternative to spiritual consciousness. Neuroscientist Stanislas Dehaene has reported that after twelve years of using every tool available to probe the brain, he and his colleagues now have a working hypothesis for how consciousness arises in the brain. Here is his summary of their results:

> In experiment after experiment, we have seen the same signatures of consciousness: physiological markers that all, simultaneously, show a massive change when a person reports becoming aware of a piece of information (say a word, a digit, or a sound).
>
> Furthermore, when we render the same information nonconscious, or "subliminal," all the signatures disappear. We have a theory about why these signatures occur, called the global neuronal workspace theory. Realistic computer simulations of neurons reproduce our main experimental findings: when the information processed exceeds a threshold for large-scale communication across many brain areas, the network ignites into a large-scale synchronous state, and all our signatures suddenly appear.[26]

If these results hold up under subsequent, independent replication, we may be able to say that we have achieved the "smoking gun" connecting consciousness and the brain.

The theory of consciousness Dehaene talks about has actually been around for almost two decades. It is called the *consciousness access hypothesis* and is presented within global workspace theory. As brain scans show, information is widely distributed within the brain. In the theory, the nervous system is viewed as a massive distributed set of specialized networks. Coordination, control, and problem solving take place by way of a central information exchange, allowing some regions, such as the sensory cortex, to distribute information to the whole. A similar architecture is used in large-scale computers. Evidence is growing that consciousness is the primary agent of such global access functions in humans and other mammals.[27]

Here is another example of how religious belief does more harm than good. If neuroscientists relied on faith and the traditional teachings of their churches to inform them that consciousness was immaterial, that is, was a part of the human system that exists in another world inaccessible to science, then they would have no motivation to understand consciousness. It would simply be beyond their ken. The result then would be that humanity would not enjoy the benefits of the research that are bound to come. Dehaene and his coworkers are already applying their ideas to noncommunicating patients in comas, vegetative states, or locked-in syndromes. A patient with locked-in syndrome is aware and awake but cannot move or communicate verbally. Unlike quadriplegia, which is caused by spinal injury, locked-in syndrome generally is caused by brain injury.

So neuroscience finally seems to be making progress toward a fully material understanding of consciousness. The reason for this is that brain imaging and other technologies have advanced to the point where direct hypothesis testing of models is available. The gap for God and the immaterial that for a long time was provided by consciousness is rapidly closing.

FREE WILL

As described above, experiments have shown that a network of high-level control areas in the brain begins to shape upcoming decisions long before they enter awareness. This would seem to challenge the whole notion of free will and the associated religious teachings about sin and redemption. If our brains are making our decisions for us subconsciously, how can we be responsible for our actions? Is free will an illusion?[28]

While "conscious will" may be an illusion, it can be argued that our material selves do still possess a kind of free will. Every decision we make is the result of a complex calculation made by our individual conscious and unconscious brains working together. That calculation relies on input from our immediate circumstances and our past experiences. So the decision is uniquely ours, based on our specific knowledge, experience, and abilities. That seems pretty free to me. While others can influence us, no one has access to all the data that went into the calculation except our unique selves. Another brain operating according to the same decision algorithms as ours would not necessarily come up with the same final decision, since the lifetime experiences leading up to that point would be different.

Let us look at free will from a physics angle. As we have already discussed in chapter 3, two centuries ago French physicist Pierre Laplace argued that Newton's mechanics implies that the motion of every particle in the universe could be predicted, in principle, from the knowledge of its position, momentum, and the force acting on it. This is the Newtonian world machine. Since, in the atomic model, everything in the universe, including human bodies and brains, is made up of particles, this would seem to do away with free will. However, as we saw in chapter 6, Heisenberg's uncertainty principle of quantum mechanics rendered the Newtonian world machine inoperable at very small distances. Let us ask if this provides a source of free will for humans.

THE QUANTUM BRAIN

The highly respected mathematician Roger Penrose has argued forcefully that the brain cannot be regarded simply as a mechanical computer. In his 1989 book *The Emperor's New Mind*, he argued, based on the incompleteness theorems of mathematician Kurt Gödel,[29] that the human thinking process cannot be reduced to computer algorithms and that some quantum process was involved.[30] Using mathematics as the exemplar, Penrose applied Gödel's theorem to demonstrate that mathematicians are aware of mathematical truths that cannot be proved by any algorithmic process. He then joined with anesthesiologist Stuart Hameroff in proposing that certain structural components of cells called *microtubules* are the seat of quantum effects in the brain, enabling humans to perform tasks no computer will every be able to do.[31]

Penrose's and Hameroff's ideas were discussed in my 1995 book *The Unconscious Quantum*.[32] It is safe to say that Penrose and Hameroff have not persuaded a consensus of experts in neuroscience, mathematics, philosophy, or computer science.[33] A 1995 issue of the journal *Psyche* contains a number of critical reviews,[34] accompanied by a detailed response from Penrose.[35] But he is not convincing. As physicist Taner Edis has shown, nonalgorithmicity can be accommodated within the mainstream of machine intelligence research.[36]

It is easy to show that the moving parts of the brain are large by microscopic standards and move around at relatively high speeds because the brain is hot. Furthermore, the distances involved are also large by *macroscopic* standards. Although you might need a microscope to see them, they are still in the realm of classical physics. I have used the term "submicroscopic" to refer to distances where quantum mechanics is important.

Let me make this quantitative. The entity that carries signals across synaptic gaps is called a "neurotransmitter." Its mass is typically 10^{-25} kilogram. Its typical speed is 358 meters per second, the average speed of a body of this mass in thermal equilibrium at body temperature, 37 degrees Celsius. Suppose that a neurotransmitter is initially located within a synaptic gap, which is about ten nanometers (10^{-8} meter) wide—about two hundred times the size of a hydrogen atom. The uncertainty in the speed of the neurotransmitter from the uncertainty principle is only 0.05 meter per second, or 0.014

percent. It follows that we can use classical Newtonian mechanics to describe the motion of the neurotransmitter with reasonable precision.

This conclusion agrees with a detailed calculation by physicist Max Tegmark, who showed that the coherence between states that is necessary to maintain a quantum system is lost in a tiny period of time in the brain, far too short for any quantum effects to have a role.[37]

Hameroff and two collaborators have challenged Tegmark's and my conclusions.[38] As mentioned, Penrose and Hameroff proposed microtubules in cells as the source of quantum effects in the brain.[39] A quantum mechanical model for microtubules has been worked out in detail by Travis John Craddock and Jack A. Tuszinski.[40] They found that while quantum effects were possible at low temperatures, below 30 degrees Kelvin (−243 degrees Celsius), thermal vibrations of the environment at ambient temperatures are more than sufficient to remove any form of collective excitation. While mechanisms have been proposed to shield the microtubules from the environment, no experiment has produced any evidence for quantum effects in microtubules or, indeed, anywhere in the brain.

Hameroff has pointed to a paper in *Nature* in which quantum effects are reported to have been observed in photosynthesis in marine algae at ambient temperature.[41] Now, nothing in physics prevents quantum effects at room or body temperatures. The warm quantum effects that are reported in photosynthesis involve photons, which are quantum objects. Hot photons are just as quantum as cold photons. The basic process of getting energy from light involves photons exiting electrons in atoms, a quantum process. So these results are not surprising or in violation of any known physics. Furthermore, the quantum coherence observed in the photosynthesis experiment lasted only on the order of 10^{-13} second. This is, as with Tegmark's calculations, far too short to produce quantum effects in the brain.[42]

A quantum brain is not required by either theory or experiment. Now, this does not mean that quantum mechanics cannot play any role in the brain. Ultimately, everything is quantum mechanical. The brain is made up of the same subatomic particles as a rock, and they all obey the rules of quantum mechanics. There simply is nothing special about the quantum mechanics of the brain that is any different from that of a rock.

However, quantum effects can still involve brain processes by another route. The brain is bathed in electrically charged particles from cosmic rays (muons) that reach Earth and beta-rays (electrons) from the radioactive potassium isotope K40 in our blood. These are energetic enough to break atomic and molecular bonds, unlike the radio waves from power lines and cell phones that people worry so much about.[43] And they are ultimately quantum mechanical.

Although the brain is a Newtonian machine, its complexity and nonlinearity put it in a category where deterministic chaos can play a role. As we saw in chapter 6, deterministic chaos is a purely classical phenomenon in which a complex system becomes extremely sensitive to initial conditions. We can imagine someone's brain carrying out a classical algorithm, like a computer, but a high-energy muon or electron breaks up a bit or two in either the code or the data and changes the outcome. This would result in the person making a random decision. But it would give the appearance of free will.

Now, that is not to say that all our decisions are random. An *Australopithecus* brain that decided at random whether to run from a leopard would not have left many descendants. So the brain must be mostly deterministic and perhaps an occasional random event is what provides us the creativity that Penrose argues is not possible if it is a purely algorithmic computer.

CHAPTER 12

METAPHOR, ATHEIST SPIRITUALITY, AND IMMANENCE

Either God is everywhere present in nature or He is present nowhere.

—Aubrey Moore[1]

The price of metaphor is eternal vigilance.

—Arturo Rosenbluth and Norbert Wiener[2]

A METAPHORICAL GOD

The Christian apologists who write about science and religion try to tell us that science flourished because of Christianity. In chapter 3, we saw that after a good start in ancient Greece and Rome, science in fact did not flourish in Europe for more than a millennium, at least in part because of Christianity. Islam preserved much of Greek and Roman science and developed it further, but then saw it fade to the point where science is hardly practiced in Muslim nations today.[3] As we have seen, science took hold in seventeenth-century Europe only after the authority of the Roman Catholic Church was challenged by the Renaissance and the Reformation.

Still, apologists argue, the great founding fathers of science—Bacon, Descartes, Copernicus, Galileo, Kepler, and Newton—were devout, albeit unconventional, believers. True enough, but they lived in an age when no clear separation existed between nature and supernature. It was not until that separation was made explicit during the following Enlightenment that atheism became associated with science and materialism.

273

While the greatest physicists of the nineteenth century, Michael Faraday and James Clerk Maxwell, were believers, most of the important physicists of the twentieth century, including Einstein, Bohr, Schrödinger, Dirac, and Feynman did not believe in a personal God. The best-known physicists alive today, Steven Weinberg and Stephen Hawking, are outspoken atheists. Notable other atheist scientists of recent times include Carl Sagan, Francis Crick, James Watson, Steven Pinker, Alan Turing, Jacques Monod, Richard Leakey, Linus Pauling, Claude Shannon, and Stephen Jay Gould.

As mentioned, Einstein and Hawking often used the word "God" metaphorically to represent the order of nature. Heisenberg, himself apparently a practicing Christian, wrote about how at the legendary 1927 Solvay Conference in Brussels, he, Dirac, Pauli, and some younger attendees stayed behind one day in the lounge of the hotel.[4] One of them said, "Einstein keeps talking about God; what are we to make of it? It is extremely difficult to imagine that a scientist like Einstein should have such strong ties with a religious tradition." Heisenberg quotes Dirac as commenting,

> I dislike religious myths on principle, if only because the myths of the different religions contradict one another. After all, it was purely by chance that I was born in Europe and not in Asia, and that is surely no criterion for judging what is true or what I ought to believe. And I can only believe what is true. As for right action, I can deduce it by reason alone from the situation in which I find myself: I live in society with others, to whom, in principle, I must grant the same rights I claim for myself. I must simply try to strike a fair balance; no more can be asked of me. All this talk about God's will, about sin and repentance, about a world beyond by which we must direct our lives, only serves to disguise the sober truth. Belief in God merely encourages us to think that God wills us to submit to a higher force, and it is this idea which helps to preserve social structures that may have been perfectly good in their day but no longer fit the modern world. All your talk of a wider context and the like strikes me as quite unacceptable. Life, when all is said and done, is just like science: we come up against difficulties and have to solve them. And we can never solve more than one difficulty at a time; your wider context is nothing but a mental superstructure added a posteriori.[5]

As mentioned, a 1998 study found that only 7 percent of the members of the American National Academy of Sciences believe in a personal God.[6] More up-to-date data were provided in 2007 by sociologists Elaine Howard Ecklund and Christopher Scheitle.[7] They asked thirty-six questions of 1,646 randomly selected natural and social scientists from twenty-one top research universities and found that 31.2 percent were atheists (do not believe in God); 31.0 percent were agnostics (no way of knowing); 15.5 percent believed but had doubts; 9.7 percent were sure there is a God, 7.2 percent believed in a higher power that is not God; 5.4 percent believed in God "sometimes." Disbelief is greatest among physicists and biologists, each with about 70 percent atheists or agnostics and only 6 or 7 percent "true believers."

SPIRITUAL ATHEISM

In 2010, Ecklund published a book, titled *Science vs. Religion: What Scientists Really Think*, that received considerable media attention because of its conclusion that scientists are more "spiritual" than we have been led to assume.[8] While based on the same study of elite universities just mentioned, Ecklund seems to contradict her own data when she writes: "Much of what we believe about the faith lives of elite scientists is wrong. The 'insurmountable hostility' between science and religion is a caricature, a thought-cliché, perhaps useful as a satire on groupthink, but hardly representative of reality." On the next page, however, she says many academic scientists practice a "*closeted faith*" because of the hostility of their colleagues. She provides no data, just a personal impression from her interviews. Many more scientists, she says, are "*spiritual atheists* who practice a new kind of individual spirituality—one that has no need for God."[9]

Ecklund asserts, "The institutional infrastructure of the academy has changed to allow more of a place of religion."[10] This is misleading because she fails to make the important distinction (although she does so later) that is made in these institutions themselves between "religion" and "religious studies." I have visited many religious studies departments around the country and find a common story. The majority of religious studies professors in secular universities are nonbelievers, to the great distress of students who enroll in these courses expecting to have their faiths strengthened—only to be taught what the Bible really says and how it really came to be written. Many atheist scholars, notably philosopher Daniel Dennett, have urged that religion should be studied scientifically as an important social phenomenon.[11] But as far as I can tell, it already is in these vibrant religious studies departments.

Ecklund's book does not spend any time dissecting the data reported in her paper with Scheitle mentioned above. Rather, Ecklund is more interested in whatever significance she can glean from her 275 anecdotal personal interviews. She concludes that scientists are more spiritual than we think while admitting that "spirituality" is a difficult term to define.

Ecklund notes, "Religion scholars think that Americans tend to link spirituality to interaction with *some* form of higher being."[12] She refers to a study reported by sociologist Robert Wuthnow in his 1998 book, *After Heaven: Spirituality in America Since the 1950s.* Wuthnow asked Americans to define spirituality: they mentioned near-death experiences, unseen spirit guides, belief in angels, meditation, and prayer groups.[13] That is, the general public associates spirituality with the supernatural.

Presenting data from a 1998 General Social Survey (she gives no exact reference), Ecklund reports that nearly 29 percent of Americans say they are "very spiritual," compared to only 9 percent of scientists. On the other hand, 32 percent of scientists consider themselves "slightly spiritual" compared to 21 percent of the general population. Ecklund calls this "thin spirituality."

The thin spirituality of scientists is clearly poles apart from the thick spirituality of the general public. As we saw above, two out of three are still atheists or agnostics and only 6–7 percent are committed believers. The spirituality Ecklund attributes to some scientists is not supernatural. A biologist's response is typical: "I get my spirituality . . . from being in nature. But I don't really believe there's a God, so I don't consider it's necessary for what I do or how I behave."[14]

I was looking in Ecklund's book for some evidence of the New Age quantum spirituality that was discussed in chapter 6. Apparently there is little among scientists. The word "quantum" does not even appear in Ecklund's index. Evidently quantum spirituality lives outside the mainstream scientific community and is found mainly on the pseudoscientific fringes under designations such as "parapsychology" or "neuroquantology."

Behavior, rather than belief, seems to be the defining factor of the spiritual atheist. Those who call themselves spiritual are engaged in helping others, caring for the environment, enjoying the outdoors, and generally spending time discussing and pondering central themes. We can't fault that.

In any case, my assertion that science and religion are fundamentally incompatible is not based on the opinions of individual scientists or even on some statistical distribution of a large sample. We have already seen that many are happy with Stephen Jay Gould's model of nonoverlapping magisteria, despite its disagreement with the observed fact that science and religion overlap considerably. Again I need to reiterate: the incompatibility being claimed here is not between the majority of religionists and scientists. It is between the worldviews and methods of science and religion as systems of thought.

THE MODERNIST VIEW OF RELIGION

Ian Barbour has made a heroic attempt to reconcile religion with science, fully justifying his £1 million Templeton Prize. We have already covered much of what he wrote about the history of the religion-science conflict in his book *Religion and Science*.[15] Now let us discuss the theological proposals that he argues reconcile the two worldviews. In what follows I will quote freely from Barbour's book.

The traditional view of God as a kind of supernatural king of the universe is still held by most believers as well as by evangelical Christian theologians such as William Lane Craig. Barbour and other contemporary liberal protestant theologians have moved theology a long way from that image. What Barbour calls the "modernist" view of God is based on evolution. He says that after Darwin, "divine creative action must be depicted not as external and once for all but as within the process and continuous in time. God's principle attribute is immanence in nature rather than transcendence."[16] That is, rather than the familiar picture of God existing outside of nature and outside of time and space, the modernist deity pervades the universe.

Barbour writes:

> Religion is rooted in human experience, and theological interpretations are secondary. [Isn't this a theological interpretation?] Human effort, not some special divine action, will bring in the Kingdom. Jesus was not the divine savior but the great teacher of high ideals. Human salvation comes through increased knowledge and noble goals, not through supernatural aid or any basic reorientation of the self.[17]

Barbour does not mention any of the "great teachers of high ideals" of many of the other religions on Earth, such as Buddha, Lao Tzu, and Muhammad. And what about others who more recently have been regarded by some as "great teachers," such as Joseph Smith, L. Ron Hubbard, Jim Jones, and David Koresh?

Barbour refers to the German theologian Friedrich Schleiermacher (died 1834), whom he calls "the father of liberal theology." According to Barbour, Schleiermacher held that

> the basis of religion is not revealed doctrine, as in traditionalism, or cognitive reason, as in natural theology, or even ethical will, as in Kant's system, but a distinctive religious awareness. Religion is a matter of living experience, not formal beliefs; it is not reducible to practical ethics or speculative philosophy but must be understood on its own terms.[18]

Barbour is keen on *process theology*, which is based on the metaphysical *process philosophy* of Alfred North Whitehead that strips the reality off matter and puts it into events.[19] Barbour views process theology as

> a theological program in which the "hard core" of the Christian tradition is taken to be belief in God as a creative love, revealed in Christ, while divine omnipotence is treated as an "auxiliary hypothesis" that can be modified to allow for the data of human freedom, evil and suffering, and evolutionary history.[20]

Thus Barbour recognizes that the only way to justify evil and suffering and maintain belief in God at the same time is to relax at least one of his traditional trinity of attributes: omnipotence, omniscience, and omnibenevolence (not to be confused with the Father, Son, and Holy Ghost). Of course, no one wants to believe in a nonomnibenevolent God, but either omnipotence or omniscience can go. If God isn't all-powerful, then he hasn't the power to alleviate all suffering. If he isn't all-knowing, then he may not know about every case of suffering. Notice, however, that science eliminated the suffering due to smallpox *without* being aware of every case and *without* being omnipotent. Certainly any benevolent god worth his salt could do a better job in easing suffering.

Barbour disagrees with any characterization, such as I have made, that faith is foolish since it is not based on evidence and has a history of failure in application. Barbour says, "*Faith* is personal trust, confidence, and loyalty. Like faith in a friend or faith in a doctor, it is not 'blind faith,' for it is closely tied to experience."[21] But is that experience reliable? Where is there the kind of evidence for trust in God that we find in doctors and friends?

Barbour bases his faith in faith on a belief in divine revelation: "The God-given encounter was experienced, interpreted, and reported by fallible human beings. . . . Revelation is recognized by its ability to illuminate *present experience*. Revelation helps us to understand our lives as individuals and as a community today."[22] But, again, where is the evidence for revelation?

He suggests these parallels between science and religion: "the interaction of data and theory (or experience and interpretation); the historical character of the interpretive community; the use of models; and the influence of paradigms."[23]

Barbour also disagrees with the notion, which I proposed in *God: The Failed Hypothesis*,[24] that God is a hypothesis and that the assumed God can be scientifically tested by looking for evidence that should be there given that God has the attributes usually assigned to him by believers. Barbour counters, "To make God a hypothesis to be tested or a conclusion of an argument (as in the argument from design) is to lose the experiential basis of religion. In my [Barbour's] view, God is known through *interpreted experience*."[25]

Barbour concludes:

Religion is indeed *a way of life*. Religious language serves diverse functions, many of which have no parallel in science. It encourages ethical attitudes and behaviors. It evokes feelings and emotions. Its typical forms are worship and meditation. Above all, its goal is to effect a personal transformation and reorientation (salvation, fulfillment, liberation, and enlightenment).[26]

Now, let me perform a religion-to-science transformation of this statement:

Science is indeed *a way of life*. Scientific language serves diverse functions, many of which have no parallel in religion. It encourages ethical attitudes and behaviors. It evokes feelings and emotions. Its typical forms are observation and model building. Above all, its goal is to effect a personal transformation and reorientation (critical thinking, fulfillment, liberation, and enlightenment).

So the big question is, which activity works better?

SCIENCE WORKS BETTER

Barbour's basic claims are (1) that good experiential reasons exist to believe in a God as informed by, but not taken literally from, Christian scriptures and traditions; and (2) that the practice of religion is beneficial for both the individual and society. Let us look at those claims and see how they fail to agree with the observed facts.

Let me begin with the notion that God is not some great king of heaven looking down on us from outside of nature, but is "immanent," that is, within space and time. If God were everywhere and everywhen, then surely the effects of his presence would have been detected by now, if not in everyday life then by scientific instruments. This is not the case. Let me list all the observations we have discussed in this book where we should see evidence for God but do not:

- The cosmos and creation
- The fine-tuning of natural laws and parameters
- The origin and evolution of life
- The structure of matter
- Answering prayers
- Revealing truths that can be tested

- Moral behavior in humans
- An immaterial component to mind
- Religious experiences
- Purpose in the universe
- Top-down causality

If God is a purely metaphysical being, meaning noninteracting with nature, then his existence can neither be verified nor disconfirmed. But, then, what does it matter if he has no effect on nature and humanity? He might as well not exist.

In any case, Barbour says theological interpretations are secondary and religion is rooted in human experience. He even dismisses the need for supernatural intervention. Rather, God is a "creative love revealed in Christ." However, as we saw in chapter 10, Jesus was not always such a "great teacher of high ideals." And we have also seen how the religion of Jesus has not had a beneficial effect on human society, as it has an almost unbroken record of every conceivable atrocity. Furthermore, as we will find in chapter 14, Christianity is a major contributor to the problems of modern society that threaten the very existence of the human race.

Barbour asks us to believe that faith in divine revelation has some informative quality, that it illuminates present experience. But, once again, where are the data? Where has divine revelation ever told us anything that wasn't already between our ears? We have "faith" in friends and doctors because it is earned. That is not faith but *trust*; the term *faith* should be reserved for unfounded beliefs. Such faith is foolish. It is a failure. It doesn't work. We have good reasons for trust in science because it works.

The parallel between science and religion—that both are based on data (experience) and theory (interpretation)—is strained. Science takes its data and formulates theories (that is, models) that can be tested against other data. When religion does that at all, it always fails the test.

CHAPTER 13
FROM CONFLICT TO INCOMPATIBILITY

Science and faith are fundamentally *incompatible*, and for precisely the same reason that irrationality and rationality are incompatible. They are different forms of inquiry, with only one, science, equipped to find real truth. And while they may have a dialogue, it's not a constructive one. Science helps religion only by disproving its claims, while religion has nothing to add to science.

—Jerry Coyne[1]

THE STATE OF RELIGION TODAY

L et us take a moment to review the state of religion in America and the world today. Religious belief in the United States is an anomaly among developed nations. In a 2007 survey by the Pew Research Center, the United States scored 1.5 on a scale measuring religiosity compared to a range from 0.25 to 0.75 in Western Europe.[2] Researcher Gregory Paul has shown a significant correlation between religiosity and income disparity, both of which are exceptionally high in America and may account for the anomaly.[3] This conclusion has been confirmed in a more recent independent study.[4]

The rich usually exploit the poor, and religion is very often a tool of that exploitation, offering promises of a better life in the hereafter. In India, for example, the poor do not revolt over their abominable living conditions because they have been told by those in higher castes that it is their karma to be where they are in society and that if they behave themselves, they will do better in the next life.

283

At the same time, despite the financial power wielded by the theocrats in the United States today, religion is experiencing a significant decline among the middle class in America and even more so in other developed nations. In *Fading Faith: The Rise of the Secular Age*, James A. Haught, editor of the *Charleston Daily Mail*, presents a range of statistical evidence that religion is waning in advanced nations.[5] Here are a few highlights (references to the original data have been added):

In France, fewer than 7 percent of adults attend worship.[6] Continent wide, only 15 percent of adults attend worship.[7]

In Denmark and Sweden, fewer than 5 percent of adults are in church on a typical Sunday. Denmark's Christian-Democratic political party attracts only 2 percent of voters.[8]

Although Ireland remains dominated by the Catholic Church, the Archdiocese of Dublin graduated only one priest in 2004.[9]

In Britain, only 4 percent of children attend Sunday school compared to 50 percent in 1900.[10]

The Anglican Church of Canada lost more than half its members between 1961 and 2001. The United Church of Canada's membership dropped 30 percent in the same period. The Presbyterian Church of Canada's membership has fallen 35 percent during those fifty years.[11]

In Japan, 77 percent of adults say they do not believe in a specific religion, although many practice a uniquely Japanese form of spirituality.[12]

In exceptionally religious America, secularism is on the march, especially among the young. The 2008 American Religious Identification Survey (ARIS) of more than fifty thousand Americans shows 15 percent of Americans identifying themselves as "nones"—as not belonging to any organized religion. This is to be compared with 8 percent in 1990.[13] Most significantly for the future, 28 percent of Americans between the ages of 18 and 29 are unaffiliated with any religion.[14]

A more recent Gallup Poll illustrates why, when interpreting polls, one has to look carefully at the wording of the question.[15] To the question, "Do you believe in God?" 92 percent of Americans said "yes." This was down from 98 percent in 1967, but it is still a high percentage. But then when respondents were asked to distinguish between God and a universal spirit, 80 percent claimed God and 12 percent chose "universal spirit." When asked about the depth of their belief, 73 percent said they were convinced God exists, while 19 percent said he probably does but they have doubts.

According to the ARIS study, since 1990, the number of Christians in the United States dropped 10 percent, mostly in the mainline churches, which fell 6 percent. Here are some other interesting conclusions from ARIS:

- Based on their stated beliefs rather than on their religious identification in 2008, 70 percent of Americans said they believed in a personal God, roughly 12 percent were atheist (no God) or agnostic (unknowable or unsure), and another 12 percent were deistic (believed in a higher power but not in a personal God). Note that these facts differ markedly from the common wisdom you often hear that 90 percent of Americans believe in God, which is generally taken to mean a personal God.

- America's religious geography has been transformed since 1990. Religion-switching, together with Hispanic immigration, has significantly changed the religious profile of some states and regions. Between 1990 and 2008, the proportion of Catholics to the general population in New England states fell from 50 percent to 36 percent, and in New York it fell from 44 percent to 37 percent, while in California it rose from 29 percent to 37 percent and in Texas from 23 percent to 32 percent.

- One sign of lack of attachment to religion by Americans is that 27 percent do not expect a religious funeral at their death.

- The challenge to Christianity in the United States does not come from other religions but rather from a rejection of all forms of organized religion.

The Catholic Church in America has remained steady in its membership at about 25 percent, but this is mainly the result of the influx of immigrants from Catholic countries. According to the 2008 Pew survey, while 31 percent of Americans were raised as Catholics, today only 24 percent identify themselves as such.[16]

Mainline churches in America are suffering steep declines. What Haught calls the "seven sisters of American Protestantism"—Presbyterians, Lutherans, Episcopalians, Methodists, American (northern) Baptists, Disciples of Christ, and United Church of Christ—lost 10 million followers while the US population doubled.[17]

In his 2011book titled *American Religion: Contemporary Trends*, Mark Chaves, Divinity School professor at Duke University, reports on the results of several surveys that show a sharp decline in interest in religion in the United States that he admits flies in the face of the perception that Americans have become more religious.[18] One particularly revealing statistic is that only about 25 percent of Americans attend religious services, while 35 to 40 percent say they do. Also, less that 25 percent of Americans express "great confidence" in religious leaders, and 44 percent strongly agree that religious leaders should avoid political involvement.

The manner in which these figures relate to the conflict between science and religion has to do with the central role religion has played in American politics, contributing to a cultural divide with extremist religion, big money, and right-wing ideology on one side and science, secularism, and the evidence-based community on the other. Due to the dominating influence of money in American politics, the political center has moved far to the right. Even the "liberal" Democratic Party has moved to the right of many European conservative parties. There is a Far Left, of course, but it has almost no influence on American society today.

As we have seen in earlier chapters, liberal theologians (that is, liberal in their theology but not necessarily in their politics), who are outside the evangelical community, are making an honest attempt to reconcile religion with science. They have failed so far, but their efforts should be acknowledged. Unfortunately, theologians have as little influence on American society as do atheists and the Far Left. So regardless of what science-friendly religious scholars come up with, most believers continue to hold to unsupportable ancient myths. Indeed, one of the interesting aspects of modern religion is the great disconnect between theology and what is preached from the pulpit, often by preachers who are fully aware of this disconnect but choose to hide it from their flocks.[19]

The other groups of believers who present little direct threat to science are those associated with the mainline Protestant churches, the so-called moderate Christians. They generally support science, particularly the teaching of evolution, but we have seen that they still profess a God-guided form of evolution that is distinctly not Darwinian but rather is a variation on intelligent design. As we saw previously, however, membership in mainline Protestant churches is declining steadily. Like theologians and atheists, mainline Protestant believers have little influence in politics.

The religious groups wielding the most influence are Catholics, comprising 24 percent of the population, and evangelical Christians, with 26 percent. These figures are for the year 2008.[20] The effectiveness of these two highly conservative religious groups is amplified by the enormous private and corporate wealth that supports their crusades and the way the government ignores their unconstitional interference in politics. Much of that wealth is directed toward electing pliable public officials who will look out for the best interests of the original donors, with the churches being used to get out the vote among their parishioners—often voting against their own best economic interests because they have been bamboozled into believing it is God's will.

IS RELIGION GOOD FOR YOU?

A widespread belief exists that, even if religion is not true, it is worth practicing because of its benefits. As we already saw in chapter 9, any health benefits of religious practice are problematical. The only definite positive correlation between religious practice and health is with church-going, and it is not proven that this is anything supernatural. It is far more likely to be simply the result of a healthier, less stressful lifestyle among churchgoers. Most nonbelievers I know also live healthy, stress-free lives, but they are never included in these studies.

The lifestyle interpretation agrees with the evidence, mentioned in chapter 10, that less religious nations are happier and healthier than religious ones. According to the study by sociologist Phil Zuckerman, the godless societies of Scandinavia rank near the top in every measure of societal and personal health.[21] Amusingly, the nation ranked the "happiest" is Denmark, which, incidentally, also pays the highest taxes.

And what about the negative impact that religion has on health? Between 1975 and 1995, at least 172 children in the United Stated died, perhaps 140 of medically curable illnesses, because their parents refused them medical treatment for religious reasons.[22] While those numbers are not large, every child is significant, and many more children are harmed by lack of immunizations and other refusals by religious parents to provide modern medical treatments and preventative measures. Parents are allowed to do this because of unconscionable religious exemptions in child-abuse prevention laws.[23] It should be noted that antivaccination affects everyone, not just the unfortunate children of religious fanatics.

Believers often bring up the famous argument called *Pascal's wager*, formulated by the French philosopher, physicist, and mathematician Blaise Pascal (died 1662). A medieval Muslim thinker, Abu Hamid al-Ghazali, may have proposed the wager earlier. Basically the argument is that you have everything to gain and nothing to lose by betting on the existence of God. On the other hand, you have nothing to gain and everything to lose in rejecting the argument.

Many people, including the great philosopher Bertrand Russell, have seen the flaw in this argument. Assuming that God is just, wouldn't he look with more favor on someone who honestly didn't believe for lack of evidence than someone who, without evidence, says he believes just so he can get his ass into heaven?

One of the great appeals of religion is the promise of eternal life. The most successful religions—Christianity, Islam, and Hinduism—thrive on that promise. But, as we have seen, it is a forlorn hope.

D'Souza has drawn up a list of benefits for belief in God and, in particular, an afterlife. I will list these systematically:[24]

Assets of Belief in an Afterlife

1. It provides us with hope at the point of death and a way to cope with our deaths.
2. It infuses life with a sense of meaning and purpose.

3. It gives us a reason to be moral and a way to transmit morality to our children.

4. Clinical evidence exists that religious people who affirm the afterlife are healthier than nonbelievers.

Let me challenge each of D'Souza's points and show why they are, in fact, liabilities:

Liabilities of Belief in an Afterlife

1. The idea that you will live forever gives a false sense of a glorious self that leads to extreme self-centeredness in this life. Furthermore, you may live in constant fear that any sin you might have committed will condemn you to an eternity of suffering in hell. Knowing you are not going to live forever restores a true sense of your place in the scheme of things and you don't have to worry about hell.

2. If you don't believe in an afterlife you will find more meaning and purpose in this world and live your life to the fullest since it is the only life you have.

3. As we have seen, morality comes from human intelligence and has nothing to do with belief or nonbelief in God or an afterlife. You may not take action to seek justice in this life if you assume it will be provided in the next. You may not exercise your own best judgment in matters, instead allowing yourself to be controlled by others who claim sacred authority.

4. As described above and in chapter 9, there is no convincing clinical evidence that religious belief improves health, and there is some indication that it has harmful effects.

Simply put, no basis exists for the claim often made by believers that religion is necessary for a person to live a healthy, happy, and moral life. Religion blinds, deafens, and numbs us to the reality around us, and though this may temporarily soothe our anxieties, like drugs or alcohol, there is a painful price to be paid down the road for such cowardly denial and self-defeating ignorance. Not only can we be both well and good without God, we can be better.

Twenty-five-hundred years ago the Buddha showed how to cope with the existence of suffering and death in the world. The individual must find a way to eliminate his or her ego, to cease being self-absorbed, and to realize that he or she is not the center of the universe. The problem is, this process is very difficult because we all have a consciousness that separates us from every other human in the world. Yet there can be no other way. The fact is that each of us will die, and all the evidence points to our consciousness ceasing upon death. Year by year, science extends human lifetimes, but it will never provide immortality. All we can do is accept that fact and learn to live with it.

SUMMARY OF CONFLICTS

At this point, let me summarize the conflicts discussed in previous pages, pointing out specifically why the worldviews and methods of science and religion are fundamentally incompatible.

The Nature of Reality

All religions, even Buddhism, teach that a reality exists that goes beyond—that transcends—the material world that presents itself to our senses and scientific instruments, one that we can access by extrasensory means. While science is willing to consider any evidence that comes along, so far none has appeared that requires any immaterial entity be added to the models that describe observations. Likewise, no information obtained by extrasensory means has ever been objectively verified.

In this regard, it is often claimed that science has nothing to say about the supernatural. But this is wrong. If the supernatural exists and has effects on the material world, then those effects are subject to scientific study.

The Origin of the Universe

Fundamental to most religions is the notion of divine creation. At one time it seemed impossible that the universe came into existence naturally. Christians saw the success of the big-bang model as a further confirmation of the biblical creation story. At least it seemed to prove that the universe had a beginning and it followed, by their reasoning, that the cause of that beginning could only be a Creator God.

However, modern cosmology has considerably dampened this hope. It has shown that the big bang need not have been the beginning of space and time and that the universe could be eternal, or at least that theological claims that an eternal universe is mathematically impossible can be shown to be false. It now seems possible, or even likely, that our universe is just one of an unlimited number of other universes. Several plausible scenarios for the natural origin of our universe have been published by reputable scholars. This is sufficient to refute any claim that a miraculous origin is required.

As with all scientific claims, these conclusions are provisional. While they do not directly conflict with the existence of a creator, science sees no need for one while religion cannot do without one.

Fine-Tuning

Theologians argue that the parameters of physics are so delicately balanced that any slight changes in their values and life would not have been possible. Therefore, a creator must have fine-tuned these parameters so that we, and our form of life, would evolve.

This claim can be refuted on several fronts. The most popular explanation among most physicists and cosmologists is that many universes exist and we just happen to live in the one suited for us.

However, even if only our universe exists, adequate explanations within existing knowledge can be found for the values of the most crucial parameters. Others can be shown to have ranges that make some form of life probable.

The Argument from Design

For centuries thinkers have argued that the observed order we see around us is evidence for divine design in the universe. However, the character Philo in David Hume's *Dialogues Concerning Natural Religion*, published in 1779, enumerated many flaws in the argument.[25] Quite simply, the universe does not look as if it were designed by a perfect, benevolent God. It is too imperfect, too filled with evil and suffering.

Today's intelligent design movement argues that complex structures require an architect and builder, and that natural processes cannot generate new information. They assert a law of conservation of information that is provably wrong. Information is the equivalent to negative entropy, and entropy is not conserved. Furthermore, the generation of complex systems from simpler systems can be seen in many physical situations, such as the phase transitions that go naturally from gas to liquid to solid in the absence of external energy.

Evolution

Life on Earth is the prime example of natural complexity. Darwinian evolution showed how in the billions of years of Earth's history, all existing species evolved from simpler forms by a process of chance mutation and natural selection. Evolution demonstrates how simplicity can lead naturally to complexity, although it does not require it.

More than 50 percent of Americans say explicitly that they do not believe in evolution. While fundamentalist Protestants in America have refused to accept evolution because of its gross conflict with the Bible, Catholics and more moderate Protestants have claimed they acknowledge evolution and say they see no conflict with their faith. However, surveys indicate that virtually no Christians accept the theory of evolution as understood by modern biology. They insist that evolution is still God-guided, while God plays no role in the conventional Darwinian theory. The mechanism of evolution is precisely *unguided* variation plus natural selection, which explains the observed complexity and variety of life and the clear absence of beneficent purpose—or, indeed, any purpose.

Attempts to teach creationism in public schools in America under the guise of "creation science" or "intelligent design" have been ruled in federal courts to be a violation of the United States Constitution since they have a clear motivation to promote religion. However, 17 percent of high school biology teachers do not cover evolution and 60 percent spend less than five hours on a subject that is the foundation of modern biology. Here the negative impact of religion on science could not be more evident.

Quantum Consciousness and a Holistic Universe

Since the 1970s, New Age gurus and promoters of psychic phenomena have been making the claim that quantum mechanics implies that human consciousness can affect reality, not just here and now but every place in space and every time in the past and future. This implies that the universe is one unified whole with the human mind tuned into a "cosmic consciousness."

This incredible notion is not supported by either empirical evidence or by the actual quantum theory itself as it is understood by experts.

The theistic claim that modern physics has undermined materialism and points toward a holistic universe is also not borne out by the facts. The standard model of particles and forces, which is based on relativity and quantum mechanics, is fully materialist and reductionist.

Reductionism and Emergence

From the time of the Greek atomists through Newtonian physics to the current standard model of particles and forces, physics has been reductionist—reducing everything to the sum of its parts. The physical world is hierarchical, going from elementary particles to atoms to molecules and so on up to mind, to human societies, and on further to the cosmos. At each level new principles are discovered that generally cannot be derived from particle physics. These new principles are said to "emerge" from the levels below.

In this process, however, the common wisdom is that the "causal arrow" still points upward from the lowest level of particles. Theologians and some scientists, however, have conjectured that another causal arrow points down so that the emergent principles can affect the behavior of lower-level entities. They then envisage God as the ultimate emergent principle acting down on the world of matter and mind.

However, the examples they give are trivial, such as a piston in a cylinder of gas moving the atoms of the gas. The atoms in the piston are simply moving the atoms of the gas by particle collisions in fully reductionist fashion. It can be shown that some emergent principles, such as those of thermodynamics, fluid dynamics, and chaos theory, can in fact be derived from particle physics. It's particles all the way up as well as down.

Information Theology

After the scientific revolution in the seventeenth century, the machine or clockwork became the popular metaphor for the universe, running deterministically according to the laws of Newtonian mechanics. After the quantum revolution and the uncertainty principle undermined determinism, many authors declared the "end of materialism," illogically conflating matter and machinery. With the advent of computers, the world as a computer, in particular, a quantum computer, has become the new metaphor for nature.

Some authors have proposed that the real stuff a computer runs on is not matter but information. Theologians have jumped on the antimatter bandwagon, but its collision with the matter in the standard model is not producing much light. Even quantum computers are still composed of matter, and a universal quantum computer made of elementary particles turns out to provide a useful metaphor for how the complex universe we live in naturally evolved 13.7 billion years from an original patch of chaos having minimal information.

The Material Mind

Belief in a disembodied soul is basic to virtually all religious faith. This soul is usually associated with the mind. Evidence for the existence of some immaterial element in humans is claimed from religious experiences, especially near-death experiences (NDEs). The data have been examined carefully and found to have plausible natural explanations.

Paranormal or psychic phenomena, which might also indicate special powers of the mind that go beyond matter, have been investigated for the better part of two centuries without a single positive claim being independently replicated.

Neuroscience has demonstrated that much of what previously was thought to be the product of an immaterial component of mind has a material basis in the brain. Quantum mechanics plays no special role in the brain. A material explanation of consciousness has not yet achieved consensus but promising models have been proposed and are being tested.

Free Will, Morality, and Justice

It is commonly assumed that if the mind is a purely material phenomenon, then humans have no free will. Of course this wreaks havoc with the elemental religious doctrine of sin and atonement. Nevertheless, a kind of free will can be conceived since our individual decisions are still based on the experiences of a lifetime, which are uniquely our own.

Those who believe in God assume without question that he is the source of morality and that society would be wicked, depraved, corrupt, and debauched if they did not have God enforcing their righteousness. This totally disagrees with the observed fact that the overwhelming majority of nonbelievers are not wicked, depraved, corrupt, or debauched while a nonnegligible minority of believers are.

We all can see that the world is not a just place, with the wicked often prospering while the good suffer. If one believes in ultimate justice, then such justice can only come about in an afterlife. Unfortunately, we have no reason to believe in ultimate justice other than our own pious wishes.

The natural origins of morality are currently a major area of study, and a considerable literature on this subject already exists. While many authors propose evolutionary models, good reasons lead us to believe that human biology is now in a postevolutionary stage in which we are able to use our intellects to develop codes of behavior that maximize everyone's well-being. That we have not yet done so completely can be laid at the feet of religion and its insistence on ancient moral codes developed in primitive societies. They may have served a purpose once; they surely do not now.

Modern Theology

While the great bulk of religious believers still hold to traditional notions of God as the ruler of the universe, modern theologians are seeking different models for God that are more consistent with scientific results in evolution, cosmology, and archeology as well as with other philosophical and theological advances.

Some have attacked the problem of divine action in the face of the absence of evidence for the miracles that would be expected to appear occasionally as the result of those actions. However, attempts to use quantum mechanics and chaos theory to provide a means for God to act in the world without being detected have not been satisfactory, especially since this perspective seems to lead to a new kind of deist god, one who plays dice with the universe. Try as they might, Christian theologians cannot reconcile this model with traditional Christian teachings. Furthermore, why would a beneficent God want to hide from his creatures so only those of limited intellect who have unquestioning faith receive the gift of an eternity of bliss?

Another view promoted by liberal Protestant theologians is that God should not be viewed as a supernatural entity outside of space and time, but as a being existing everywhere and everyplace in space and time. They claim their methods of arriving at this model closely follow those of science. In particular, the model is based on data, namely human experience, and theory, namely interpretation of the data. However, while that may be the case, this procedure differs from scientific method in that wherever it has been tested, as in prayer experiments or religious experiences, it has uniformly failed. When that happens in science, the hypothesis being tested is discarded. This never happens in religion, at least as the direct result of testing.

While some scientists are religious, they have compartmentalized and compromised their thinking. Religion is based on faith—beliefs that have no basis in evidence or reason. Faith is foolish. Faith is a failure. It is the very antithesis of science. And, ultimately, this is why science and religion are forever fundamentally incompatible.

CONFRONTATION OR ACCOMMODATION?

The court decisions on evolution and creationism have not addressed the more fundamental issues at the interface between science and religion. Science leaders in America generally have avoided confronting religious leaders with the irrationality of many of their positions. Let me quote from the National Academy of Sciences' position paper on the evolution controversy:

> *Can a person believe in God and still accept evolution?*
>
> Many do. Most religions of the world do not have any direct conflict with the idea of evolution. Within the Judeo-Christian religions, many people believe that God works through the process of evolution. That is, God has created both a world that is ever-changing and a mechanism through which creatures can adapt to environmental change over time.
>
> At the root of the apparent conflict between some religions and evolution is a misunderstanding of the critical difference between religious and scientific ways of knowing. Religions and science answer different questions about the world. Whether there is a purpose to the universe or a purpose for human existence are not questions for science. Religious and scientific ways of knowing have played, and will continue to play, significant roles in human history.
>
> No one way of knowing can provide all of the answers to the questions that humans ask. Consequently, many people, including many scientists, hold strong religious beliefs and simultaneously accept the occurrence of evolution.[26]

However, we have seen that those Catholics and moderate Christians who say they accept evolution really do not. They view evolution as the process by which God achieves his purpose, and they believe that humans are central to that purpose. This is not evolution as it is understood by science. The Darwinian scheme has no role for God. In this scheme, humanity is an accident. The National Academy of Sciences has simply swept this fact under the rug, just as it has ignored the fact, mentioned in chapter 9, that scientists are, in principle, able to discover the supernatural and some very competent researchers are pursuing that end. The failure to find any sign of the supernatural so far is piling up evidence that it is nonexistent.

Those who hold views similar to the National Academy have come to be called "accommodationists." In 2006 and 2007, the term "new atheism" entered into the lexicon of the discourse on science and religion, referring to those who are taking a more confrontational stand in dealing with religion.[27] A series of books appeared that immediately hit the bestseller lists: *The End of Faith*[28] and *Letter to a Christian Nation*[29] by neuroscientist and philosopher Sam Harris; *Breaking the Spell*[30] by philosopher Daniel Dennett; *The God Delusion*[31] by biologist Richard Dawkins; *God Is Not Great*[32] by journalist and literary critic Christopher Hitchens; and my own *God: The Failed Hypothesis*—the only one by a physicist.[33]

While many other authors, bloggers, and speakers have eloquently portrayed the atheist position, this particular group was singled out by the media as the "new atheists" because of the much harder line against religion they advocated compared to what had been the widespread convention. Of course, these five are certainly not the only ones out there who are confronting rather than accommodating religion. Several popular bloggers[34] and a rapidly growing, well-organized collection of university students[35] support and promote the new atheist position.

Prior to the new atheism, the prevalent view among nonbelievers was that religion is generally a benign force in the world and that only a small minority of fanatics presented any serious threat to the public well-being. Also, those scientists and science organizations particularly concerned with the financial support for and teaching of science—especially evolution—felt that it was strategically a bad idea to alienate believers, especially since they constitute such a large majority in America and completely dominate government at all levels. Journalist Chris Mooney exemplifies the accommodationist position:

> If the goal is to create an America more friendly toward science and reason, the combativeness of the New Atheists is strongly counterproductive. . . . America is a very religious nation, and if forced to choose between faith and science, a vast number of Americans will select faith.[36]

Although individual new atheists (there is no central new atheism dogma) fully appreciated the merit of this argument and have thought about it carefully, they decided that religion is too destructive a force in society to just sit back and allow it to proceed unopposed. And this opposition was not to be limited to religious extremists. To the great dismay of accommodationists, moderate believers also found themselves criticized by new atheists. Despite moderate believers' sincere public endorsements of science and reason, they still hold to ancient mythologies. In our final chapter we will see why this, in the long run, can only impede humanity's progress toward a more rational world.

CHAPTER 14
WHY DOES IT MATTER?

The delusional is no longer marginal.

—Kevin Phillips[1]

Those who control what young people are taught, and what they experience—what they see, hear, think, and believe—will determine the future course of the nation.

—James Dobson[2]

THE AMERICAN ANOMALY

I f the conflict between science and religion were just a matter of intellectual debate, a battle between eggheads in theology and eggheads in science and philosophy, the stakes would not be very high. There is a famous quotation that is variously attributed to C. P. Snow, Henry Kissinger, Daniel Patrick Moynihan, Woodrow Wilson, and others: "Academic politics are so vicious precisely because the stakes are so small."[3]

However, in the case of the incompatibility of science and religion, the stakes are high—not because of academic politics but because of American and, indeed, world politics. The role of religion in today's political and social scene is ubiquitous, from Islamic terrorism to attempts by the Christian Right in America to replace democracy with theocracy. A theocracy? In America? Most of my scientific colleagues in the cozy comfort of their cluttered campus offices would scoff at the notion. But they need to be good scientists and look at the data. As I write this, potential Republican candidates to oppose

Democrat Obama in the 2012 presidential election are busy campaigning and debating. Most are injecting religion into the political dialogue as never before, some even claiming divine guidance. At this writing, all the front-runners, even those who have been relatively moderate in the past, are making strong antiscientific statements that are obviously intended to appeal to those who vote in Republican primaries.

In 2003, columnist Kimberly Blaker edited a series of essays titled *The Fundamentals of Extremism: The Christian Right in America* that documented the insidious role of Christian fundamentalism in American family and political life.[4] While this book is almost a decade old, the problems described have, if anything, become more critical.

In his 2006 book *American Theocracy*, former Republican strategist and bestselling author Kevin Phillips wrote, "The rapture, end times, and Armageddon hucksters in the United States rank with any Shiite ayatollahs, and the last two presidential elections [2000 and 2004] make the transformation of the GOP into the first religious party in U.S. history."[5]

In her 2006 *New York Times* bestseller *Kingdom Coming: The Rise of Christian Nationalism*, journalist Michelle Goldberg chronicled the dominant role in American politics played by extremist Christians.[6] In the same year, Damon Linker, former editor of the Catholic magazine *First Things*, wrote in his book *The Theocons* about how, over the course of three decades, a few determined men have succeeded in injecting their radical religious ideas into the nation's politics.[7]

Also in 2006, in *American Fascists: The Christian Right and the War on America*, Pulitzer Prize–winning author and theist Chris Hedges noted:

Democratic and Christian [values] are being dismantled, often with stealth, by a radical Christian movement known as dominionism, which seeks to cloak itself in the mantle of Christian faith and American patriotism. Dominionism takes its name from Genesis 1:26–31, in which God gives human beings "dominion" over creation. This movement, small in number but influential, departs from traditional evangelicalism. Dominionists now control at least six national television networks, each reaching tens of millions of homes, and virtually all of the nation's more than 2,000 religious radio stations, as well as denominations such as the Southern Baptist

Convention. Dominionism seeks to redefine traditional democratic and Christian terms and concepts to fit an ideology that calls on the radical church to take political power.[8]

In *The Family: The Secret Fundamentalism at the Heart of American Power*, published in 2009, journalist Jeff Sharlet revealed how a small, secretive group of extremely conservative Christians called "The Family" have wielded increasing national and international political power. They organize the yearly Washington Prayer Breakfasts attended by presidents and foreign diplomats, provide prayerful retreats for congressmen, senators, and Supreme Court justices, and preach a gospel of biblical capitalism, military might, and American empire.[9] In a followup volume published in 2010, *C Street: The Fundamentalist Threat to American Democracy*, Sharlet told how The Family also aided several prominent senators and congressmen in covering up their extramarital affairs.[10]

In a 2011 essay in *Religion Dispatches*, Peter Montgomery, senior fellow at People for the American Way, summarized how "religious right leaders and activists have spent decades creating fertile soil for anti-union campaigns through the promotion of 'biblical capitalism.'"[11] These leaders have proclaimed that Jesus and the Bible oppose progressive taxes, capital gains taxes, estate taxes, and minimum wage laws. They also enlist Jesus in a war against unions, which they regard as unbiblical.

The founder of The Family was a Norwegian immigrant, Abraham Vereide. According to Sharlet, also writing in *Religion Dispatches*:

In 1942 [Vereide] moved to the capital where the National Association of Manufacturers staked him to a meeting of congressmen who would become students of his spiritual politics, among them Virginia senator Absalom Willis Robertson—Pat Robertson's father. Vereide returned the manufacturers' favor by telling his new congressional followers that God wanted them to break the spine of organized labor. They did.[12]

Montgomery writes that the 1990 Christian Coalition leadership manual coauthored by coalition founder Ralph Reed cites four biblical passages instructing slaves to be obedient to their masters. For example:

Slaves, submit yourselves to your masters with all respect, not only to those who are good and considerate, but also to those who are harsh. For it is commendable if a man bears up under the pain of unjust suffering because he is conscious of God. (1 Peter 2:18–19)

Reed interprets this to mean, "Christians have a responsibility to submit to the authority of their employers, since they are designated as part of God's plan for the exercise of authority on earth by man."[13] We see here the same argument that has been used for millennia to justify the right of monarchs, no matter how incompetent or brutal, to rule over everyone else: God put them there so it must be his will.

In 2005, journalist Chris Mooney documented how the antiscience attitudes of the George W. Bush administration, motivated in part by the heavy representation of Catholics and evangelical Christians in virtually all federal offices from the president on down and dominating most advisory panels, suppressed reports by government scientists on issues such as birth control, global warming, and stem-cell research.[14]

Antiscience is implicit, or even explicit, throughout this movement. Scientists have to stop sitting back and start stepping up to challenge religion. The welfare and, indeed, the survival of our species is at stake.

THE UPS AND DOWNS OF SCIENCE IN THE PUBLIC MIND

Mooney and marine biologist Sheril Kirshenbaum have chronicled how science has fallen in and out of favor with the American public in the years since World War II.[15] Prior to that war, fundamental science in the United States was respectable but lagged behind that in Europe, especially England and Germany, and had little public attention. That changed dramatically when, with the help of expatriate scientists from the continent, the United States developed the nuclear bomb, along with radar and other dramatic military technologies. Then, public and politicians alike recognized the power and importance of basic research, especially physics.

The result was a greatly increased expenditure of federal funds on research and scientific education in the postwar period, spurred on by the Cold War. Then, in 1957 came the Soviet Union's launch of Sputnik, which generated an even greater anxiety in Americans that our very way of life was threatened. The result was further expansion of spending on science. Mooney and Kirshenbaum note that in 1940 the total federal expenditure on research and development was $74 million. By 1962–63 it was $12.3 *billion*.[16] (Inflation played a small role in this increase: $1 in 1940 had the same buying power as $2.14 in 1962.)

I had the good fortune of getting my PhD in physics in 1963. I did not have to seek a job; jobs sought me. I didn't need to spend the usual several years' apprenticeship as a postdoc but moved directly into the professorate. Two colleagues and I were able to set up from scratch a research group in elementary particle physics at the University of Hawaii, funded by the US Atomic Energy Commission. Our research had no application to nuclear bombs or reactors, but the message of the war was that it was imperative for our national defense to always be at the forefront of research into the fundamental nature of matter and energy.

In just a few years, however, by the mid 1970s, the trend of ever-increasing funding for scientific research began to reverse. This resulted from the cultural shift brought about by the Vietnam War. The growing antiwar, antinuclear, environmental, and feminist movements on the Left were matched by the antievolution, antireproductive rights, and antienvironmental movements on the Right. The culture war began and science was caught in the middle.

Science also became more specialized and more remote from public understanding; the number of scientific journals expanded from five thousand to one hundred thousand—each filled with articles that were unreadable not only by the general public but also by most scientists outside each specialty. Unsurprisingly, many people could not see the benefit of much of this effort, unless it directly affected them, such as with medical research.

That is not to say science was shut off. We landed men on the moon, a robot vehicle on Mars, and discovered the structure of DNA. In my field, we developed the standard model of particles and forces that has stood empirically nonfalsified for a generation. But in 1993, Congress cancelled the

Superconducting Supercollider that would have probed to the next level of matter, leaving that job to be done in Europe with the Large Hadron Collider, only now going into operation almost two decades later.

A brief period of renewed interest in science was brought about in the 1980s, largely by the brilliant efforts of one man, astronomer Carl Sagan, with his books, speeches, and the extraordinary TV series *Cosmos*. The first of *Cosmos*'s thirteen one-hour episodes premiered in 1980. However, Sagan soon got in trouble with the establishment for his criticism of Ronald Reagan's inane Strategic Defense Initiative ("Star Wars") and his promotion of the "nuclear winter" scenario that Sagan and others claimed would make any nuclear war unwinnable.

The scientific community did not win any awards for courage and statesmanship in its treatment of Sagan, who was refused tenure at Harvard and denied membership in the National Academy of Sciences because he was not a "pure" scientist but a "popularizer," as if popularizing science was bad for science. In fact, Sagan was not just a great popularizer of science; he had made significant contributions in research that normally would have warranted these honors.

Thus, with the Reagan administration began the conservative antipathy toward science that reached its peak in the administration of G. W. Bush.

THE CULTURAL DIVIDE

In 2008, science found a more congenial president in Barack Obama. However, Obama and his opponents in the primary and general elections demonstrated their priorities by ignoring invitations to speak about science issues during the campaigns. By contrast, Obama and Hillary Clinton had no qualms about attending a "compassion forum" on faith and values at Messiah College during the primary, and Obama and John McCain appeared at a forum at pastor Rick Warren's Saddleback Church during the general election.[17] Warren would give the invocation at Obama's inaugural (another obvious violation of the separation of church and state).

Despite his commitment to bipartisanship, Obama's election only served to deepen the cultural divide between two irreconcilable ideologies. Progressives are generally for more government involvement in society, with social programs and regulations on industry, while the conservatives fight for minimal government and special benefits for the rich. Conservatives exhibit the inconsistency of being against all forms of government intervention on business while seeking laws to control individual sexual behavior. The actual sexual behavior of many priests, preachers, and politicians perhaps indicates that it is not so much inconsistency as hypocrisy.

A clear tendency exists for evangelicals to be conservative Republicans and nonbelievers to be progressive Democrats. That is not to say that Democratic politicians are all atheists. Hardly. Politicians from both parties profess belief in a higher power in greater proportion than the general public, as a necessity for being elected. At this writing, only one member of Congress is an avowed atheist, and President Obama has continued the unconstitutional faith-based initiative programs instituted by President Bush. We can't count on the heavily Catholic Supreme Court, including the recent liberals appointed by Obama, to strike these programs down.

Nevertheless, the Republican Party today is completely dominated by the Christian Right while the Democrats at least exhibit some cautious independence from Christian ideology. Following is a typical example of religious intrusion in American politics. In 2011, Senator Jim DeMint (Republican from South Carolina) appeared on the Family Research Council's weekly radio show and said:

> Some are trying to separate the social, cultural issues from fiscal issues, but you really can't do that. America works, freedom works, when people have that internal gyroscope that comes from a belief in God and biblical faith. Once we push that out, you no longer have the capacity to live as a free person without the external controls of an authoritarian government. I've said it often and I believe it—the bigger government gets, the smaller God gets, as people become more dependent on government and less dependent on God.[18]

DeMint then added:

We've found we can't set up free societies around the world because they
don't have the moral underpinnings that come from biblical faith. I don't
think Christians should cower from this debate, should be told that their
views and their values should be separate from government policies, because
America doesn't work without the faith that created it.

A poll taken by the Pew Forum on Religion and Public Life in 2004 found that
70 percent of traditional evangelicals were Republican, 20 percent Democratic,
and 10 percent independent. On the other hand, 54 percent of atheists and
agnostics in the poll were Democratic, 27 percent independent, and only 19
percent Republican.[19] A more recent Gallup poll reported that the very religious
were 19 percentage points more likely to be Republican than the nonreligious.[20]
As I write this today, almost every vote in Congress and state legislatures is falling
along party lines. And the split is religious as well as political.

Consider the recent phenomenon called the Tea Party. A 2011 poll from
the Pew Forum showed a strong religious Right influence on the Tea Party.[21]
Among registered voters, 69 percent of those who agreed with the religious
Right said they also agreed with the Tea Party. Only 4 percent disagreed.
Tea Party members support the social agenda of religious conservatives by
large fractions: 64 percent oppose same-sex marriage; 59 percent say abortion
should be illegal in most or all cases.

The religious beliefs of Americans who consider themselves part of the
Tea Party were reported in a survey by Robert P. Jones and Daniel Cox as part
of the Third Biennial American Values Survey.[22] They found that 47 percent
believe the Bible is the literal word of God compared to one-third of the
general population and 64 percent of white evangelicals.

An interesting anomaly has developed in the demographics of support for
the two major political parties. Since the 1950s, Republicans have increasingly
drawn their support from the higher income levels of society while Democrats
have obtained theirs from the lower levels. However, when it comes to evangelical Christians and Tea Party members, those of all income levels are solidly
Republican.[23]

In America today we have megachurches with thousands of members
who are taught the "prosperity gospel": God wants everyone to be wealthy.

Since the preacher is obviously the chosen of God, he is entitled to his fleet of Mercedes, private jets, and mansions in Texas and Bermuda.

Still, income disparity in America today is so great as to be almost unbelievable for a nation that prides itself on equality. In a 2010 book titled *Winner-Take-All Politics*, political scientists Jacob S. Hacker and Paul Pierson showed that in the thirty-year period ending in 2006, those among the top 1 percent income group in American took in almost $700,000 a year more than they would have had they not received special treatment over that time. By contrast, those among the bottom fifth had an annual income almost $6,000 lower.[24] In 1960, the richest 1 percent of Americans held 10 percent of the nation's income; in 2007 they held 23 percent.[25] Extensive statistics on income distribution can also be found in a 2011 report by sociologist G. William Domhoff.[26]

It is beyond contempt the way the conservative politicians in America bring down their wrath on the low-paid members of our society, such as schoolteachers and janitors, while the CEOs and other top executives of major corporations and Wall Street financial operators receive outrageous salaries and benefits. And it is beyond foolishness that other low-paid members of our society, including many Tea Party members, 19 percent of whom have annual incomes of less than $30,000,[27] vote for these monsters because they have been told it is the will of God.

I have thought about what conservatives might have against schoolteachers. I surmise that they want to undermine public schools where students are taught evolution (in some places), tolerance of other races and religions, female equality, the facts of history, and critical thinking. In the theocracy sought by the Far Right, children will be shielded from these dangers in Christian madrassas and home schooling.

In my experience, scientists in the academy line up pretty much on the moderate center of the political spectrum while those working for industry tend to position themselves on the not-too-far Right. There are few anymore on the extreme Left, which is usually associated with socialism. Of course, many academic scientists consult for industry, and they also tend to toe the conservative line. This ideological split would not be serious if it did not have real consequences for issues that affect public health and welfare.

TOBACCO AND POLLUTION

An eye-opening, meticulously documented exposé, *Merchants of Doubt: How a Handful of Scientists Obscured the Truth on Issues from Tobacco Smoke to Global Warming*, written by Naomi Oreskes and Erik M. Conway, was published in 2010.[28] The authors tell the story of prominent senior physicists Frederic Seitz, S. Fred Singer, William Nierenberg, and Robert Jastrow writing and giving "expert" testimony to presidents, congressional committees, and the media on a wide range of issues where they came down hard against the prevailing scientific consensus whenever that consensus threatened the economic interests of their corporate backers.

Seitz was a solid-state physicist who had won a National Medal of Science and many other awards, including honorary doctorates from thirty-one universities. He was president of the National Academy of Sciences and of Rockefeller University. He died in 2008.

Singer is an Austrian-born atmospheric and space physicist and was a leading figure in early space research. He has held a variety of military, government, and academic positions. Singer received a Gold Medal Award for Distinguished Federal Service in 1954.

Jastrow was an astronomer and physicist who was the first chairman of NASA's Lunar Exploration Committee and founding director of the Goddard Institute for Space Studies. He was also a popular author who was quoted in chapter 7 as claiming that the big bang is evidence for a creator. He died in 2008.

Nierenberg was a physicist whom I remember as the well-liked director of the Scripps Institution of Oceanography in La Jolla, California, when I visited there frequently in the 1980s. He served on several important federal panels including the White House Office of Science and Technology Policy and the National Aeronautics and Space Administration's Advisory Council. He was on Ronald Reagan's transition team. Like the others, Nierenberg was conservative and a cold-war hawk. He died in 2000.

Serious conflict between scientists of differing political views began in 1979 when the tobacco industry launched a massive campaign against the discovery that smoking caused cancer and other deadly ailments. Seitz was particularly central in spearheading this counterattack by big tobacco. Eventually tobacco lost the battle and the industry was found guilty of hiding facts about the dangers of smoking that it had known about since 1953.

Later, Singer aided the tobacco industry in unsuccessfully challenging the evidence for the dangers of secondhand smoke. In 1994 he coauthored a report attacking the findings of the US Environmental Protection Agency (EPA) that secondhand smoking was hazardous. The report was funded by the Tobacco Institute.[29] Today, few people question that smoking is a life-threatening habit, both to the smoker and to those unfortunate enough to be in his or her vicinity.

Despite the tobacco losses, the strategy of throwing doubt on science continued to be applied to prevent government actions to curb atmospheric pollutants. Fred Singer played a key role, questioning the claims of atmospheric scientists that chlorofluorocarbons (CFCs) were responsible for the growing ozone hole over Antarctica.[30] He also took the lead in arguing that acid rain from burning fossil fuels was not a serious problem.[31]

Bill Nierenberg chaired the Acid Rain Peer Review Panel, which in 1982 was charged with studying the consequences of acid rain. The panel included Sherwood Rowland, who would share the 1995 Nobel Prize in Chemistry with two colleagues for their work in the formation and composition of the ozone hole. The panel concluded 8–1 that acid raid was a serious problem, and it was well documented that acid rain resulted from "man-made sulfur dioxide (SO_2)."[32] Nierenberg himself is quoted as saying, "You just know in your heart that you can't throw 25 million tons of sulfates into the Northeast and not expect some . . . consequences."[33] However, the Reagan White House, with Singer's help (the one negative vote and the only one on the committee appointed by Reagan) and Nierenberg's compliance, was able to obfuscate this finding.[34] At the time, Singer was working for the conservative Heritage Foundation that vociferously opposes environmental regulation.

Conservative physicists also moved to downplay the "nuclear winter" scenario, developed by Carl Sagan and other scientists, which projected disastrous climate effects that would result from nuclear war. In 1985, Singer wrote an attack on the scenario that was published in the prestigious journal *Science*.[35] In a related matter, Jastrow was prominent in supporting Ronald Reagan's

Strategic Defense Initiative ("Star Wars"), which most scientists said would not work and furthermore was destabilizing. All four physicists have been involved in the huge battle over global warming, which continues to this day.

In all these cases, the strategy is to cast doubt on scientific findings that, for the most part, are considered conclusive by a large majority of experts on the subject. The doubters are generally not experts, but men (rarely women) of sufficient stature that they command attention. They are able to take advantage of the way scientists properly avoid claiming "certainty" in any conclusion but always leave the door open a little crack for future developments. The politicians, the media, and the public generally do not appreciate this fine distinction. As we have seen, while scientific conclusions are technically never 100 percent final, they often get close enough for all practical purposes.

A specific example that supports Oreskes and Conway's thesis can be found in a 2007 report issued by the highly respected Union of Concerned Scientists on "How ExxonMobil Uses Big Tobacco Tactics to 'Manufacture Uncertainty' on Climate Change."[36] This is from the executive summary:

> In an effort to deceive the public about the reality of global warming, ExxonMobil has underwritten the most sophisticated and most successful disinformation campaign since the tobacco industry misled the public about the scientific evidence linking smoking to lung cancer and heart disease. As this report documents, the two disinformation campaigns are strikingly similar. ExxonMobil has drawn upon the tactics and even some of the organizations and actors involved in the callous disinformation campaign the tobacco industry waged for 40 years. Like the tobacco industry, ExxonMobil has:
>
> - *Manufactured uncertainty* by raising doubts about even the most indisputable scientific evidence.
> - Adopted a strategy of *information laundering* by using seemingly independent front organizations to publicly further its desired message and thereby confuse the public.
> - *Promoted scientific spokespeople* who misrepresent peer-reviewed scientific findings or cherry-pick facts in their attempts to persuade the media and the public that there is still serious debate among scientists that burning fossil fuels has contributed to global warming and that human-caused warming will have serious consequences.

- *Attempted to shift the focus* away from meaningful action on global warming with misleading charges about the need for "sound science."
- *Used its extraordinary access to the Bush administration* to block federal policies and shape government communications on global warming.

One technique is to argue that "both sides" of an issue are entitled to equal time, even when one side has all the data and successful models while the other side has nothing but smoke and mirrors.

Although Oreskes and Conway do not discuss it, we can see the same strategies being used by the opponents of evolution. Right-wing politicians in several states have tried and in some cases succeeded in making it mandatory to teach "criticisms" of Darwinian evolution, which usually means sneaking divine creationism into schools (as we saw in chapter 4, this has been declared unconstitutional by the courts).[37]

Currently the big battle between industry-funded scientists and the majority of experts is over the existence and possible consequences of anthropogenic (human-caused) global warming (AGW). Since those who denied that smoking causes cancer, that CFCs generate the ozone hole, and that pollution from fossil fuels causes serious acid rain have been proven wrong, it seems a good guess that the global warming deniers, many of them the same people, are wrong too.

GLOBAL WARMING

There is no doubt that Earth is getting warmer. The overwhelming majority of climate scientists claim both that the magnitude of the trend cannot be explained as part of the normal cycles of warming and cooling that Earth has experienced in the past and that humans are a major cause.[38] A 2009 survey showed that 97.4 percent of active publishers on climate change agreed that "human activity is a significant contributing factor in changing mean global temperatures."[39]

The Intergovernmental Panel on Climate Change (IPCC) is a huge effort that has produced numerous technical reports and disseminated public information on the climate problem.[40] Its Fourth Assessment Report, published in

2007, declared, "Warming of the climate system is unequivocal" and "anthropogenic warming over the last three decades *likely* had a discernable influence at the global scale on observed changes in many physical and biological systems."[41] The IPCC received the 2007 Nobel Peace Prize for its efforts.

Increased warming is said to be the result of the carbon dioxide and other greenhouse gases emitted by humanity's massive use of fossil fuels. Climate scientists warn that if the nations of the world do not take drastic action soon to reduce these emissions, we are heading for drastic climate changes and flooding of vast coastal areas now occupied by millions of people. Already polar ice is melting dramatically.[42] A 2011 report by the National Research Council states, "Climate change is occurring, is very likely caused by human activities, and poses significant risks for a broad range of human and natural systems."[43]

Fierce political opposition has arisen, especially among conservatives in the United States, who object to placing any limits on fossil fuels because they say it will have significant impact on the economy (and their own well-oiled pocketbooks). A number of scientists have joined in the fray by expressing skepticism that human activity plays a significant role in global warming, suggesting instead that the current warming trend is part of natural climate fluctuations. They argue that the models used by climate scientists in their calculations cannot be trusted because the models have too many uncertainties and the climate scientists make too many unjustified assumptions.

It is beyond the scope of this book to delve deeply and definitively into either the science or politics of global warming. Let me just consider the role of religion in what should be primarily a scientific question. Both sides of the issue claim the other is letting their "religion" rule over their science. Perhaps the best-known scientist who is skeptical of AGW is the brilliant physicist Freeman Dyson, who in the 1940s was one of the founders of the highly successful theory of quantum electrodynamics described in chapter 6.[44] He writes: "There is a worldwide secular religion, which we may call environmentalism, holding that we are stewards of the earth, that despoiling the planet with waste products of our luxurious living is a sin, and that the path of righteousness is to live as frugally as possible. The ethics of environmentalism are being taught to children in kindergartens, schools, and colleges all over the world."[45]

Certainly extreme environmentalism, as well as nazism, fascism, and communism, share some of the trappings of religion, namely irrational ideologies. Indeed, defining religion to encompass the latter three serves to defuse the argument so often heard that religion had nothing to do with the mass murders of the twentieth century (see chapter 3).

For our purposes, however, let us use "religion" to imply belief in a transcendent world beyond matter. No doubt many environmentalists have such a belief, but a spirit world is not a part of the doctrines of environmentalism.

Evidence exists that many who deny the dangers of global warming do so out of religious conviction. Yet another Pew Research Center survey asked the following question: "Is there solid evidence the earth is warming?" Let me just give the percentages of thise who said yes and agreed that it is the result of human activity:[46]

Total US population	47%
Unaffiliated with any church	58%
White mainline Protestants	48%
White, non-Hispanic Catholics	44%
Black Protestants	39%
White evangelical Protestants	34%

Also interesting was the result that 21 percent of all Americans, 18 percent of the people unaffiliated with religion, and 31 percent of white evangelicals said there was no global warming at all. While mainline Protestants and Catholics are close to the national average, they still are below that of the unaffiliated. Surely the fact that 58 percent who are unaffiliated support the scientific consensus while less than 50 percent of believers do is evidence for some correlation between religion and the denial of global warming.

The role of religion in global warming denialism can be seen in the political battles over the teaching of evolution.[47] In 2010, the Kentucky Legislature introduced a bill encouraging teachers to discuss "the advantages and disadvantages of scientific theories," including "evolution, the origins of life, global warming and human cloning." A similar bill was passed in Louisiana in 2008, and in 2009 the Texas Board of Education required that teachers present all sides of the evidence on evolution and global warming.[48]

Demanding equal time for opposing views on evolution and global warming is like demanding equal time for phlogiston and flat-Earth theories. But what is the connection between global warming and evolution? The only one I can think of is religion.

Those who consider the Bible to be the literal world of God and who reject evolution also take seriously the last book of the New Testament, *Revelation*, which describes the end of times. What's more, the Jesus of the Gospels predicted that the Son of Man would return in a generation to set up the Kingdom of God on Earth (Matt. 16:28, Matt. 24:34, Mark 9:1, Mark 13:30, and Luke 9:27). Of course he didn't, but for two-thousand-plus years Christians have always thought the end was right around the corner.[49] Why worry about global warming if the Kingdom of God is at hand?

At this writing, John Shimkus, Republican of Illinois, is a member of the Energy and Commerce Committee of the US House of Representatives. He has argued that climate change is a myth because God told Noah he would never again destroy Earth by flood (Gen 8:21–22). Shimkus can be seen on video saying: "The earth will end only when God declares it's time to be over. Man will not destroy this earth. This earth will not be destroyed by a flood. . . . I do believe God's word is infallible, unchanging, perfect."[50]

In 2009, Representative "Smokey Joe" Barton, Republican from Texas, told C-Span:

> I would also point out that CO_2, carbon dioxide, is not a pollutant in any normal definition of the term. . . . I am creating it [CO_2] as I talk to you. [Carbon dioxide is] in your Coca-Cola, your Dr Pepper, your Perrier water. It is necessary for human life. It is odorless, colorless, tasteless, does not cause cancer, does not cause asthma.
>
> A lot of the CO_2 that is created in the United States is naturally created. You can't regulate God. Not even the Democratic majority in the U.S. Congress can regulate God.
>
> If you think greenhouse gases are bad, life couldn't exist without greenhouse gases. . . . So, there is a, there is a climate theory—and it's a theory, it's not a fact, it's never been proven—that increasing concentrations of CO_2 in the upper atmosphere somehow interact to trap more heat than the atmosphere would otherwise.[51]

Personally, I can't see how pumping back into the atmosphere, over only a century or two, carbon that took millions of years to accumulate in the Earth can be harmless.

Perhaps the most vocal person denying global warming is Republican senator James Inhofe. In a speech on the Senate floor on July 28, 2003, Inhofe called catastrophic global warming "a hoax." He said that "satellite data, confirmed by NOAA [National Oceanic and Atmospheric Administration] balloon data, confirms that no meaningful warming has occurred over the last century."[52] This is false. The satellite data in fact corroborated the warming trend reported from surface measurements. Here's the summary of the NOAA report:

> Global temperatures in 2004 were 0.54°C (0.97°F) above the long-term (1880–2003) average, ranking 2004 the fourth warmest year on record. The warmest year on record is 1998, having an anomaly of 0.63°C (1.13°F), followed by 2002 and 2003 both having an anomaly of 0.56°C (1.01°F). Land temperatures in 2004 were 0.83°C (1.50°F) above average, ranking fourth in the period of record while ocean temperatures were third warmest with 0.42°C (0.76°F) above the 1880–2003 mean.[53]

Inhofe is one of the most conservative members of the Senate and, characteristically, also promotes evangelical causes. He has used government funds at least twenty times to travel to Africa on missions that he himself has referred to publically as "Jesus things." There he has "played an active role in the faith-based aspect of our anti-AIDS campaign," according to a Ugandan diplomat.[54]

The Cornwall Alliance for the Stewardship of Creation has issued what it calls "An Evangelical Declaration on Global Warming." Here's what it says (quoted exactly):

WHAT WE BELIEVE

1. We believe Earth and its ecosystems—created by God's intelligent design and infinite power and sustained by His faithful providence—are robust, resilient, self-regulating, and self-correcting, admirably suited for human flourishing, and displaying His glory. Earth's climate system is no exception. Recent global warming is one of many natural cycles of warming and cooling in geologic history.

2. We believe abundant, affordable energy is indispensable to human flourishing, particularly to societies which are rising out of abject poverty and the high rates of disease and premature death that accompany it. With present technologies, fossil and nuclear fuels are indispensable if energy is to be abundant and affordable.

3. We believe mandatory reductions in carbon dioxide and other greenhouse gas emissions, achievable mainly by greatly reduced use of fossil fuels, will greatly increase the price of energy and harm economies.

4. We believe such policies will harm the poor more than others because the poor spend a higher percentage of their income on energy and desperately need economic growth to rise out of poverty and overcome its miseries.

WHAT WE DENY

1. We deny that Earth and its ecosystems are the fragile and unstable products of chance, and particularly that Earth's climate system is vulnerable to dangerous alteration because of minuscule changes in atmospheric chemistry. Recent warming was neither abnormally large nor abnormally rapid. There is no convincing scientific evidence that human contribution to greenhouse gases is causing dangerous global warming.

2. We deny that alternative, renewable fuels can, with present or near-term technology, replace fossil and nuclear fuels, either wholly or in significant part, to provide the abundant, affordable energy necessary to sustain prosperous economies or overcome poverty.

3. We deny that carbon dioxide—essential to all plant growth—is a pollutant. Reducing greenhouse gases cannot achieve significant reductions in future global temperatures, and the costs of the policies would far exceed the benefits.

4. We deny that such policies, which amount to a regressive tax, comply with the Biblical requirement of protecting the poor from harm and oppression.[55]

This statement has been endorsed by approximately five hundred people, including a large number of scientists and other academics.[56] It is not simply the view of a small fringe group but that of the large majority of American evangelicals.

The Cornwall Alliance has also produced a DVD series titled *Resisting the Green Dragon: A Biblical Response to One of the Greatest Deceptions of Our Day* that attacks environmentalism as a "new religion" that "threatens the sanctity of life. If you buy the series you will "learn how the Bible powerfully confronts environmental fears and how—in God's wise design—people and nature can thrive together.[57]

Nevertheless, it must be reported that some believers, including a number of prominent evangelical Christians, see their "stewardship" of Earth, ordained by God, as requiring that they pay attention to the warnings of climate scientists. The Rev. Jim Ball, senior director for climate programs at the Evangelical Environmental Network, which supports the science of global warming, said many of the deniers feel that "it is hubris to think that human beings could disrupt something that God created." However, he adds, "This group already feels like scientists are attacking their faith and calling them idiots so they are likely to be skeptical" about global warming.[58]

Those evangelicals who have gone against the grain merit our notice. In 2006, some eighty-six evangelical leaders signed a statement saying, "Millions of people could die in this century because of climate change, most of them our poorest global neighbors." The list includes Rick Warren, author of the blockbuster bestseller *The Purpose-Driven Life.*[59] However, other leaders, including Watergate felon Charles Colson, founder of the Prison Fellowship Ministries, and James C. Dobson, founder of Focus on the Family, objected to the statement.[60]

The Catholic Church is becoming increasingly green. In 2007, Pope Benedict told a Vatican conference on climate change to "respect creation" while "focusing on the needs of sustainable development."[61] Still, more than 50 percent of white, non-Hispanic Catholics do not believe in anthropogenic global warming. Besides, the Church's continuing refusal to accept birth control (another Church teaching most educated Catholics ignore) negates any positive role it might play in providing for a livable future world.

Global warming denialism is a part of a growing distrust of science in America, described above, that is not limited to evangelicals or conservatives. In a 2009 book, *Denialism: How Irrational Thinking Hinders Scientific Progress, Harms the Planet, and Threatens Our Lives,*[62] journalist Michael Specter discusses how science is viewed by many as a special-interest group that does not always serve the public's best interest. Because of this much harm is being done as a result of the avoidance of technologies such as vaccination and genetically modified foods. Furthermore, billions of dollars are spent on "natural" foods and alternative remedies that scientific studies have found to be worthless. At the same time, the examples Specter gives, and those covered in this chapter, make it clear that elements in the scientific community have not always acted for the commonweal. We scientists have to work harder to regain the public's *trust* (not faith).

Corporate greed is the force behind global warming denial. Antiscience, fueled by religion, is being exploited to prevent the US government from taking actions that might be essential for everyone's welfare, including the grandchildren of those industrialists, preachers, politicians, and scientists who now so vehemently oppose any action.[63] For industrialists there is a double irony to their denialism. Not only will their grandchildren benefit from reducing our dependence on fossil fuels, but who else is going to make money from the development of new energy sources?

THE IMPENDING CRASH

In the nineteenth century, petroleum was discovered to be an incredibly cheap source of energy and the building material for countless products, from plastics to pesticides. And petroleum came inexpensively from the ground. It was used to replace human and animal muscle power with machines that could do the work of thousands for a tiny fraction of the cost. One barrel of oil, which costs about a dollar to extract from a Saudi well, can be used to accomplish a year's worth of physical labor from eleven men. A new world economy was developed based on petroleum and other fossil fuels such as coal and natural gas. Food production soared from the use of farm machinery and petrochemicals on crops. World population exploded, and lifestyles in developed societies reached unprecedented heights of comfort and wealth.

But this period will be seen in the future as a tiny blip in world history. Although most people cannot imagine otherwise, the society that we now live in cannot last. While the end of the supply of fossil fuels is still not in sight, all of Earth's natural resources are finite. All the fossil fuels, which took 100 million years to accumulate in the Earth, inevitably must be used up. Even before this happens, Earth may become uninhabitable.

A large effort is going into developing alternative energy sources, but the political will to put them to use is still not there. Some of these alternatives, notably solar energy, may already be competitive with fossil fuels if the pricing of energy were to reflect its true economic cost in terms of environmental damage, human health, and the huge military expenditure required to guard sources and transport systems.

Nuclear fission based on uranium also has a limited lifetime. Unfortunately, the 2011 disaster in Japan underlined the problems with older reactor designs. Advanced designs, particularly those that use liquid thorium instead of uranium offer longer-term prospects, but public fears have made developments of new designs politically untenable.

Nuclear fusion offers the promise of an endless, pollution-free source of energy. However, despite billions of dollars expended by several nations for more than fifty years, the practicality of nuclear fusion remains unproved.

A personal note here should underscore this fact. When I entered graduate school in physics at UCLA in 1956, I planned to work on controlled fusion, thinking it was the fuel of the (near) future. Luckily I found better opportunities in particle physics.

No one can predict exactly when we will no longer be able to rely on fossil fuels. If fusion proves impractical and fission is discarded, human society will experience an economic crash greater than any in history, as energy and food production dwindle while population continues to grow. Steps could be taken to ameliorate this crash, but they seem beyond our political capability.

The public and its political representatives seem to think that we can continue to carry on living as we have. Many have been fooled into believing that Earth and its ecosystems were created by God and so will be sustained by "His faithful providence." Most just refuse to think about it.

But at least there will be one positive outcome when there is no longer sufficient oil to run our engines. It will bring to an end all the conflicts of the last hundred years or so that were directly the result of the world's reliance on oil. And our atmosphere, rivers, and oceans should become clean again. Let's hope someone is around to enjoy it.

A FINAL CONCLUSION

Religious faith would not be such a negative force in society if it were just about religion. However, the magical thinking that becomes deeply ingrained whenever faith rules over facts warps all areas of life. It produces a frame of mind in which concepts are formulated with deep passion but without the slightest attention paid to the evidence that bears on the concept. Nowhere is this more evident than in America today, where the large majority of the public hold onto a whole set of beliefs despite a total lack of evidence to support these beliefs and, indeed, strong evidence that denies them. These beliefs are not just limited to religion but extend to the occult (often condemned by churches), economics, politics, and health. It is not that the public lacks information. In fact, today we all are inundated with information, especially on the Internet. However, much of that information is untrustworthy, and it takes a trained thinker to filter out the good from the bad. Magical thinking and blind faith are the worst mental systems we can apply under these circumstances. They allow the most outrageous lies to be accepted as facts.

From its very beginning, religion has been a tool used by those in power to retain that power and keep the masses in line. This continues today, with religious groups being manipulated to work against believers' own best interests in health and economic well-being in order to cast doubt on well-established scientific findings. This would not be possible except for the diametrically opposed worldviews of science and religion. Science is not going to change its commitment to the truth. We can only hope religion will change its commitment to nonsense.

I have an urgent plea to scientists and all thinking people. We need to focus our attention on one goal, which will not be reached in the lifetime of the youngest among us, but which has to be achieved someday if humanity is to survive: the eradication of foolish faith from the face of this planet.

NOTES

FOREWORD

1. Thomas Jefferson, letter to Rev. James Smith, December 8, 1822.

2. I am not conceding that Jesus actually existed or that if he did exist, he actually spoke the words attributed to him in the New Testament. Since Christians do believe all of this, I quote the words "as if" they were historical and true, which is the framework within which Christians like the Neumanns operate.

3. In one of my debates with Dinesh D'Souza, he countered that the Neumanns had foolishly misinterpreted James 5:15, which says only that the prayer of faith will "save" the sick, not "heal" the sick. But we have to take into account not only the context of illness in which this verse appears, but also the fact that a direct distinction was made between "heal" and "save" in the words that follow: "and the Lord shall raise him up; and if he have committed sins, they shall be forgiven him." The word "if" separates the two issues. The "raising up" was related to the physical illness; the rest was salvific. Besides, other translations indeed consider "save" as "heal," such as the New International Version (NIV, popular with evangelicals, if not scholars), which says: "And the prayer offered in faith will make the sick person well; the Lord will raise him up." The Amplified Bible renders "raise up" as "restore," which clearly deals with a physical healing because no Christian believes that salvation "restores" a person to a previously sinless state.

4. Notice that "keeping the commandments" here has nothing to do with morality or right living. The commandment is to have faith: "believe on . . . Jesus."

5. Luke Skywalker: "All right, I'll give it a try." Yoda: "No. Try not. Do . . . or do not. There is no try."

6. Thomas Aquinas, *Summa theologica* 2.2.q2.a.9.

7. "An Introduction to St. Paul's Letter to the Romans," Luther's German Bible of 1522.

8. John Calvin, *Institutes of the Christian Religion*, vol. 3, chap. 2, sec.7.

9. I know there are some "rational apologists" who claim to have reasoned their way to belief. Depending on the assumptions in the premises, you can use an if-then argument to conclude almost anything you want. When I questioned the premises themselves, I found that they basically assume what they are aiming to conclude, which is begging the question. For example, it seems reasonable to conclude that if "everything needs a cause," then there must have been a first cause (or Prime Mover), until you realize that the premise is really saying

"everything (except God) needs a cause" (a statement of faith), which is circular reasoning because it brings the conclusion into the premise.

10. *Freethought Radio*, June 11, 2011.

11. That is actually a biblical defense: "Truly you are a God who hides himself, O God and Savior of Israel" (Isa. 45:15). This, however, sounds more like a rationalization than a defense: if God should be detected, but isn't, it must be because he chooses not to be found, not because he doesn't exist. Just seven verses earlier, by the way, the biblical God admitted that he is responsible for creating evil: "I form the light, and create darkness: I make peace, and create evil: I the LORD do all these things" (Isa. 45:7). He's not just playing Hide and Seek: he's playing Hit and Run.

12. *Freethought Radio*, March 21, 2009.

13. http://ffrf.org/out/?billboard/2078 (accessed August 26, 2011).

PREFACE

1. Bertrand Russell, *Has Religion Made Useful Contributions to Civilization? An Examination and a Criticism* (London: Watts & Co., 1930).

2. Bart D. Ehrman, *Jesus, Interrupted: Revealing the Hidden Contradictions in the Bible (and Why We Don't Know about Them)* (New York: HarperOne, 2009).

3. I thank Brent Meeker for suggesting this scenario.

1. INTRODUCTION

1. Gili S. Drori et al., *Science in the Modern World Polity: Institutionalization and Globalization* (Stanford, CA: Stanford University Press, 2003).

2. John William Draper, *History of the Conflict between Science and Religion*, 7th ed. (London: Harry S. King, 1876).

3. Andrew Dickson White, *A History of the Warfare of Science with Theology in Christendom: Two Volumes in One* (Amherst, NY: Prometheus Books, 1993; first published in 1886).

4. Draper, *History of the Conflict between Religion and Science*, pp. 51–52, as discussed in David C. Lindberg and Ronald L. Numbers, *God and Nature: Historical Essays on the Encounter between Christianity and Science* (Berkeley, CA: University of California Press, 1986), p. 1.

5. As quoted in Lindberg and Numbers, *God and Nature*, p. 2.

6. Ronald L. Numbers, "Aggressors, Victims, and Peacemakers: Historical Actors in the Drama of Science and Religion," in *The Science and Religion Debate: Why Does It Continue?* ed. Harold W. Attridge (New Haven: Yale University Press, 2009), p. 31.

7. Ian G. Barbour, *Religion and Science: Historical and Contemporary Issues* (San Francisco: HarperSanFrancisco, 1997); Lindberg and Numbers, *God and Nature*; David C. Lindberg, *The Beginnings of Western Science: The European Scientific Tradition in Philosophical, Religious, and Institutional Context, Prehistory to AD 1450*, 2nd ed. (Chicago: University of Chicago Press, 2007).

8. John Hedley Brooke, *Science and Religion: Some Historical Perspectives* (Cambridge, UK; New York: Cambridge University Press, 1991) pp. 15–16.

9. Lawrence M. Principe, *Science and Religion*, Philosophy & Intellectual History (Chantilly, VA: The Teaching Company, 2006).

10. These quotations are from the Principe Course Guidebook, pp. 9–10.

11. White, *A History of the Warfare of Science with Theology in Christendom*, p. 97.

12. Jim al-Khalili, *The House of Wisdom: How Arabic Science Saved Ancient Knowledge and Gave Us the Renaissance* (New York: Penguin, 2011).

13. Stephen Jay Gould, *Rocks of Ages: Science and Religion in the Fullness of Life* (New York: Ballantine, 1999).

14. Francis S. Collins, *The Language of God: A Scientist Presents Evidence for Belief* (New York: Free Press, 2006).

15. Sam Harris, *The Moral Landscape: How Science Can Determine Human Values* (New York: Free Press, 2010), pp. 5–6.

16. Ibid., p. 23.

17. Editorial in *Nature* 432 (2004): 657.

18. Barbour, *Religion and Science*.

19. Michael Ruse, *Science and Spirituality: Making Room for Faith in the Age of Science* (Cambridge, UK; New York: Cambridge University Press, 2010), p. 234.

20. John Templeton Foundation, "Mission," http://www.templeton.org/who-we-are/about-the-foundation/mission (accessed February 7, 2011).

21. William Grassie, *The New Sciences of Religion: Exploring Spirituality from the Outside In and Bottom Up* (New York: Palgrave Macmillan, 2010).

22. See, for example, Barbour, *Religion and Science*, p. 25.

23. Carl Zimmer, *Soul Made Flesh: The Discovery of the Brain—and How It Changed the World* (London: Heinemann, 2004).

24. Ibid., p. 26.

25. Charles Webster, "Puritanism, Separatism, and Science," in Lindberg and Numbers, *God and Nature*, pp. 192–217.

26. Barbour, *Religion and Science*, p. 27.

27. Harris, *The Moral Landscape*, p. 10.

28. Barbour, *Religion and Science*, p. 28.

29. Ibid., p. 29.

2. THE EARLIEST SKIRMISHES

1. Paul-Henri Thiry, Baron d'Holbach, *The System of Nature* (New York: Garland, 1984).

2. Edward Gibbon, *The History of the Decline and Fall of the Roman Empire* (London: W. Strahan, 1783), vol. 1.

3. Robert Roy Britt, "Startling Discovery: The First Human Ritual," *Live Science*, November 30, 2006, http://www.livescience.com/history/061130_oldest_ritual.html (accessed November 4, 2010).

4. David C. Lindberg, *The Beginnings of Western Science: The European Scientific Tradition in Philosophical, Religious, and Institutional Context, Prehistory to AD 1450*, 2nd ed. (Chicago: University of Chicago Press, 2007).

5. Pascal Boyer, *Religion Explained: The Evolutionary Origins of Religious Thought* (New York: Basic Books, 2001).

6. Scott Atran, *In Gods We Trust: The Evolutionary Landscape of Religion* (Oxford, UK; New York: Oxford University Press, 2002).

7. Daniel C. Dennett, *Breaking the Spell: Religion as a Natural Phenomenon* (New York: Viking, 2006), p. 107.

8. Justin L. Barrett, "Exploring the Natural Foundations of Religion," *Trends in Cognitive Science* 4 (2000): pp. 29–34.

9. Dennett, *Breaking the Spell*, pp. 108–09.

10. Ibid., p. 51.

11. Michael Shermer, *The Believing Brain: From Ghosts and Gods to Politics and Conspiracies—How We Construct Beliefs and Reinforce Them as Truths* (New York: Times Books, 2011).

12. Hank Davis, *Caveman Logic: The Persistence of Primitive Thinking in a Modern World* (Amherst, NY: Prometheus Books, 2009), pp. 31–32.

13. Ibid.

14. Ibid., p. 47.

15. Ibid., p. 55.

16. It wasn't until 1833 that the word "scientist" was used for what until then were called "natural philosophers." However, they are known now as scientists, so why should I be pedantic and use an antiquated term?

17. Lindberg, *The Beginnings of Western Science*, p. 11.

18. Bertrand Russell, *A History of Western Philosophy, and Its Connection with Political and Social Circumstances from the Earliest Times to the Present Day* (New York: Simon & Schuster, 1945).

19. Whenever the historicity of Jesus is doubted, Christians will often claim that the evidence for Jesus being a real person is as good as that for Socrates. However, if Socrates did not exist, this would not change anything. Plato's philosophy, attributed to Socrates, would still be the same. This is not the case for Jesus, whose existence as a man and God is central to Christian belief.

20. Victor J. Stenger, *Quantum Gods: Creation, Chaos, and the Search for Cosmic Consciousness* (Amherst, NY: Prometheus Books, 2009).

21. D. N. Sedley, *Creationism and Its Critics in Antiquity* (Berkeley: University of California Press, 2007), pp. 31–32.

22. Ibid., p. 25.

23. Ibid., pp. 78–79.

24. Will Durant, *The Story of Philosophy: The Lives and Opinions of the Greater Philosophers* (New York: Simon & Schuster, 1953), p. 9.

25. See my discussion in *Quantum Gods*, pp. 227–37.

26. Sedley, *Creationism and Its Critics in Antiquity*, pp. 82–85.

27. Lindberg, *The Beginnings of Western Science*, p. 37.

28. Ibid., p. 34, and references therein.

29. Roger Penrose, *The Road to Reality: A Complete Guide to the Laws of the Universe* (London: Jonathan Cape, 2004), p. 13.

30. However, both Penrose and Nobel laureate physicist Steven Weinberg have admitted they are Platonists, at least in some sense of the term.

31. Eugene O'Connor, trans., *The Essential Epicurus: Letters, Principle Doctrines, Vatican Sayings, and Fragments* (Amherst, NY: Prometheus Books, 1993).

32. Stephen Greenblatt, "The Answer Man: An Ancient Poem Was Rediscovered and the World Swerved," *New Yorker* (August 8, 2011): 28–33.

33. Stephen Greenblatt, *The Swerve: How the World Became Modern* (New York: W. W. Norton, 2011).

34. As quoted in Greenblatt, "The Answer Man."

35. Alison Brown, *The Return of Lucretius to Renaissance Florence* (Cambridge, MA: Harvard University Press, 2010).

36. Greenblatt, "The Answer Man."

37. Lindberg, *The Beginnings of Western Science*, p. 77.

38. Sedley, *Creationism and Its Critics in Antiquity*, p. 134.

39. Stephen Hawking and Leonard Mlodinow, *The Grand Design* (New York: Bantam Books, 2010).

40. Sedley, *Creationism and Its Critics in Antiquity*, pp. 136–8.

41. As translated by Sedley, *Creationism and Its Critics in Antiquity*, p. 143.

42. Richard R. LaCroix, "Unjustified Evil and God's Choice," *Sophia* 13, no. 1 (1974): 20–28.

43. Nicholas Everitt, "The Argument from Imperfection: A New Proof of the Non-existence of God," *Philo* 9, no. 2 (2006): 113–30.

44. As translated by Sedley, *Creationism and Its Critics in Antiquity*, p. 144.

45. James T. Cushing, *Philosophical Concepts in Physics: The Historical Relation between Philosophy and Scientific Theories* (Cambridge, UK; New York: Cambridge University Press, 1998), p. 4.

46. Sedley, *Creationism and Its Critics in Antiquity*, p. 170.

47. *Internet Encyclopedia of Philosophy*, "Aristotle (384–322 BCE)," http://www.iep.utm.edu/aristotl/ (accessed January 30, 2011).

48. As translated by Sedley in *Creationism and Its Critics in Antiquity*, p. 195.

49. Lindberg, *The Beginnings of Western Science*, pp. 148–50.

50. Ibid., p. 133.

51. S. P. Scott, *History of the Moorish Empire in Europe* (Philadelphia and London: J. B. Lippincott Co., 1904), vol. 3, pp. 461–2.

52. Jim al-Khalili, *The House of Wisdom: How Arabic Science Saved Ancient Knowledge and Gave Us the Renaissance* (New York: Penguin, 2011), p. 208.

53. Lucretius, *De Rerum Natura* (Cambridge, MA: Harvard University Press, 1992), pp. 87–93 and 177–81.

54. Richard Carrier, "Attitudes Toward the Natural Philosopher in the Early Roman Empire (100 BC to 313 AD)" (PhD dissertation, Columbia University, 2008), p. 323.

55. Vern L. Bullough, "Science and Religion in Historical Perspective," in *Science and Religion: Are They Compatible?* ed. Paul Kurtz (Amherst, NY: Prometheus Books, 2003), pp. 129–38.

56. Lindberg, *The Beginnings of Western Science*, p. 133.

3. THE REBIRTH AND TRIUMPH OF SCIENCE

1. Benjamin Silliman, "Address before the Association of American Geologists and Naturalists, Assembled in Boston, April 24, 1842," *American Journal of Science* 43 (1842): 217–50.

2. James Read Eckard, "The Logical Relations of Religion and Natural Science," *Biblical Reporatory and Princeton Review* 32 (1860): 577–608.

3. David C. Lindberg, *The Beginnings of Western Science: The European Scientific Tradition in Philosophical, Religious, and Institutional Context, Prehistory to AD 1450*, 2nd ed. (Chicago: University of Chicago Press, 2007), pp. 166–70.

4. Reviel Netz and William Noel, *The Archimedes Codex: How a Medieval Prayer Book Is Revealing the True Genius of Antiquity's Greatest Scientist* (Philadelphia, PA: Da Capo Press, 2007).

5. Ibid., p. 181.

6. Ibid., p. 177.

7. Jim al-Khalili, *The House of Wisdom: How Arabic Science Saved Ancient Knowledge and Gave Us the Renaissance* (New York: Penguin, 2011).

8. Ibid., p. xxviii.

9. Quoted in Eric John Holmyard, *Makers of Chemistry* (Oxford, UK: Clarendon, 1931).

10. Edward Grant, *Science and Religion, 400 BC to AD 1550: From Aristotle to Copernicus* (Westport, CT: Greenwood Press, 2004).

11. Lindberg, *The Beginnings of Western Science*, p. 201.

12. Ibid., pp. 194–96.

13. Ibid., p. 210.

14. al-Khalili, *The House of Wisdom*, pp. 232–34.

15. Ibid., pp. 243–44.

16. Taner Edis, *An Illusion of Harmony: Science and Religion in Islam* (Amherst, NY: Prometheus Books, 2007).

17. al-Khalili, *The House of Wisdom*, p. 246.

18. Ibid., p. 247.

19. Ibid., p. 221.

20. Ibid., pp. 223–24.

21. Ibid., pp. 226–28.

22. Ibid., pp. 229–30.

23. William Lane Craig and James D. Sinclair, "The *Kalâm* Cosmological Argument," in *The Blackwell Companion to Natural Theology*, ed. William Lane Craig and James Porter Moreland (Chichester, UK; Malden, MA: Wiley-Blackwell, 2009), pp. 101–201.

24. Lindberg, *The Beginnings of Western Science*, pp. 233–34.

25. Ibid., pp. 236.

26. Ibid., p. 244.

27. Ibid., p. 246.

28. Ibid., p. 248.

29. Ibid., pp. 250–253.

30. Ian G. Barbour, *Religion and Science: Historical and Contemporary Issues* (San Francisco: HarperSanFrancisco, 1997), pp. 4–17.

31. Victor J. Stenger, *Quantum Gods: Creation, Chaos, and the Search for Cosmic Consciousness* (Amherst, NY: Prometheus Books, 2009).

32. James Hannam, *God's Philosophers: How the Medieval World Laid the Foundations of Modern Science* (London: Icon, 2009).

33. Ian Sample, "Stephen Hawking: 'There's No Heaven; It's a Fairy Story,'" *Guardian* (May 16, 2011).

34. Cardinal Paul Poupard, ed. *Galileo Galilei: Toward a Resolution of 350 Years of Debate, 1633–1983* (Pittsburgh, PA: Duquesne University Press, 1987).

35. Edward Rosen, "Calvin's Attitude Toward Copernicus," *Journal of the History of Ideas* (July 1960): 431.

36. Thomas Dixon, *Science and Religion: A Very Short Introduction* (New York: Oxford University Press, 2008), p. 31.

37. Ibid.

38. As quoted in Michael J. Crowe, *Theories of the World from Antiquity to the Copernican Revolution*, 2nd rev. ed. (Mineola, NY: Dover, 2001), pp. 74–75.

39. P. B. Scheurer and G. Debrock, *Newton's Scientific and Philosophical Legacy* (Dordrecht, Netherlands; Boston, MA: Kluwer Academic, 1988), p. 28.

40. Barbour, *Religion and Science*, p. 23.

41. Ibid.

42. Roger Hahn, "Laplace and the Mechanistic Universe," in *God and Nature*, ed. David Lindberg and Ronald Numbers (Berkeley and Los Angeles: University of California Press, 1986).

43. Richard C. Vitzthum, *Materialism: An Affirmative History and Definition* (Amherst, NY: Prometheus Books, 1995).

44. Barbour, *Religion and Science*, pp. 34–39.

45. Philipp Blom, *A Wicked Company: The Forgotten Radicalism of the European Enlightenment* (New York: Basic Books, 2010), pp. 32–33.

46. Julien O. de La Mettrie, *L'Homme Machine* (Lyde: 1748). An English translation can be found at a University of Michigan website, http://cscs.umich.edu/~crshalizi/LaMettrie/Machine/ (accessed December 9, 2010).

47. Ibid., p. 34.

48. Vitzthum, *Materialism*, pp. 66–103.

49. Blom, *A Wicked Company*.

50. Barbour, *Religion and Science*, p. 36, as stated by Dampier in William C. Dampier's *A History of Science and Its Relations with Philosophy and Religion* (Cambridge, UK: Cambridge University Press, 4th ed., 1948), p. 196. [First ed. 1930].

51. Barbour, *Religion and Science*, p. 40.

52. For a good summary, see "Locke, Science, Morality, and Knowledge," on the Oregon State University website at http://oregonstate.edu/instruct/phl302/distance_arc/locke/locke-science-lec.html (accessed December 9, 2010).

53. Barbour, *Religion and Science*, p. 44.

54. Ibid., p. 45.

55. Michael Shermer, *How We Believe: The Search for God in an Age of Science* (New York: W. H. Freeman, 2000), p. 78.

56. See, for example, Richard Swinburne and Alan G. Padgett, *Reason and the Christian Religion: Essays in Honour of Richard Swinburne* (Oxford, UK; New York: Clarendon Press, Oxford University Press, 1994); William Lane Craig, *Reasonable Faith: Christian Truth and Apologetics* (Wheaton, IL: Crossway Books, 1994).

57. Stephen Jay Gould, *Rocks of Ages: Science and Religion in the Fullness of Life* (New York: Ballantine, 1999).

58. See, for example, Hannam, *God's Philosophers*.

59. Robert J. Hutchinson, *The Politically Incorrect Guide to the Bible* (Washington, DC: Regnery, 2007), p. 139.

60. Alvin J. Schmidt, *How Christianity Changed the World* (Zondervan, 2004), p. 222.

61. Richard Carrier, "Christianity Was Not Responsible for Modern Science," in *The Christian Delusion: Why Faith Fails*, ed. John Loftus (Amherst, NY: Prometheus Books, 2010), pp. 396–419.

62. Ibid., p. 397.

63. Ibid., p. 414.

4. DARWIN, DESIGN, AND DEITY

1. Charles Darwin, manuscript outline for *On the Origin of Species by Means of Natural Selection* (London: J. Murray, 1859).

2. Ronald L. Numbers, "Aggressors, Victims, and Peacemakers: Historical Actors in the Drama of Science and Religion," in *The Science and Religion Debate: Why Does It Continue?* ed. Harold W. Attridge (New Haven: Yale University Press, 2009), pp. 15–53.

3. William Paley, *Natural Theology* (London: Printed for R. Faulder by Wilks and Taylor, 1802).

4. Darwin, *On the Origin of Species by Means of Natural Selection*.

5. Charles Lyell and G. P. Deshayes, *Principles of Geology: Being an Attempt to Explain the Former Changes of the Earth's Surface, by Reference to Causes Now in Operation* (London: J. Murray, 1830).

6. Michael Shermer, *In Darwin's Shadow: The Life and Science of Alfred Russel Wallace: A Biographical Study on the Psychology of History* (Oxford, UK; New York: Oxford University Press, 2002).

7. Charles Darwin, *The Descent of Man, and Selection in Relation to Sex* (New York: D. Appleton & Co., 1871).

8. Letter to Miss Gerard from Adam Sedgwick, January 2, 1860, in *The Life and Letters of the Rev. Adam Sedgwick*, vol. 2 (1890), pp. 359–60.

9. Philip C. England et al., "Kelvin, Perry, and the Age of the Earth," *American Scientist* 95 (2007): 342–49. Online at http://www.es.ucsc.edu/~pkoch/EART_206/09-0108/Supplemental/Englandpercent20etpercent2007percent20AmScipercent2095-342.pdf (accessed December 18, 2010).

10. See Numbers, "Aggressors, Victims, and Peacemakers," p. 34, and references therein.

11. Keith Thomson, "Huxley, Wilberforce and the Oxford Museum," *American Scientist* 88, no. 5 (2000): 210. Online at http://www.americanscientist.org/issues/pub/2000/5/huxley-wilberforce-and-the-oxford-museum/3 (accessed December 18, 2010).

12. Peter Nichols, *Evolution's Captain: The Dark Fate of the Man Who Sailed Charles Darwin around the World* (New York: HarperCollins 2003).

13. Ibid., p. 311.

14. Thomas Dixon, *Science and Religion: A Very Short Introduction* (New York: Oxford University Press, 2008), p. 62.

15. Jerry A. Coyne, *Why Evolution Is True* (Oxford, UK; New York: Oxford University Press, 2009).

16. Richard Dawkins, *The Greatest Show on Earth: The Evidence for Evolution* (New York: Free Press, 2010).

17. Stuart A. Kauffman, *Reinventing the Sacred: A New View of Science, Reason and Religion* (New York: Basic Books, 2008).

18. Ian G. Barbour, *Religion and Science: Historical and Contemporary Issues* (San Francisco: HarperSanFrancisco, 1997).

19. John Paul II, message to Pontifical Academy of Sciences, October 22, 1996, Catholic Information Network. Online at http://www.cin.org/jp2evolu.html (accessed December 20, 2008).

20. Pope Benedict XVI, "Easter Vigil Homily: We Celebrate the First Day of the New Creation," April 25, 2011. Online at http://www.catholic.org/clife/lent/story.php ?id=41157&page=1 (accessed May 8, 2011).

21. Gallup Poll, "Republicans, Democrats Differ on Creationism," 2010 http://www.gallup.com/poll/108226/Republicans-Democrats-Differ-Creationism.aspx 2010 (accessed January 22, 2011).

22. Alvin Plantinga in *Science and Religion: Are They Compatible?* (New York: Oxford University Press, 2011), p. 4.

23. Daniel Dennett in *Science and Religion: Are They Compatible?* p. 28.

24. Alvin Plantinga in *Science and Religion: Are They Compatible?* p. 41.

25. Daniel Dennett in *Science and Religion: Are They Compatible?* p 46.

26. John D. Miller et al., "Public Acceptance of Evolution," *Science* 11 (2006): 765.

27. Tom Peck and Jerome Taylor, "Scientist Imam Threatened over Darwinist Views," *The Independent* (March 5, 2011).

28. For a comprehensive account of science in the Muslim world, see Taner Edis, *An Illusion of Harmony: Science and Religion in Islam* (Amherst, NY: Prometheus Books, 2007)

29. Numbers, "Aggressors, Victims, and Peacemakers," p. 38.

30. Edward J. Larson, *Summer for the Gods: The Scopes Trial and America's Continuing Debate over Science and Religion* (New York: Basic Books, 1997).

31. Ronald L. Numbers, *The Creationists: From Scientific Creationism to Intelligent Design*, expanded ed., first Harvard University Press pbk. ed. (Cambridge, MA: Harvard University Press, 2006).

32. George McCready Price: *Outlines of Modern Christianity and Modern Science* (Oakland, CA: Pacific Press, 1902); *Illogical Geology, the Weakest Point in the Evolution Theory* (Los Angeles: Modern Heretic Co., 1906); *The Fundamentals of Geology and Their Bearings on the Doctrine of a Literal Creation* (Mountain View, CA: Pacific Press, 1913); *The New Geology: A Textbook for Colleges, Normal Schools, and Training Schools; and for the General Reader* (Mountain View, CA: Pacific Press, 1923).

33. Numbers, "Aggressors, Victims, and Peacemakers," p. 37.

34. Ronald L. Numbers, "The Creationists," in *God and Nature: Historical Essays on the Encounter between Christianity and Science*, ed. David C. Lindberg and Ronald L. Numbers (Berkeley, CA: University of California Press, 1986), pp. 399–410.

35. Ibid., p. 402.

36. Numbers, "Aggressors, Victims, and Peacemakers," p. 40.

37. John Clement Whitcomb Jr. and Henry M. Morris, *The Genesis Flood: The Biblical Record and Its Scientific Implications* (Philadelphia, PA: Presbyterian and Reformed Pub. Co., 1961).

38. Massimo Pigliucci, *Denying Evolution: Creationism, Scientism, and the Nature of Science* (Sunderland, MA: Sinauer Associates, 2002).

39. William R. Overton, U. S. District Court Opinion, *McLean v. Arkansas*, 1982. Reprinted in *But Is It Science? The Philosophical Question in the Creation/Evolution Controversy*, ed. Michael Ruse (Amherst, NY: Prometheus Books, 1988), pp. 307–31.

40. *Lemon v. Kurtzman*, 403 U.S. 602 (1971), available at http://supreme.justia.com/us/403/602/case.html (accessed January 6, 2011).

41. For a full discussion of the philosophical issues, see Ruse, *But Is It Science?*

42. Michael J. Behe, *Darwin's Black Box: The Biochemical Challenge to Evolution* (New York: Free Press, 1996).

43. David Ussery, "Darwin's Transparent Box: The Biochemical Evidence for Evolution," in *Why Intelligent Design Fails: A Scientific Critique of the New Creationism*, ed. Matt Young and Taner Edis (New Brunswick, NJ: Rutgers University Press, 2004), pp. 48–57; Alan D. Gishlick, "Evolutionary Paths to Irreducible Systems," ibid., pp 58–71; Ian Musgrave, "Evolution of the Bacterial Flagellum," ibid., pp. 72–84.

44. William A. Dembski, *Intelligent Design: The Bridge between Science and Theology* (Downers Grove, IL: InterVarsity Press, 1999), p. 160.

45. Victor J. Stenger, *Has Science Found God? The Latest Results in the Search for Purpose in the Universe* (Amherst, NY: Prometheus Books, 2003), pp. 102–10.

46. Claude Elwood Shannon and Warren Weaver, *The Mathematical Theory of Communication* (Urbana, IL: University of Illinois Press, 1949).

47. Barbara Forrest and Paul R. Gross, *Creationism's Trojan Horse: The Wedge of Intelligent Design* (Oxford, UK; New York: Oxford University Press, 2004).

48. Phillip E. Johnson, *Defeating Darwinism by Opening Minds* (Downers Grove, IL: InterVarsity Press, 1997), p. 16.

49. Phillip E. Johnson, *Reason in the Balance: The Case Against Naturalism in Science, Law & Education* (Downers Grove, IL: InterVarsity Press, 1995).

50. Phillip E. Johnson, *The Wedge of Truth: Splitting the Foundations of Naturalism* (Downers Grove, IL: InterVarsity Press, 2000).

51. "The Wedge Strategy," Center for the Renewal of Science and Culture, http://www.

antievolution.org/features/wedge.html (accessed January 6, 2011). For more details, see Forrest and Gross, *Creationism's Trojan Horse*.

52. Steve Benen, "From Genesis to Dominion: Fat-Cat Theocrat Funds Creationism Crusade," *Church & State* (July/August 2000). Online at http://www.texscience.org/files/discovery.htm (accessed March 4, 2011).

53. *Kitzmiller v. Dover Area School District* trial transcript: Day 12 (October 1, 2005), AM session, part 1, http://www.talkorigins.org/faqs/dover/day12am.html (accessed January 11, 2012).

54. For his account of the Dover trial, see Kenneth R. Miller, "Darwin, God, and Dover: What the Collapse of 'Intelligent Design' Means for Science and Faith in America," in *The Religion and Science Debate: Why Does It Continue?* ed. Harold W. Attridge (New Haven: Yale University Press, 2009), pp. 55–92.

55. Kenneth R. Miller, *Finding Darwin's God: A Scientist's Search for Common Ground between God and Evolution* (New York: Cliff Street Books, 1999).

56. Ibid., chapter 8.

57. Victor J. Stenger, *Quantum Gods: Creation, Chaos, and the Search for Cosmic Consciousness* (Amherst, NY: Prometheus Books, 2009).

58. Jones, John E. III, "*Kitzmiller v. Dover Area School District et al.*, Case No. 04cv2688," December 20, 2005, http://www.pamd.uscourts.gov/kitzmiller/kitzmiller_342.pdf (accessed January 5, 2011).

59. Bradley John Monton, *Seeking God in Science: An Atheist Defends Intelligent Design* (Peterborough, ON: Broadview Press, 2009).

60. "U.S. Religious Knowledge Survey," Pew Forum on Religion and Public Life, September 28, 2010. http://pewforum.org/other-beliefs-and-practices/u-s-religious-knowledge-survey.aspx (accessed January 5, 2011).

61. For example, see Dan Barker, *Losing Faith in Faith: From Preacher to Atheist* (Madison, WI: Freedom from Religion Foundation, 2006); John W. Loftus, *Why I Became an Atheist: A Former Preacher Rejects Christianity* (Amherst, NY: Prometheus Books, 2008); Bart D. Ehrman, *Jesus, Interrupted: Revealing the Hidden Contradictions in the Bible (and Why We Don't Know about Them)* (New York: HarperOne, 2009).

62. Michael Berkman and Eric Plutzer, *Evolution, Creationism, and the Battle to Control America's Classrooms* (Cambridge, UK; New York: Cambridge University Press), p. 123.

63. Ibid., p. 81.

64. Barbour, *Religion and Science*, pp. 72–74.

65. Henry Drummond, *The Lowell Lectures on the Ascent of Man* (London: Hodder & Stoughton, 1894).

66. Arthur Peacocke, "The Sciences of Complexity: A New Theological Resource?" in *Information and the Nature of Reality: From Physics to Metaphysics*, ed. Paul Davies and Niels Henrik Gregersen (Cambridge, UK; New York: Cambridge University Press, 2010), pp. 249–81.

67. Michael Dowd, *Thank God for Evolution! How the Marriage of Science and Religion Will Transform Your Life and Our World* (Tulsa, OK: Council Oak Books, 2007).

68. David J. Bartholomew, *God, Chance, and Purpose: Can God Have It Both Ways?* (Cambridge, UK; New York: Cambridge University Press, 2008).

69. Barbour, *Religion and Science*, pp. 237–247.

70. Stenger, *Quantum Gods*.

71. See, for example, the 2008 documentary film *Expelled: No Intelligence Allowed*, directed by Nathan Frankowski and narrated by Ben Stein.

72. Herbert Spencer, *The Principles of Biology* (London: William and Norgate, 1864).

73. Herbert Spencer, *The Principles of Sociology* (New York: D. Appleton & Co., 1880).

74. Herbert Spencer, *System of Synthetic Philosophy*, 10 vols. See vol. 1, *First Principles* (London: Williams and Norgate, 1863).

75. Richard Dawkins, *The Selfish Gene*, new ed. (Oxford, UK; New York: Oxford University Press, 1989), p. 192.

76. Edward O. Wilson, *Sociobiology: The New Synthesis* (Cambridge, MA: Belknap Press of Harvard University Press, 1975).

77. Dawkins, *The Selfish Gene,* p. 198.

78. Richard Dawkins, *The God Delusion* (Boston, MA: Houghton Mifflin, 2006), pp. 199–200.

79. John Gray, "The Atheist Delusion," *Guardian* (March 15, 2008), http://www.guardian.co.uk/books/2008/mar/15/society (accessed May 30, 2011).

80. Craig A. James, *The Religion Virus: Why We Believe in God—An Evolutionist Explains Religion's Incredible Hold on Humanity* (n.p.: O Books, 2010), p. 58.

81. Daniel C. Dennett, *Breaking the Spell: Religion as a Natural Phenomenon* (New York: Viking, 2006), p. 84.

82. Darrel W. Ray, *The God Virus: How Religion Infects Our Lives and Culture* (Bonner Springs, KS: IPC Press, 2009), p. 23.

83. Ibid., pp. 23–29.

84. James, *The Religion Virus*, p. 58.

85. Ibid., p. 25.

86. Becky Garrison, *The New Atheist Crusaders and Their Unholy Grail: The Misguided Quest to Destroy Your Faith* (Nashville: Thomas Nelson, 2007), pp. 104–105.

87. James, *The Religion Virus*, p. 172.

5. TOWARD THE NEW PHYSICS

1. Sam Harris, *The Moral Landscape: How Science Can Determine Human Values* (New York: Free Press, 2010), p. 25.

2. Helge S. Kragh, *Entropic Creation: Religious Contexts of Thermodynamics and Cosmology* (Aldershot, Hampshire, England; Burlington, VT: Ashgate, 2008).

3. Ibid., p. 11.

4. Ibid., p. 13.

5. Here I use "particles" generically to refer to the fundamental objects, which may turn out to be strings, membranes, or whatever.

6. Victor J. Stenger, *Quantum Gods: Creation, Chaos, and the Search for Cosmic Consciousness* (Amherst, NY: Prometheus Books, 2009), pp. 55–62.

7. Victor J. Stenger, "Bioenergetic Fields," *Scientific Review of Alternative Medicine* 13, no. 1 (1999): 26–30, http://www.colorado.edu/philosophy/vstenger/Medicine/Biofield.html (accessed December 22, 2010).

8. For discussions on the arrow of time, see Huw Price, *Time's Arrow and Archimedes' Point: New Directions for the Physics of Time* (New York: Oxford University Press, 1996); Victor J. Stenger, *Timeless Reality: Symmetry, Simplicity, and Multiple Universes* (Amherst, NY: Prometheus Books, 2000); Sean M. Carroll, *From Eternity to Here: The Quest for the Ultimate Theory of Time* (New York: Dutton, 2010).

9. An English translation can be found in H. A. Lorentz et al., *The Principle of Relativity: A Collection of Original Memoirs on the Special and General Theory of Relativity* (London: Methuen & Co., 1923).

10. Tom Roberts and Siegmar Schleif, "What Is the Experimental Basis of Special Relativity?" http://www.xs4all.nl/~johanw/PhysFAQ/Relativity/SR/experiments.html#Tests_of_the_poR (accessed December 21, 2010).

11. Clifford M. Will, *Was Einstein Right? Putting General Relativity to the Test*, 2nd ed. (New York, NY: Basic Books, 1993).

6. PARTICLES AND WAVES

1. Steven Weinberg, *Dreams of a Final Theory* (New York: Pantheon Books, 1992), p. 64.

2. Diana Lutz, "Physicist Disputes 'Quantum Mind' in Debate Hosted by Deepak Chopra," *Skeptical Inquirer* 35, no. 3 (2011): 8.

3. William Grassie, *The New Sciences of Religion: Exploring Spirituality from the Outside In and Bottom Up* (New York: Palgrave Macmillan, 2010)0, p. 169.

4. Ernan McMullin, "From Matter to Materialism and (Almost) Back," in *Information and the Nature of Reality: From Physics to Metaphysics*, ed. Paul Davies and Niels Henrik Gregersen (Cambridge, UK; New York: Cambridge University Press, 2010). pp. 13–37.

5. Philip Clayton, "Unsolved Dilemmas: The Concept of Matter in the History of Philosophy and in Contemporary Physics," in *Information and the Nature of Reality: From Physics to Metaphysics*, ed. Paul Davies and Neils Henrik Gregersen (Cambridge, UK; New York: Cambridge University Press, 2010), pp. 38–62.

6. Ibid., pp. 53–54.

7. Victor J. Stenger, *Quantum Gods: Creation, Chaos, and the Search for Cosmic Consciousness* (Amherst, NY: Prometheus Books, 2009).

8. Of course "perfect" is an idealization used in the model, just as in any model. While a perfect black body does not exist, the approximation does describe what is observed in many instances.

9. Larry Laudan, "A Confutation of Convergent Realism," *Philosophy of Science* 48, no. 1 (1981): 19–49.

10. P. A. M. Dirac, *The Principles of Quantum Mechanics* (Oxford, UK: Clarendon Press, 1930).

11. Victor J. Stenger, *The Comprehensible Cosmos: Where Do the Laws of Physics Come From?* (Amherst, NY: Prometheus Books, 2006), appendix D.

12. I use "submicroscopic" rather than "microscopic" to refer to the quantum regime since all of the objects one observes with a familiar laboratory microscope, such as microbes, still obey classical mechanics.

13. Spins are conventionally expressed in units of $\hbar$.

14. Not all quantum mechanics follows in this fashion. Some quantum effects have no classical counterpart.

15. For a complete history of QED, see S. S. Schweber, *QED and the Men Who Made It: Dyson, Feynman, Schwinger, and Tomonaga* (Princeton, NJ: Princeton University Press, 1994).

16. Fritjof Capra, *The Tao of Physics: An Exploration of the Parallels between Modern Physics and Eastern Mysticism* (Boston, MA [New York]: Shambhala [distributed in the United States by Random House], 1975).

17. George Zweig, a graduate student at Cal Tech (California Institute of Technology) where Gell-Mann taught, curiously arrived at the same idea independently.

18. The reason for the "chromo" in the name is that the quantity in the theory that is analagous to electric charge is called "color charge" because it has mathematical properties likened to the primary colors.

19. F. Englert and R. Brout, "Broken Symmetry and the Mass of Gauge Vector Bosons," *Physical Review Letters* 13 (1964): 321; G. S. Gurainik, C. R. Hagen, and T. W. Kibble, "Global Conservation Laws and Massless Particles," *Physical Review Letters* 13 (1964): 585; Peter W. Higgs, "Broken Symmetries and the Masses of Gauge Bosons," *Physical Review Letters* 13 (1964): 508.

20. Clayton, "Unsolved Dilemmas," pp. 54–55.

21. Ibid., p. 55.

22. Deepak Chopra, *Quantum Healing: Exploring the Frontiers of Mind/Body Medicine* (New York: Bantam Books, 1989); Deepak Chopra, *Ageless Body, Timeless Mind: The Quantum Alternative to Growing Old* (New York: Harmony Books, 1993).

23. Rhonda Byrne, *The Secret* (New York: Atria Books; and Hillsboro, OR: Beyond Words, 2006).

24. Nicholas Saunders, *Divine Action and Modern Science* (Cambridge, UK; New York: Cambridge University Press, 2002).

25. Philip Clayton et al., eds. *Quantum Mechanics: Scientific Perspectives on Divine Action* (Vatican City: Vatican Observatory Publications, 2001).

26. Ibid.

27. James Gleick, *Chaos: Making a New Science*, 20th anniversary ed. (New York: Penguin Books, 2008).

28. Victor J. Stenger, *The Fallacy of Fine-Tuning: How the Universe Is Not Designed for Us* (Amherst, NY: Prometheus Books, 2011), pp. 274–76.

29. Victor J. Stenger, *The Unconscious Quantum: Metaphysics in Modern Physics and Cosmology* (Amherst, NY: Prometheus Books, 1995).

30. Peter Godfrey-Smith, *Theory and Reality: An Introduction to the Philosophy of Science* (Chicago: University of Chicago Press, 2003), p. 176.

31. Ibid.

32. Ibid., p. 237.

33. Roger Penrose, *The Road to Reality: A Complete Guide to the Laws of the Universe* (New York: Alfred A. Knopf, 2005), pp. 12–13.

34. Victor J. Stenger, *Physics and Psychics: The Search for a World beyond the Senses* (Amherst, NY: Prometheus Books, 1990).

35. The report by T. Adam, N. Agafonova, and A. Aleksandrov et al. in their article "Measurement of the Neutrino Velocity with the Opera Detector in the CNGS Beam," preprint submitted to *Journal of High Energy Physics*, November 17, 2011, http://xxx.lanl.gov/pdf/1109.4897v2 (accessed December 22, 2011), that neutrinos have been observed traveling faster than the speed of light is yet to be independently confirmed and, in fact, disagrees with earlier observations.

36. A. Einstein et al., "Can the Quantum Mechanical Description of Physical Reality Be Considered Complete?" *Physical Review* 47 (1935): 777.

37. David Bohm, "A Suggested Interpretation of Quantum Theory in Terms of 'Hidden Variables,' I and II," *Physical Review* 85 (1952): 166.

38. Alain Aspect et al., "Experimental Realization of the Einstein-Podolosky-Rosen *Gedankenexperiment*: A New Violation of Bell's Inequalities," *Physical Review Letters* 49 (1982): 91.

39. David. Bohm and B. J. Hiley, *The Undivided Universe: An Ontological Interpretation of Quantum Theory* (London; New York: Routledge, 1993).

40. David Bohm, *Wholeness and the Implicate Order* (London: Routledge & Kegan Paul, 1981).

41. Clayton, "Unsolved Dilemmas," p. 56.

42. Phillipe H. Eberhard and Ronald R. Ross, "Quantum Field Theory Cannot Provide Faster-Than-Light Communication," *Foundations of Physics Letters* 2 (1989): 127–79.

43. For more details, see Stenger, *The Unconscious Quantum*.

7. COSMOS AND CREATOR

1. Barbara C. Sproul, *Primal Myths: Creating the World* (San Francisco: Harper & Row, 1979).

2. For a philosophical perspective on theism and cosmology, see Hans Halvorson and Helge S. Kragh, "Theism and Physical Cosmology," in *The Routledge Companion to Theism*, ed. Stewart Goetz et al. (scheduled for release June 12, 2012).

3. Francis S. Collins, *The Language of God: A Scientist Presents Evidence for Belief* (New York: Free Press, 2006), p. 67.

4. Georges-Henri Lemaître, "Un Univers homogène de masse constante et de rayon croissant rendant compte de la vitesse radiale des nébuleuses extragalactiques (A homogeneous Universe of constant mass and growing radius accounting for the radial velocity of extragalactic nebulae)," *Annales de la Société Scientifique de Bruxelles* (Annals of the Scientific Society of Brussels), 47 (April 1927): 49.

5. A. Deprit, "Monsignor Georges Lemaître," in *The Big Bang and Georges Lemaître*, ed. A. Barger (Dordrecht, Germany: Reidel, 1984), p. 370.

6. A. S. Eddington, "On the Instability of Einstein's Spherical World," *Monthly Notices of the Royal Society* 90 (1930): 668–88.

7. Humason had started as a mule driver bringing supplies up the mountain.

8. Marcia Bartusiak, *The Day We Found the Universe* (New York: Pantheon Books, 2009); Harry Nussbaumer and Lydia Bieri, *Discovering the Expanding Universe* (Cambridge, UK; New York: Cambridge University Press, 2009).

9. Our closest galaxy, Andromeda, is actually heading toward us and may collide with the Milky Way some day.

10. The redshift is analogous to the decrease in pitch of an ambulance siren as it moves away from us. When moving toward us, the siren pitch is higher.

11. I discuss some possible reasons for a zero cosmological constant in *The Fallacy of Fine-Tuning: How the Universe Is Not Designed for Us* (Amherst, NY: Prometheus Books, 2011).

12. Pope Pius XII, "The Proofs for the Existence of God in the Light of Modern Natural Science," address by Pope Pius XII to the Pontifical Academy of Sciences, November 22, 1951, reprinted as "Modern Science and the Existence of God," *Catholic Mind* 49 (1972): 182–92.

13. Joseph Silk, *The Big Bang*, 3rd ed. (New York: W.H. Freeman, 2001).

14. Robert Jastrow, *God and the Astronomers*, 2nd ed. (New York: W. W. Norton, 2000), p. 107.

15. Stephen W. Hawking and Roger Penrose, "The Singularities of Gravitational Collapse," *Proceedings of the Royal Society of London*, series A, 314 (1970): 529–48.

16. William Lane Craig and James D. Sinclair, "The *Kalâm* Cosmological Argument," in *The Blackwell Companion to Natural Theology*, ed. William Lane Craig and James Porter Moreland (Chichester, UK; Malden, MA: Wiley-Blackwell, 2009), pp. 101–201.

17. David Hilbert, "On the Infinite," in *Philosophy of Mathematics*, ed. Paul Benacerraf and Hillary Putnam (Englewood Cliffs, NJ: Prentice-Hall, 1964), pp. 139–41.

18. Craig and Sinclair, "The *Kalâm* Cosmological Argument," p. 103.

19. John Hedley Brooke, *Science and Religion: Some Historical Perspectives* (Cambridge, UK; New York: Cambridge University Press, 1991), p. 28.

20. Arvind Borde, Alan H. Guth, and Alexander Vilenkin, "Inflationary Spacetimes Are Not Past-Complete," *Physical Review Letters* 90 (2003): 151301.

21. Ravi K. Zacharias, *The End of Reason: A Response to the New Atheists* (Grand Rapids, MI: Zondervan, 2008), p. 31.

22. Dinesh D'Souza, *What's So Great About Christianity?* (Washington, DC: Regnery, 2007), p. 116.

23. Tony Rothman, "A 'What You See Is What You Beget' Theory," *Discover* (May 1987), p. 99.

24. Collins, *The Language of God*, p. 75.

25. Stenger, *The Fallacy of Fine-Tuning*, chapter 2.

26. Brandon Carter, "Large Number Coincidences and the Anthropic Principle in Cosmology" (paper presented at the IAU [International Astronomical Union] Symposium 63: Confrontation of Cosmological Theories with Observational Data, Dordrecht, Germany, 1974).

27. John D. Barrow and Frank J. Tipler, *The Anthropic Cosmological Principle* (Oxford, UK; New York: Oxford University Press, 1986).

28. Martin J. Rees, *Just Six Numbers: The Deep Forces That Shape the Universe* (New York: Basic Books, 2000; first published in 1999); John D. Barrow, *The Constants of Nature: From Alpha to Omega* (London: Jonathan Cape, 2002); Paul Davies, *The Goldilocks Enigma: Why Is the Universe Just Right for Life?* (London: Allen Lane, 2006).

29. Hugh Ross, *The Creator and the Cosmos: How the Greatest Scientific Discoveries of the Century Reveal God* (Colorado Springs, CO: NavPress, 1995); Richard Swinburne, "Argument from the Fine-Tuning of the Universe," in *Modern Cosmology & Philosophy*, ed. John Leslie (Amherst, NY: Prometheus Books, 1988), pp. 160–79.

30. Hugh Ross, "Big Bang Model Refined by Fire," in *Mere Creation: Science, Faith & Intelligent Design*, ed. William A. Dembski (Downers Grove, IL: InterVarsity Press, 1998), pp. 363–83.

31. Rich Deem, "Evidence for the Fine Tuning of the Universe," Evidence for God, http://www.godandscience.org/apologetics/designun.html (accessed January 10, 2011).

32. Stephen W. Hawking, *A Brief History of Time: From the Big Bang to Black Holes* (Toronto, ON; New York: Bantam Books, 1988), pp. 121–22.

33. Deem says "mass of the universe," but referring to his source, Ross, it is clear that Deem should have said mass density of the universe.

34. Ross, "Big Bang Model Refined by Fire," p. 373.

35. Leonard Susskind, *The Cosmic Landscape: String Theory and the Illusion of Intelligent Design* (New York: Little, Brown, 2005), pp. 65–78.

36. Demos Kazanas, "Dynamics of the Universe and Spontaneous Symmetry Breaking," *Astrophysical Journal* 241 (1980): L59–L63; Alan H. Guth, "Inflationary Universe: A Possible Solution to the Horizon and Flatness Problems," *Physical Review D* 23, no. 2 (1981): 347–56; Andrei D. Linde, "A New Inflationary Universe Scenario: A Possible Solution of the Horizon, Flatness, Homogeneity, Isotropy, and Primordial Monopole Problems," *Physics Letters B* 108 (1982): 389.

37. See Hawking, *A Brief History of Time*, p. 128.

38. Ned Wright, "Errors in the Steady State and Quasi-SS Models," http://www.astro.ucla.edu/~wright/stdystat.htm (accessed January 11, 2011).

39. A. D. Linde, "Eternally Existing Self-Reproducing Chaotic Inflationary Universe," *Physics Letters B* 175, no. 4 (1986): 395–400; A. Vilenkin, *Many Worlds in One: The Search for Other Universes* (New York: Hill and Wang, 2006); Bernard Carr, ed., *Universe or Multiverse?* (Cambridge, UK: Cambridge University Press, 2007).

40. Richard Swinburne, *Is There a God?* (Oxford, UK; New York: Oxford University Press, 1996).

41. Don N. Page, "Evidence against Fine Tuning for Life," January 12, 2011, http://xxx.lanl.gov/abs/1101.2444v1 (accessed January 20, 2011); Don Page, "Does God So Love the Multiverse?" in *Science and Religion in Dialogue*, ed. Melville Y. Stewart (Chichester, UK: Wiley-Blackwell, 2010), pp. 380–95.

42. Hawking, *A Brief History of Time*, p. 50.

43. Ibid.

44. Anthony N. Aguirre and Steven Gratton, "Inflation without a Beginning: A Null Boundary Proposal," *Physical Review D* 67 (2003): 083516.

45. Alexander Vilenkin, e-mail communication, May 21, 2010.

46. Page, "Evidence against Fine Tuning for Life."

47. Linde, "Eternally Existing Self-Reproducing Chaotic Inflationary Universe"; Vilenkin, *Many Worlds in One*; Carr, *Universe or Multiverse?*

48. Vilenkin, *Many Worlds in One*, pp. 141–51; Stephen Feeney et al., "First Observational Tests of Eternal Inflation," December 10, 2010, http://arxiv.org/pdf/1012.1995v1 (accessed January 14, 2011).

49. Susskind, *The Cosmic Landscape*.

50. Steven Weinberg, "Living in the Multiverse," in *Universe or Multiverse?* ed. Bernard Carr (New York: Cambridge University Press, 2007), pp. 29–42.

51. For a detailed demonstration, see Stenger, *The Fallacy of Fine-Tuning*.

52. Hawking, *A Brief History of Time*, p. 128.

53. Gerard 'tHooft, "Dimensional Reduction in Quantum Gravity," last revised March 2009, http://lanl.arxiv.org/abs/gr-qc/9310026 (accessed January 11, 2011); Raphael Buosso, "The Holographic Principle," *Reviews of Modern Physics* 74 (2002): 825–75.

54. F. Hoyle et al., "A State in C12 Predicted from Astronomical Evidence," *Physical Review Letters* 92 (1953): 1095; Barrow and Tipler, *The Anthropic Cosmological Principle*, p. 252.

55. M. Livio et al., "The Anthropic Significance of the Existence of an Excited State of C12," *Nature* 340 (1989): 281–84.

56. D. I. Kazakov, "Beyond the Standard Model (In Search of Supersymmetry)," lectures given at the European School for High Energy Physics, Caramulo, Portugal, August/September 2000.

57. Ross, *The Creator and the Cosmos*.

58. André Linde, "Quantum Creation of the Inflationary Universe," *Physics Letters* 108B (1982): 389–92; James B. Hartle and Stephen W. Hawking, "Wave Function of the Universe," *Physical Review D* 28 (1983): 2960–75; Alexander Vilenkin, "Boundary Conditions in Quantum Cosmology," *Physical Review D* 33 (1986): 3560–69; David Atkatz, "Quantum Cosmology for Pedestrians," *American Journal of Physics* 62, no. 7 (1994): 619–27; David Atkatz and Heinz Pagels, "Origin of the Universe as a Quantum Tunneling Event," *Physical Review Letters D* 25 (2001): 083508.

59. Vilenkin, Alexander, "Creation of Universes from Nothing," *Physics Letters B* 117B (1982): 25–28.

60. A. Vilenkin, *Many Worlds in One*, chapter 17.

61. James B. Hartle and Stephen W. Hawking, "Wave Function of the Universe," *Physical Review D* 28 (1983): 2960–75.

62. This scenario is worked out at an undergraduate mathematical level in Victor J. Stenger, *The Comprehensible Cosmos: Where Do the Laws of Physics Come From?* (Amherst, NY: Prometheus Books, 2006), pp: 312–19.

63. Stephen W. Hawking and Leonard Mlodinow, *The Grand Design* (New York: Bantam Books, 2010).

64. Hawking, *A Brief History of Time*, p. 175.

65. Carl Sagan, in Hawking, *A Brief History of Time*, p. x.

66. Hartle and Hawking, "Wave Function of the Universe," 2960–75.

67. Hawking and Mlodinow, *The Grand Design*, p. 180.

68. Ibid., p. 5.

69. Ibid., p. 42.

70. Ibid., p. 46.

71. Ibid., p. 136.

72. Ibid., p. 164.

73. Robert Wright, *The Evolution of God* (New York: Little, Brown, 2009).

74. Paul Davies and John R. Gribbin, *The Matter Myth: Dramatic Discoveries That Challenge Our Understanding of Physical Reality* (New York: Simon & Schuster, 1992).

75. Seth Lloyd, *Programming the Universe: A Quantum Computer Scientist Takes on the Cosmos* (New York: Knopf, 2006).

76. Niels Henrik Gregersen, "God, Matter, and Information: Towards a Stoicizing Logos Christianity," in *Information and the Nature of Reality: From Physics to Metaphysics*, ed. Paul Davies and Niels Henrik Gregersen (Cambridge: Cambridge University Press, 2010), p. 339.

77. Wright, *The Evolution of God*, p, 216.

78. Gregersen, "God, Matter, and Information," p. 325.

79. John F. Haught, "Information, Theology, and the Universe," in *Information and the Nature of Reality: From Physics to Metaphysics*, ed. Paul Davies and Niels Henrik Gregersen (Cambridge, UK; New York: Cambridge University Press, 2010), pp. 301–38.

80. Keith Ward, "God as the Ultimate Informational Principle," in *Information and the Nature of Reality: From Physics to Metaphysics* , ed. Paul Davies and Niels Henrik Gregersen (Cambridge, UK; New York: Cambridge University Press, 2010), pp. 282–299.

81. Claude Elwood Shannon and Warren Weaver, *The Mathematical Theory of Communication* (Urbana, IL: University of Illinois Press, 1949).

82. Paul Young, *The Nature of Information* (New York: Praeger, 1987).

83. Gregersen, "God, Matter, and Information," pp. 319–48.

84. John A. Wheeler, "Information, Physics, Quantum: The Search for Links," in *Complexity, Entropy, and the Physics of Information*, ed. W. Zurek (Redwood City, CA: Addison-Wesley, 1990).

85. Lloyd, *Programming the Universe*, p. 151.

86. Ibid., chapters 7 and 8.

8. PURPOSE

1. David Hume and Henry David Aiken, *Dialogues Concerning Natural Religion* (New York: Hafner, 1972).

2. Freeman J. Dyson, *A Many-Colored Glass: Reflections on the Place of Life in the Universe* (Charlottesville: University of Virginia Press, 2007), p. 76.

3. Dinesh D'Souza, *Life after Death: The Evidence* (Washington, DC: Regnery, 2009), p. 103.

4. Ibid., p. 104.

5. Simon Conway Morris, *Life's Solution: Inevitable Humans in a Lonely Universe* (Cambridge, UK; New York: Cambridge University Press, 2003).

6. Elliott Sober, "It Had to Happen," *New York Times* (November 30, 2003).

7. Simon Conway Morris, "Darwin Was Right. Up to a Point," *Manchester Guardian* (February 12, 2009), available at http://www.guardian.co.uk/global/2009/feb/12/simon-conway-morris-darwin (accessed December 20, 2009).

8. Jerry A. Coyne, *Why Evolution Is True* (Oxford, UK; New York: Oxford University Press, 2009), pp. 92–94.

9. Jerry A. Coyne, "Simon Conway Morris Becomes a Creationist," *Why Evolution Is True*, February 14, 2009, http://whyevolutionistrue.wordpress.com/2009/02/14/simon-conway-morris-becomes-a-creationist/ (accessed December 18, 2009).

10. H. James Birx, *Interpreting Evolution: Darwin & Teilhard de Chardin* (Amherst, NY: Prometheus Books, 1991).

11. For example, Jean Telemond in *Shoes of the Fisherman* and Father Lankester Merrin in *The Exorcist*.

12. Pierre Teilhard de Chardin, *The Phenomenon of Man* (New York: Harper, 1959).

13. Ian G. Barbour, *Religion and Science: Historical and Contemporary Issues* (San Francisco: HarperSanFrancisco, 1997), p. 247.

14. Birx, *Interpreting Evolution*, p. 193–200.

15. John Maynard Smith, *On Evolution* (Edinburgh, UK: Edinburgh University Press, 1972), p. 89.

16. Stephen Jay Gould, *Ever Since Darwin: Reflections in Natural History* (New York: W. W. Norton, 1977).

17. William B. Provine, "Darwinism: Science or Naturalistic Philosophy?" *Origins Research Archives* 16, no. 1 (1995): 9.

18. Frank J. Tipler, *The Physics of Immortality: Modern Cosmology, God, and the Resurrection of the Dead* (New York: Anchor Books, 1994).

19. Victor J. Stenger, "Scientist Nitwit Atheist Proves Existence of God," *Free Inquiry* 15, no. 2(1995): 54–55.

20. William Grassie, *The New Sciences of Religion: Exploring Spirituality from the Outside In and Bottom Up* (New York: Palgrave Macmillan, 2010), p. 166.

21. Ibid., p. 168.

22. Philip W. Anderson, "More Is Different," *Science* 177, no. 4047 (1972): 393–96.

23. Philip Clayton and Paul Davies, eds. *The Re-Emergence of Emergence: The Emergentist Hypothesis from Science to Religion* (Oxford, UK; New York: Oxford University Press, 2006).

24. Stuart A. Kauffman, *Reinventing the Sacred: A New View of Science, Reason, and Religion* (New York: Basic Books, 2008), p. 230.

25. Grassie, *The New Sciences of Religion*, p. 107.

26. James Gleick, *Chaos: Making a New Science*, 20th anniversary ed. (New York: Penguin Books, 2008).

27. Philip Ball, *The Self-Made Tapestry: Pattern Formation in Nature* (Oxford, UK; New York: Oxford University Press, 1999).

28. Philip Ball, *Critical Mass: How One Thing Leads to Another* (New York: Farrar, Straus, and Giroux, 2004).

29. Friedrich Wöhler, "Über künstliche Bildung des Harnstoffs" ("On theArtificial Formation of Urea")," *Annalen der Physik und Chemie* 88, no. 2 (1828): 253–56.

30. See, for example, Thomas Crean, *God Is No Delusion: A Refutation of Richard Dawkins* (San Francisco: Ignatius Press, 2007), p. 10.

31. Victor J. Stenger, *God: The Failed Hypothesis* (Amherst, NY: Prometheus Books, 2007).

32. Richard Dawkins, *The God Delusion* (Boston, MA: Houghton Mifflin, 2006), p. 109.

33. William A. Dembski, *Intelligent Design: The Bridge between Science and Theology* (Downers Grove, IL: InterVarsity Press, 1999).

34. Michael J. Behe, *Darwin's Black Box: The Biochemical Challenge to Evolution* (New York: Free Press, 1996).

35. Paul Davies, *The Cosmic Blueprint: New Discoveries in Nature's Creative Ability to Order the Universe* (Philadelphia, PA: Templeton Foundation Press, 2004), p. 142.

36. George F. R. Ellis, "True Complexity and Its Associated Ontology," in *Science and Ultimate Reality: Quantum Theory, Cosmology, and Complexity*, ed. John D. Barrow et al. (Cambridge, UK: Cambridge University Press, 2004), p. 613.

37. Philip D. Clayton, "Emergence: Us from It," in *Science and Ultimate Reality: Quantum Theory, Cosmology, and Complexity*, ed. John D. Barrow et al. (Cambridge, UK: Cambridge University Press, 2004), pp. 577–606.

38. Arthur Peacocke, "The Sciences of Complexity: A New Theological Resource?" in *Information and the Nature of Reality: From Physics to Metaphysics*, ed. Paul Davies, and Niels Henrik Gregersen (Cambridge, UK; New York: Cambridge University Press, 2010).

39. Ibid., p. 265.

40. Ibid., p. 253.

41. Paul Davies, *The Physics of Time Asymmetry* (Berkeley, CA: University of California Press, 1974).

42. Heinz-Dieter Zeh, *The Physical Basis of the Direction of Time*, 5th ed. (Berlin; New York: Springer, 2007).

43. Peacocke, "The Sciences of Complexity," p. 255.

44. Stephen K. Scott, *Oscillations, Waves, and Chaos in Chemical Kinetics* (Oxford, UK; New York: Oxford University Press, 1994).

45. Alexander K. Dewdney, "The Hodge-Podge Machine Makes Waves," *Scientific American* 259, no. 2 (1998): 104–107; Stephen Wolfram, *A New Kind of Science* (Champaign, IL: Wolfram Media, 2002).

46. Ontario Consultants on Religious Tolerance, "Roman Catholicism and Abortion Access: Pagan & Christian Beliefs 400 BCE—1980 CE," http://www.religioustolerance.org/abo_hist.htm (accessed January 13, 2012).

47. Carol E. Cleland and Christopher F. Chyba, "Defining 'Life,'" *Origins of Life and Evolution of the Biosphere* 32, no. 4 (2002): 387–93.

48. Jeff Schweitzer and Giuseppe Notarbartolo-di-Sciara, *Beyond Cosmic Dice: Moral Life in a Random World* (Los Angeles: Jacquie Jordan, 2009), p. 47.

49. Stanley L. Miller, "A Production of Amino Acids under Possible Primitive Earth Conditions," *Science* 117 (1953): 528–29.

50. Paul F. Deisler Jr., "How Did Life Begin? A Perspective on the Nature and Origin of Life," *Skeptic* 16, no. 2 (2011): 34–40 and references therein.

51. Albrecht Moritz, "The Origin of Life," *TalkOrigins Archive*, updated July 1, 2010, http://www.talkorigins.org/faqs/abioprob/originoflife.html (accessed December 17, 2009).

52. Alonzo Richardo and Jack W. Szostak, "Life on Earth: Fresh Clues Hint at How the

First Living Organisms Arose from Inanimate Matter," *Scientific American* (September 2009): 54–61.

9. TRANSCENDENCE

1. Some of the material in this chapter has previously appeared in Victor J. Stenger, "Life after Death: Examining the Evidence," in *The End of Christianity*, ed. John W. Loftus (Amherst, NY: Prometheus Books, 2011), pp. 305–32.

2. Deepak Chopra, *Life after Death: The Burden of Proof* (New York: Harmony Books, 2006).

3. Dinesh D'Souza, *Life after Death: The Evidence* (Washington, DC: Regnery, 2009).

4. Corliss Lamont, *The Illusion of Immortality*, 4th ed. (New York: F. Ungar, 1965).

5. Alan F. Segal, *Life after Death: A History of the Afterlife in the Religions of the West* (New York: Doubleday, 2004).

6. Most experts agree that the pagan Zoroastrians introduced this idea to the Jews. See Segal, *Life after Death*, pp. 173–203, and discussion and sources in Richard Carrier, *Not the Impossible Faith: Why Christianity Didn't Need a Miracle to Succeed* (Lulu.com, 2009), pp. 85–86, 90–99.

7. Paul Edwards, *Reincarnation: A Critical Examination* (Amherst, NY: Prometheus Books, 2002).

8. Karen Armstrong, *The Great Transformation: The Beginning of Our Religious Traditions* (New York: Knopf, 2006).

9. Actually, the speed of light can't change, by definition. See chapter 7.

10. National Academy of Sciences, *Teaching about Evolution and the Nature of Science* (Washington, DC: National Academies Press, 1998), p. 58. Free copy available at http://www. nap.edu/catalog/5787.html (accessed January 13, 2012).

11. H. J. Benson et al., "Study of the Therapeutic Effects of Intercessory Prayer (STEP) in Cardiac Bypass Patients: A Multicenter Randomized Trial of Uncertainty and Certainty of Receiving Intercessory Prayer," *American Heart Journal* 154, no. 4 (2007): 934–42.

12. For a nice summary of the STEP results, see the Templeton Foundation press release, Mechal Weiss, "Largest Study of Third-Party Prayer Suggests Such Prayer Not Effective in Reducing Complications Following Heart Surgery," June 4, 2007, http://www.templeton.org/ pdfs/press_releases/060407STEP.pdf (accessed February 12, 2011).

13. J. M. Aviles et al., "Intercessory Prayer and Cardiovascular Disease Progression in a Coronary Care Unit Population: A Randomized Controlled Trial," *Mayo Clinic Proceedings* 76, no. 12 (2001): 1192–98.

14. M. W. Krucoff et al., "Music, Imagery, Touch, and Prayer as Adjuncts to Interventional Cardiac Care: The Monitoring and Actualization of Noetic Trainings (Mantra) II Randomized Study," *Lancet* 366 (2005): 211–17.

15. C. G. Ellison and J. S. Levin, "The Religion-Health Connection: Evidence, Theory, and Future Directions," *Health Education and Behavior* 24 (1998): 700–20; D. B. Larson et al., eds. *Scientific Research on Spirituality and Health: A Report Based on the Scientific Progress in Spirituality Conferences* (Rockville, MD: National Institutes for Healthcare Research, 1998).

16. Richard P. Sloan and Emila Bagiella, "Claims about Religious Involvement in Health Outcomes," *Annals of Behavioral Medicine* 24, no. 1 (2002): 14–21.

17. Harold G. Koenig, *The Healing Power of Faith: Science Explores Medicine's Last Great Frontier* (New York: Simon & Schuster, 1999); Chester L. Tolson and Harold G. Koenig, *The Healing Power of Prayer: The Surprising Connection between Prayer and Your Health* (Grand Rapids, MI: Baker Books, 2003); Harold G. Koenig et al., *Handbook of Religion and Health*, 2nd ed. (Oxford, UK; New York: Oxford University Press, 2011).

18. Harold Koenig, "What Religion Can Do for Your Health," Beliefnet, http://www.beliefnet.com/Health/2006/05/What-Religion-Can-Do-For-Your-Health.aspx (accessed February 11, 2011).

19. L. H. Powell et al., "Religion and Spirituality: Linkages to Physical Health," *American Psychologist* 58, no. 1 (2003): 36–52.

20. Ibid., p. 36.

21. Elissa Patterson, "The Philosophy and Physics of Holistic Health Care: Spiritual Healing as a Workable Interpretation," *British Journal of Advanced Nursing* 27 (1998): 287–93.

22. Béla Scheiber and Carla Selby, *Therapeutic Touch* (Amherst, NY: Prometheus Books, 2000).

23. Joanne Stefanatos, "Introduction to Bioenergetic Medicine," in *Complementary and Alternative Veterinary Medicine: Principles and Practice*, ed. Allen M. Schoen and Susan G. Wynn (St. Louis: Mosby Year Book, 1997 [Mosby is now an Elsevier Health Sciences imprint]), p. 270.

24. David Ramey, ed. *Consumer's Guide to Alternate Therapies in the Horse* (New York: Howell Book House, 1999).

25. Victor J. Stenger, "Bioenergetic Fields," *Scientific Review of Alternative Medicine* 3, no. 1 (1999): 26–30.

26. Some of the following material is taken from Victor J. Stenger, "Life after Death: Examining the Evidence," in *The End of Christianity*, ed. John Loftus (Amherst, NY: Prometheus Books, 2011), pp. 305–58.

27. Janice Minor Holden, Bruce Greyson, and Debbie James, eds. *The Handbook of Near-Death Experiences: Thirty Years of Investigation* (Santa Barbara, CA: Praeger, 2009), chap. 1.

28. Raymond A. Moody Jr., MD, *Life after Life: The Investigation of a Phenomenon—Survival of Bodily Death* (New York: Bantam Books; HarperCollins, 1976).

29. Susan J. Blackmore, *Dying to Live: Near-Death Experiences* (Amherst, NY: Prometheus Books, 1993).

30. Gerald Woerlee, *Mortal Minds: The Biology of Near-Death Experiences* (Amherst, NY: Prometheus Books, 2005).

31. Kevin Nelson, *The Spiritual Doorway in the Brain: A Neurologist's Search for the God Experience* (New York: Dutton, 2011), pp. 128–31.

32. Holden et al., *The Handbook of Near-Death Experiences*, p. 16.

33. Ibid., p. 27.

34. Ibid., p. 186.

35. Ibid.

36. Ibid., p. 209.

37. Ibid., p. 210.

38. Dinesh D'Souza, *Life after Death: The Evidence* (Washington, DC: Regnery, 2009), p. 64.

39. Kimberly Clark, "Clinical Interventions with NDEs," in Bruce Greyson and Charles P. Flynn, eds., *The Near-Death Experience: Problems, Prospects, Perspectives* (Springfield, IL: C. C. Thomas, 1984), pp. 242–55.

40. Hayden Ebbem et al., "Maria's Near-Death Experience: Waiting for the other Shoe to Drop," *Skeptical Inquirer* 20, no. 4 (1996): 27–33.

41. Jeffrey P. Bishop and Victor J. Stenger, "Retroactive Prayer: Lots of History, Not Much Mystery, No Science," *British Medical Journal* 329 (2004): 1444–46; "Religion and Health," exchange with Andrew Weaver and Larry Dossey, http://www.colorado.edu/philosophy/vstenger/relig.html (accessed June 9, 2011).

42. Larry Dossey, *Recovering the Soul: A Scientific and Spiritual Search* (New York: Bantam Books, 1989).

43. Kenneth Ring and Sharon Cooper, *Mindsight: Near-Death and Out-of-Body Experiences in the Blind* (Palo Alto, CA: William James Center for Consciousness Studies, Institute of Transpersonal Psychology, 1999).

44. Blackmore, *Dying to Live*, pp. 131–32.

45. Ring and Cooper, *Mindsight*.

46. Jeffrey Long and Paul Perry, *Evidence of the Afterlife: The Science of Near-Death Experiences* (New York: HarperCollins, 2010).

47. Ibid.

48. Mark Fox, *Religion, Spirituality, and the Near-Death Experience* (New York: Routledge, 2003).

49. Internet Infidels, Secular Web, http://www.infidels.org/infidels/ (accessed December 11, 2009).

50. Keith Augustine, "Does Paranormal Perception Occur in Near-Death Experiences?" *Journal of Near-Death Studies* 25, no. 4 (2007): 203–36; "Near-Death Experiences with Hallucinatory Features," *Journal of Near-Death Studies* 26, no. 1 (2007): 3–31; "Psychophysiological and Cultural Correlates Undermining a Survivalist Interpretation of Near-Death Experiences," *Journal of Near-Death Studies* 26, no. 2 (2007): 89–125. See following papers in each volume, which present criticisms and Augustine's responses to these criticisms.

51. Keith Augustine, "Hallucinatory Near-Death Experiences," Secular Web, http://www.infidels.org/library/modern/keith_augustine/HNDEs.html#veridical (accessed January 26, 2011).

52. Dannion Brinkley and Paul Perry, *Saved by the Light: The True Story of a Man Who Died Twice and the Profound Revelations He Received* (New York: Villard Books, 1994).

53. *Wikipedia*, http://en.wikipedia.org/wiki/Saved_by_the_Light (accessed December 9, 2009).

54. Morey Bernstein, *The Search for Bridey Murphy* (Garden City, NY: Doubleday, 1956).

55. Terence Hines, ed. *Pseudoscience and the Paranormal: A Critical Examination of the Evidence* (Amherst, NY: Prometheus Books, 1988); Martin Gardner, *Fads and Fallacies in the Name of Science* (New York: Dover, 1957); James Alcock, "Psychology and Near-Death Experiences," *Skeptical Inquirer* 3, no. 3 (1978): 25–41.

56. Chopra, *Life after Death: The Burden of Proof*, pp. 72–73.

57. Ian Stevenson, *Twenty Cases Suggestive of Reincarnation*, 2nd ed. (Charlottesville: University Press of Virginia, 1974).

58. Ian Stevenson, *Reincarnation and Biology: A Contribution to the Etiology of Birthmarks and Birth Defects*, 2 vols. (Westport, CT: Praeger, 1997).

59. Leonard Angel, *"Reincarnation and Biology* (Book Review)," *Skeptic* 9, no. 3 (2002): 86–90.

60. Victor J. Stenger, *Physics and Psychics: The Search for a World beyond the Senses* (Amherst, NY: Prometheus Books, 1990), pp. 276–305.

61. Paul Kurtz, *The Transcendental Temptation: A Critique of Religion and the Paranormal* (Amherst, NY: Prometheus Books, 1986).

62. Daryl J. Bem, "Feeling the Future: Experimental Evidence for Anomalous Retroactive Influences on Cognition and Affect," *Journal of Personality and Social Psychology* 100, no. 3 (2011): 407–25.

63. Victor J. Stenger, *Timeless Reality: Symmetry, Simplicity, and Multiple Universes* (Amherst, NY: Prometheus Books, 2000).

64. Erik-Jan Wagenmakers et al., "Why Psychologists Must Change the Way They Analyze Their Data: The Case of Psi: Comment on Bem," *Journal of Personality and Social Psychology* 100, no. 3 (2011): 426–32.

65. Jeffrey N. Rouder and Richard D. Morey, "A Bayes Factor Meta-Analysis of Bem's ESP Claim," *Psychonomic Bulletin and Review* 18, no. 4 (May 15, 2011): 682–89, http://www.springerlink.com/content/q113m61462793241/ (accessed June 22, 2011).

66. Ben Goldacre quoted in Kendrick Frazier, "Journal That Published Bem's Paper Rejects Attempt to Replicate It," *Skeptical Inquirer* 35, no. 4 (2011): 7.

67. James E. Alcock, "Back from the Future: Parapsychology and the Bem Affair," *Skeptical Inquirer* 35, no. 2 (2011): 31–39.

68. Joseph Banks Rhine, *Extra-Sensory Perception* (Boston, MA: Boston Society for Psychic Research, 1934), p. 210.

69. Alcock, "Back from the Future," p. 31.

70. Even more often, considering the "file-drawer effect" in which negative results are

often not published because they aren't interesting. In fact, negative results are just as important as positive results in science.

71. David H. Freedman, "Lies, Damned Lies, and Medical Science," *Atlantic* (November 2010): 76–86.

72. Dean I. Radin, *The Conscious Universe: The Scientific Truth of Psychic Phenomena* (New York: HarperEdge, 1997).

73. I. J. Good, "Where Has the Billion Trillion Gone?" *Nature* 389, no. 6653 (1997): 806–807; Douglas M. Stokes, "The Shrinking Filedrawer: On the Validity of Statistical Meta-Analysis in Parapsychology," *Skeptical Inquirer* 35, no. 3 (2001): 22–25.

74. Frank J. Tipler, *The Physics of Christianity* (New York: Doubleday, 2007).

75. Joe Nickell, *Inquest on the Shroud of Turin: Latest Scientific Findings* (Amherst, NY: Prometheus Books, 1998), pp. 149–50.

76. Normally an electron and proton can interact, producing a neutrino and neutron. However, here Tipler is proposing complete annihilation into neutrinos alone.

77. Ernan McMullin, "From Matter to Materialism and (Almost) Back," in *Information and the Nature of Reality: From Physics to Metaphysics*, ed. Paul Davies and Niels Henrik Gregersen (Cambridge, UK; New York: Cambridge University Press, 2010), p. 23.

78. Clifford M. Will, *Was Einstein Right? Putting General Relativity to the Test* (New York: Basic Books, 1986), chapter 4.

79. Muslims believe that God eventually releases everyone from hell.

80. *Catechism of the Catholic Church*, rev. ed. (London: Geoffrey Chapman, 1999), pp. 1030–32.

81. As I write this, the latest prediction for the end of the world—May 12, 2011, by Oakland, California, preacher Harold Camping—had not materialized.

82. Richard H. Jones, *Piercing the Veil: Comparing Science and Mysticism as Ways of Knowing Reality* (New York: Jackson Square Books, 2010).

83. Ibid., pp. 63–76.

84. Ibid., p. 23.

85. Ibid., p. 28.

10. BEYOND EVOLUTION

1. Dinesh D'Souza, *Life after Death: The Evidence* (Washington, DC: Regnery, 2009), pp. 165–83.

2. Ibid., pp. 167–68.

3. Ibid., pp. 171–72.

4. Charles Darwin, *The Descent of Man, and Selection in Relation to Sex* (New York: D. Appleton & Co., 1871).

5. Thomas Henry Huxley, *Evolution and Ethics* (London and New York: Macmillan, 1893).

6. Matt Ridley, *The Origins of Virtue: Human Instincts and the Evolution of Cooperation* (New York: Viking, 1997); Elliott Sober and David Sloan Wilson, *Unto Others: The Evolution and Psychology of Unselfish Behavior* (Cambridge, MA: Harvard University Press, 1998); Leonard D. Katz, *Evolutionary Origins of Morality: Cross-Disciplinary Perspectives* (Thorverton, UK; Bowling Green, OH: Imprint Academic, 2000); Michael Shermer, *The Science of Good and Evil: Why People Cheat, Gossip, Care, Share, and Follow the Golden Rule* (New York: Times Books, 2004); M. Rutherford, "The Evolution of Morality," *Goundings* 1 (2007). For more information and references see the website Evolution of Morality, http://www1.umn.edu/ships/evolutionofmorality/ (accessed February 6, 2011).

7. Richard Dawkins, *The Selfish Gene*, new ed. (Oxford, UK; New York: Oxford University Press, 1989).

8. George C. Williams, *Adaption and Natural Selection* (Princeton, NJ: Princeton University Press, 1966).

9. Robert Trivers, "The Evolution of Reciprocal Altruism," *Quarterly Review of Biology* 46 (1971): 35–57.

10. Francis S. Collins, *The Language of God: A Scientist Presents Evidence for Belief* (New York: Free Press, 2006), p. 28.

11. D'Souza, *Life after Death: The Evidence*, pp. 176–77.

12. Ibid., p. 180.

13. Ibid., p. 172.

14. Michael Shermer, *The Science of Good and Evil*, pp. 235–36.

15. A. N. Franzblau, "Religious Belief and Character among Jewish Adolescents," *Teachers College Contributions to Education*, no. 634 (1934).

16. Murray G. Ross, *Religious Beliefs of Youth* (New York: Association Press, 1950).

17. Travis Hirschi and Rodney Stark, "Hellfire and Delinquency," *Social Problems* 17 (1969): 202–13.

18. R. E. Smith et al., "Faith without Works: Jesus People, Resistance to Temptation, and Altruism," *Journal of Applied Social Psychology* 5 (1975): 320–30.

19. David M. Wulff, *Psychology of Religion: Classic and Contemporary Views* (New York: Wiley, 1991).

20. Gregory S. Paul, "Cross-National Correlations of Quantifiable Societal Health with Popular Religiosity and Secularism in the Prosperous Democracies: A First Look," *Journal of Religion & Society* 7 (2005), online at http://moses.creighton.edu/JRS/pdf/2005-11.pdf (accessed February 5, 2011).

21. Pippa Norris and Ronald Inglehart, *Sacred and Secular: Religion and Politics Worldwide*, 2nd ed. (Cambridge, UK: Cambridge University Press, 2011).

22. Phil Zuckerman, *Society without God: What the Least Religious Nations Can Tell Us about Contentment* (New York: New York University Press, 2008).

23. Kimberly Blaker, "The Social Implications of Armageddon," in *The Fundamentals of*

Extremism: The Christian Right in America, ed. Kimberly Blaker (New Boston, MI: New Boston Books, 2003), pp. 114–53.

24. Dwight M. Donaldson, *Studies in Muslim Ethics* (London: SPCK, 1953), p. 25. W. A. Spooner, "The Golden Rule," in *Encyclopedia of Religion and Ethics*, ed. James Hastings (New York: Charles Scribner's Sons, 1914), pp. 310–12.

26. Diogenes Laërtus, *Lives and Opinions of Eminent Philosophers*, vol. 1, p. 39.

27. Richard Jasnow, *A Late Period Hieratic Wisdom Text: P. Brooklyn P. 47.218.135* (Chicago: University of Chicago Press, 1992), p. 85.

28. Elizabeth Anderson, "If God Is Dead, Is Everything Permitted?" in *Philosophers without Gods: Meditations on Atheism and the Secular Life*, ed. Louise M. Antony (Oxford, UK; New York: Oxford University Press, 2007), pp. 215–30.

29. For a well-documented recent example in Mormonism, see Jon Krakauer, *Under the Banner of Heaven: A Story of Violent Faith* (New York: Doubleday, 2003).

30. Ibid., p. 222.

31. D'Souza, *Life after Death: The Evidence*, p. 189.

32. Barbara Hagerty, "Nun Excommunicated for Allowing Abortion," National Public Radio, May 19, 2010.

33. Hector Avalos, *Fighting Words: The Origins of Religious Violence* (Amherst, NY: Prometheus Books, 2005), p. 326.

34. Jared Diamond, "Vengence Is Ours," *New Yorker* (April 21, 2008): 74–87.

35. Dan Barker, *Godless: How an Evangelical Preacher Became One of America's Leading Atheists* (Berkeley, CA: Ulysses Press, 2008).

36. Rodney Stark, *For the Glory of God: How Monotheism Led to Reformations, Science, Witch Hunts, and the End of Slavery* (Princeton, NJ: Princeton University Press, 2003), p. 365; Dinesh D'Souza, *What's So Great about Christianity?* (Washington, DC: Regnery, 2007), pp. 70–72; D'Souza, *Life after Death: The Evidence*, pp. 195–98.

37. Hector Avalos, *Slavery, Abolitionism, and the Ethics of Biblical Scholarship* (Sheffield, UK: Sheffield Phoenix Press, 2011).

38. Ibid., pp. 286–87.

39. Collins, *The Language of God*, p. 218.

40. C. S. Lewis, *Mere Christianity* (Westwood, NJ: Barbour, 1952 [New York: HarperCollins, 2001]), p. 21.

41. Collins, *The Language of God*, p. 29.

42. Sam Harris, *The Moral Landscape: How Science Can Determine Human Values* (New York: Free Press, 2010), p. 13.

43. Steven Pinker, *The Blank Slate: The Modern Denial of Human Nature* (New York: Viking, 2002).

44. William Grassie, *The New Sciences of Religion: Exploring Spirituality from the Outside In and Bottom Up* (New York: Palgrave Macmillan, 2010), p. 91.

45. Shermer, *The Science of Good and Evil*, p. 31.

46. Collins, *The Language of God*, p. 23.

47. Harris, *The Moral Landscape*, p. 170.

48. J. M. Nunez et al., "Intentional False Responding Shares Neural Substrates with Response Conflict and Cognitive Control," *Neuroimage* 25, no. 1 (2005): 267–77.

49. Harris, *The Moral Landscape*, p. 97; K. A. Kiel et al., "Limbic Abnormalities in Affective Processing by Criminal Psychopaths as Revealed by Functional Magnetic Resonance Imaging," *Biological Psychiatry* 50, no. 9 (2001): 677–84.

50. Dawkins, *The Selfish Gene*, pp. 3, 201.

51. D'Souza, *Life after Death: The Evidence*, p. 181.

52. For a nice review of the history of morality and an outline of a proposal for a natural morality, see Jeff Schweitzer and Giuseppe Notarbartolo-di-Sciara, *Beyond Cosmic Dice: Moral Life in a Random World* (Los Angeles: Jacquie Jordan, 2009).

11. MATTER AND MIND

1. George Lakoff and Mark Johnson, *Philosophy in the Flesh: The Embodied Mind and Its Challenge to Western Thought* (New York: Basic Books, 1999).

2. Dinesh D'Souza, *Life after Death: The Evidence* (Washington, DC: Regnery, 2009), p. 106.

3. Ibid., p. 107.

4. Ibid., p. 110.

5. Steven Pinker, *How the Mind Works* (New York: W. W. Norton, 1997), p. 21.

6. Ibid., p. 121.

7. Frank J. Tipler, *The Physics of Immortality: Modern Cosmology, God, and the Resurrection of the Dead* (New York: Anchor Books, 1994).

8. Patricia Smith Churchland, *Neurophilosophy: Toward a Unified Science of the Mind-Brain* (Cambridge, MA: MIT Press, 1986).

9. D. Kapogiannis et al., "Cognitude and Neural Foundations of Religious Belief," *Proceedings of the National Academy of Sciences USA* 106, no. 12 (2009): 4876–81.

10. William Penfield and T. Rasmussen, *The Cerebral Cortex of Man: A Clinical Study of Localization of Function* (New York: MacMillan, 1950).

11. David E. Comings, *Did Man Create God? Is Your Spiritual Brain at Peace with Your Thinking Brain?* (Duarte, CA: Hope Press, 2008), pp. 348–54.

12. K. Dewhurst and A. W. Beard, "Sudden Religious Conversions in Temporal Lobe Epilepsy," *British Journal of Psychiatry* 117 (1970): 497–507; Comings, *Did Man Create God?* pp. 355–66.

13. V. S. Ramachandran and Sandra Blakeslee, *Phantoms in the Brain: Probing the Mysteries of the Human Mind* (New York: HarperCollins, 1998), p. 179.

14. Comings, *Did Man Create God?* pp. 362–66. See Comings for a full discussion of the connection between religious experiences and the brain.

15. William Grassie, *The New Sciences of Religion: Exploring Spirituality from the Outside In and Bottom Up* (New York: Palgrave Macmillan, 2010), p. 104.

16. Andrew B. Newberg and Eugene G. D'Aquili, *The Mystical Mind: Probing the Biology of Religious Experience* (Minneapolis, MN: Fortress Press, 1999); Andrew B. Newberg et al., *Why God Won't Go Away: Brain Science and the Biology of Belief* (New York: Ballantine Books, 2001); Andrew B. Newberg and Mark Robert Waldman, *Born to Believe: God, Science, and the Origin of Ordinary and Extraordinary Beliefs* (New York: Free Press, 2007); Andrew B. Newberg and Mark Robert Waldman, *How God Changes Your Brain: Breakthrough Findings from a Leading Neuroscientist* (New York: Ballantine Books, 2009); Andrew B. Newberg, *Principles of Neurotheology* (Farnham, Surrey, England; Burlington, VT: Ashgate, 2010).

17. Newberg et al., *Why God Won't Go Away*.

18. M. A. Persinger, "Religious and Mystical Experiences as Artifacts of Temporal Lobe Function: A General Hypothesis," *Perceptual and Motor Skills* 57 (1983): 1255–62.

19. Pehr Granqvist et al., "Sensed Presence and Mystical Experiences Are Predicted by Suggestibility, Not by the Application of Transcranial Weak Complex Magnetic Fields," *Neuroscience Letters* 379, no. 1 (2005): 1–6.

20. L. S. St-Pierre and M. A. Persinger, "Experimental Facilitation of the Sensed Presence Is Predicted by the Specific Patterns of the Applied Magnetic Fields, Not by Suggestibility: Reanalyses of 19 Experiments," *International Journal of Neurosciences* 116, no. 9 (2006): 1079–96.

21. A. T. Barker et al., "Non-Invasive Magnetic Stimulation of Human Motor Cortex," *Lancet* 1, no. 8437 (1985): 1106–1107; Alvaro Pascual-Leone et al., "Transcranial Magnetic Stimulation in Cognitive Neuroscience—Virtual Lesion, Chronometry, and Functional Connectivity," *Current Opinion in Neurobiology* 10 (2000): 232–37; Mark Hallett, "Transcranial Magnetic Stimulation and the Human Brain," *Nature* 406 (2000): 147–50; Vincent Walsh and Alvaro Pascual-Leone, *Transcranial Magnetic Stimulation: A Neurochronometrics of Mind* (Cambridge, MA: MIT Press, 2003).

22. See Vincent Walsh and Alan Cowey, "Transcranial Magnetic Stimulation and Cognitive Neuroscience," *Nature Reviews Neuroscience* 1 (2000): 73–78 and references therein.

23. Harris, *The Moral Landscape*, pp. 158–59.

24. B. Libet et al., "Time of Conscious Intention to Act in Relation to Onset of Cerebral Activity (Readiness-Potential): The Unconscious Initiation of a Freely Voluntary Act," *Brain* 106 (1983): 623–42.

25. Chun Siong Soon et al., "Unconscious Determinants of Free Decisions in the Human Brain," *Nature Neuroscience* 11, no. 5 (2008): 545.

26. Stanislas Dehaene, "Signatures of Consciousness," *The Reality Club*, November 24, 2009, http://www.edge.org/3rd_culture/dehaene09/dehaene09_index.html (accessed December 27, 2009).

27. Bernard J. Baars, "The Conscious Access Hypothesis: Origins and Recent Evidence," *Trends in Cognitive Science* 6, no. 1 (2002): 47–52.

28. C. M. Fischer, "If There Were No Free Will," *Med Hypotheses* 56, no. 3 (2001): 364–66; Daniel M. Wegner, *The Illusion of Conscious Will* (Cambridge, MA: MIT Press, 2002); Daniel M. Wegner, "Précis of the Illusion of Conscious Will," *Behavioral and Brain Sciences* 27, no. 5 (2004): 659–92; Harris, *The Moral Landscape,* pp. 102–106.

29. Kurt Gödel, "Some Basic Theorems on the Foundations of Mathematics and Their Implications," in *Collected Works: Kurt Gödel, vol. 3,* ed. Solomon Feferman (Oxford, UK: Oxford University Press, 1995), pp. 304–23.

30. Roger Penrose, *The Emperor's New Mind: Concerning Computers, Minds, and the Laws of Physics* (Oxford, UK; New York: Oxford University Press, 1989).

31. Roger Penrose, *Shadows of the Mind: A Search for the Missing Science of Consciousness* (Oxford, UK; New York: Oxford University Press, 1994).

32. Victor J. Stenger, *The Unconscious Quantum: Metaphysics in Modern Physics and Cosmology* (Amherst, NY: Prometheus Books, 1995), pp. 273–80.

33. Rick Grush and Patricia S. Churchland, "Gaps in Penrose's Toilings," in *On the Contrary: Critical Essays, 1987–1997,* ed. Paul M. Churchland and Patricia S. Churchland (Cambridge, MA: MIT Press, 1998), pp. 205–30.

34. Solomon Feferman, "Penrose's Gödelian Argument: A Review of *Shadows of the Mind* by Roger Penrose," *Psyche* 2, no. 7 (1995); David J. Chalmers, "Can Physics Provide a Theory of Consciousness? A Review of *Shadows of the Mind* by Roger Penrose," *Psyche* 2, no. 9 (1995); Stanley A. Klein, "Is Quantum Mechanics Relevant to Understanding Consciousness? A Review of *Shadows of the Mind* by Roger Penrose," *Psyche* 2, no. 3 (1995); Tim Maudlin, "Between the Motion and the Act: A Review of *Shadows of the Mind* by Roger Penrose," *Psyche* 2, no. 2 (1995); John McCarthy, "Awareness and Understanding in Computer Programs: A Review of *Shadows of the Mind* by Roger Penrose," *Psyche* 2, no. 11 (1995); Daryl McCullough, "Can Humans Escape Gödel? A Review of *Shadows of the Mind* by Roger Penrose," *Psyche* 2, no. 4 (1995); Drew McDermott, "[Star] Penrose Is Wrong," *Psyche* 2, no. 17 (1995); Hans Moravec, "Roger Penrose's Gravitonic Brains: A Review of *Shadows of the Mind* by Roger Penrose," *Psyche* 2, no. 6 (1995); Gregory R. Mulhauser, "On the End of a Quantum Mechanical Romance," *Psyche* 2, no. 5 (1995).

35. Roger Penrose, "Beyond the Doubting of a Shadow: A Reply to Commentaries on *Shadows of the Mind*," *Psyche* 2, no. 23 (1995).

36. Taner Edis, "How Gödel's Theorem Supports the Possibility of Machine Intelligence," *Minds and Machines* 8 (1998): 251–62.

37. Max Tegmark, "The Importance of Quantum Decoherence in Brain Processes," *Physical Review E* 61, no. 4 (2000): 4194–206.

38. S. Hagan et al., "Quantum Computation in Brain Microtubules? Decoherence and Biological Feasibility," *Physical Review E* 65 (2002): 061901.

39. Penrose, *Shadows of the Mind.*

40. Travis John Craddock and Jack A. Tuszinski, "A Critical Assessment of the Information Processing Capabilities of Neuronal Microtubules Using Coherent Excitations," *Journal of Biological Physics* 26 (2010): 53–70.

41. Elisabetta Collini et al., "Coherently Wired Light-Harvesting in Photosynthetic Marine Algae at Ambient Temperature," *Nature* 463 (2010): 644–47.

42. Tegmark, "The Importance of Quantum Decoherence in Brain Processes."

43. Brain scans have shown that some structural changes in the brain take place near the antenna of cell phones, but while the mechanism is still unknown, it cannot be the direct result of the photons emitted by the antenna breaking chemical bonds. See Nora D. Volkow et al., "Effects of Cell Phone Radiofrequency Signal Exposure on Brain Glucose Metabolism," *Journal of the American Medical Association* 305, no. 8 (2011): 808–12.

12. METAPHOR, ATHEIST SPIRITUALITY, AND IMMANENCE

1. Aubrey Moore, "The Christian Doctrine of God," in *Lux Mundi*, ed. C. Gore (London: Murray, 1891), pp. 41–81.

2. As quoted in R. C. Lewontin, "In the Beginning Was the Word," *Science* 291, no. 5507 (2001): 1263–64.

3. Taner Edis, *An Illusion of Harmony: Science and Religion in Islam* (Amherst, NY: Prometheus Books, 2007).

4. In 2009, my wife and I stayed at the same hotel, the Metropole. A group photo of the participants in the 1927 conference, which includes all the greatest physicists of the time, from Einstein on down, is prominently displayed in the lobby.

5. Werner Heisenberg, *Physics and Beyond: Encounters and Conversations* (London: G. Allen & Unwin, 1971).

6. Edward J. Larson, "Leading Scientists Still Reject God," *Nature* 294, no. 6691 (1998): 313.

7. Elaine Howard Ecklund and Christopher P. Scheitle, "Religion among American Scientists: Distinctions, Disciplines, and Demographics," *Social Problems* 54, no. 2 (2007): 289–307.

8. Elaine Howard Ecklund, *Science vs. Religion: What Scientists Really Think* (New York: Oxford University Press, 2010), p. 5.

9. Ibid., p. 6.

10. Ibid., p. 7.

11. Daniel C. Dennett, *Breaking the Spell: Religion as a Natural Phenomenon* (New York: Viking, 2006).

12. Ecklund, *Science vs. Religion*, p. 51.

13. Robert Wuthnow, *After Heaven: Spirituality in America Since the 1950s* (Berkeley, CA: University of California Press, 1998).

14. Ecklund, *Science vs. Religion*, p. 59.

15. Ian G. Barbour, *Religion and Science: Historical and Contemporary Issues* (San Francisco: HarperSanFrancisco, 1997).

16. Ibid., p. 66.

17. Ibid., p. 67.

18. Ibid., p. 69.

19. Alfred North Whitehead et al., *Process and Reality: An Essay in Cosmology*, corrected ed. (New York: Free Press, 1978).

20. Ibid., p. 154.

21. Ibid., p. 135.

22. Ibid.

23. Ibid., p.136.

24. Victor J. Stenger, *God: The Failed Hypothesis* (Amherst, NY: Prometheus Books, 2007).

25. Barbour, *Religion and Science*, p. 153.

26. Ibid., p. 159.

13. FROM CONFLICT TO INCOMPATIBILITY

1. Jerry A. Coyne, "Science and Religion Aren't Friends," *USA Today* (October 11, 2010).

2. Pew Research Center, "World Publics Welcome Global Trade—But Not Immigration: October 4, 2007," http://pewglobal.org/files/pdf/258.pdf (accessed March 8, 2011); see fig. 1.1 on p. 24 of Victor J. Stenger, *Quantum Gods: Creation, Chaos, and the Search for Cosmic Consciousness* (Amherst, NY: Prometheus Books, 2009).

3. Gregory Paul, "The Chronic Dependence of Popular Religiosity upon Dysfunctional Psychosociological Conditions," *Evolutionary Psychology* 7, no. 3 (2009): 398–441.

4. Frederick Solt, Philip Habel, and J. Tobin Grant, "Economic Inequality, Relative Power, and Religiosity," *Social Science Quarterly* 92, no. 2 (2011): 447–65.

5. James A. Haught, *Fading Faith: The Rise of the Secular Age* (Cranford, NJ: Gustav Broukal Press, 2010).

6. "Church Struggles with Attendance in Europe," *NBC News* (April 20, 2005).

7. Christopher Dickey, "Near 'The Edge of the Abyss,'" *Newsweek* (August 15, 2005).

8. "Who Is Secular in the World Today? A Symposium" *Religion in the News* 9, no. 2 (Fall 2006).

9. Charles M. Sennott and *Boston Globe* staff, "Catholic Church Withers in Europe," *Boston Globe* (May 2, 2005).

10. Steve Bruce, *God Is Dead: Secularization in the West* (Malden, MA: Blackwell, 2002).

11. Haught gives this reference as *Spero News* (December 6, 2005). However, I have not been able to confirm it.

12. Hiroshi Matsubara, "Western Eyes Blind to Spirituality in Japan," *Japan Times* (January 1, 2002).

13. Barry A. Kosmin and Ariela Keysar, "American Religious Identification Survey (ARIS) 2008," http://commons.trincoll.edu/aris/ (accessed February 21, 2010).

14. Pew Forum on Religion and Public Life, "Religion among the Millennials," February 17, 2010, http://pewforum.org/Age/Religion-Among-the-Millennials.aspx (accessed March 22, 2010).

15. Frank Newport, "More Than 9 in 10 Americans Continue to Believe in God," Gallup Poll (June 3, 2001), http://www.gallup.com/poll/147887/Americans-Continue-Believe-God. aspx (accessed June 8, 2011).

16. Luis Lugo et al., "U.S. Religious Landscape Survey," Pew Forum on Religion and Public Life, February 2008, http://religions.pewforum.org/pdf/report-religious-landscape -study-full.pdf (accessed March 8, 2011).

17. Haught, *Fading Faith*, pp. 25–26.

18. Mark Chaves, *American Religion: Contemporary Trends* (Princeton, NJ: Princeton University Press, 2011).

19. Bart D. Ehrman, *Jesus, Interrupted: Revealing the Hidden Contradictions in the Bible (and Why We Don't Know about Them)* (New York: HarperOne, 2009), pp. 12–15.

20. Lugo et al., "U.S. Religious Landscape Survey."

21. Phil Zuckerman, *Society without God: What the Least Religious Nations Can Tell Us about Contentment* (New York: New York University Press, 2008).

22. E. Gunn, "Death by Prayer," *Freethought Today* (September 2008): 6–7.

23. George C. Cunningham, *Decoding the Language of God: Can a Scientist Really Be a Believer? A Geneticist Responds to Francis Collins* (Amherst, NY: Prometheus Books, 2010), p. 56.

24. Dinesh D'Souza, *Life after Death: The Evidence* (Washington, DC: Regnery, 2009), pp. 214–15.

25. David Hume, *Dialogues Concerning Natural Religion*, ed. Henry David Aiken (New York: Hafner, 1972).

26. National Academy of Sciences, *Teaching about Evolution and the Nature of Science* (Washington, DC: National Academies Press, 1998), p. 58. Free copy available at http://www. nap.edu/catalog/5787.html.

27. Victor J. Stenger, *The New Atheism: Taking a Stand for Science and Reason* (Amherst, NY: Prometheus Books, 2009).

28. Sam Harris, *The End of Faith: Religion, Terror, and the Future of Reason* (New York: W. W. Norton, 2004).

29. Sam Harris, *Letter to a Christian Nation*, 1st Vintage Books ed. (New York: Vintage Books, 2008).

30. Daniel C. Dennett, *Breaking the Spell: Religion as a Natural Phenomenon* (New York: Viking, 2006).

31. Richard Dawkins, *The God Delusion* (Boston: Houghton Mifflin, 2006).

32. Christopher Hitchens, *God Is Not Great: How Religion Poisons Everything* (New York: Twelve, Hachette Book Group, 2007).

33. Victor J. Stenger, *God: The Failed Hypothesis* (Amherst, NY: Prometheus Books, 2007).

34. Jerry Coyne, http://whyevolutionistrue.wordpress.com/; Hemant Mehta, http://friendlyatheist.com/; P. Z. Myers, http://scienceblogs.com/pharyngula/; Rebecca Watson, http://skepchick.org/.

35. Secular Student Alliance, http://www.secularstudents.org/.

36. Chris Mooney and Sheril Kirshenbaum, *Unscientific America: How Scientific Illiteracy Threatens Our Future* (New York: Basic Books, 2009), pp. 97–98.

14. WHY DOES IT MATTER?

1. Kevin Phillips, *American Theocracy: The Peril and Politics of Radical Religion, Oil, and Borrowed Money in the 21st Century* (New York: Viking, 2006), p. xv.

2. James C. Dobson et al., *Children at Risk: The Battle for the Hearts and Minds of Our Kids* (Prince Frederick, MD: Recorded Books, 2003), p. 27.

3. Ask MetaFilter, "Academic Politics Are Vicious because the Stakes Are So Low?" http://ask.metafilter.com/80812/Academic-politics-are-vicious-because-the-stakes-are-so-low (accessed February 27, 2011).

4. Kimberly Blaker, ed. *The Fundamentals of Extremism: The Christian Right in America* (New Boston, MI: New Boston Books, 2003).

5. Phillips, *American Theocracy*, p. vii.

6. Michelle Goldberg, *Kingdom Coming: The Rise of Christian Nationalism* (New York: W. W. Norton, 2006).

7. Damon Linker, *The Theocons: Secular America Under Siege* (New York: Doubleday, 2006).

8. Chris Hedges, *American Fascists: The Christian Right and the War on America* (New York: Free Press, 2007), p. 10.

9. Jeff Sharlet, *The Family: The Secret Fundamentalism at the Heart of American Power* (New York: Harper Perennial, 2009).

10. Jeff Sharlet, *C Street: The Fundamentalist Threat to American Democracy* (New York: Little, Brown, 2010).

11. Peter Montgomery, "Jesus Hates Taxes: Biblical Capitalism Created Fertile Anti-Union Soil," *Religion Dispatches* (March 14, 2011).

12. Jeff Sharlet, "This Is Not a Religion Column: Biblical Capitalism," *Religion Dispatches*, October 1, 2008, http://www.religiondispatches.org/archive/politics/562/ (accessed April 1, 2011).

13. William L. Fisher, "Christian Responsibility for Government," in *Christian Coalition Leadership Manual*, ed. William L. Fisher et al. (Christian Coalition, 1990), secs. 2.5–2.8.

14. Chris Mooney, *The Republican War on Science* (New York: Basic Books, 2005).

15. Mooney and Kirshenbaum, *Unscientific America*.

16. Ibid., p. 27.

17. Ibid., p. 55.

18. Scott Keyes, "Jim DeMint's Theory of Relativity: 'The Bigger Government Gets, the Smaller God Gets,'" *Think Progress* (March 15, 2011). http://thinkprogress.org/2011/03/15/demint-big-govt/ (accessed March 29, 2011).

19. Pew Forum on Religion and Public Life, "Religion and Public Life: A Faith-Based Partisan Divide," January 2005, http://pewforum.org/uploadedfiles/Topics/Issues/Politics_and_Elections/religion-and-politics-report.pdf (accessed March 17, 2011).

20. Frank Newport, "In U.S., Very Religious Americans Still Align More with GOP," Gallup Poll (June 27, 2011), http://www.gallup.com/poll/148274/Religious-Americans-Align-GOP.aspx (accessed June 28, 2011).

21. Pew Forum on Religion and Public Life, "The Tea Party and Religion," February 23, 2011, http://pewforum.org/Politics-and-Elections/Tea-Party-and-Religion.aspx (accessed February 23, 2011).

22. Robert P. Jones and Daniel Cox, "Religion and the Tea Party in the 2010 Elections: An Analysis of the Third Biennial American Values Survey," Public Religion Research Institute, http://www.publicreligion.org/objects/uploads/fck/file/AVS%202010%20Report%20FINAL.pdf (accessed August 26, 2011).

23. Jacob S. Hacker and Paul Pierson, *Winner-Take-All Politics: How Washington Made the Rich Richer—and Turned Its Back on the Middle Class* (New York: Simon & Schuster, 2010), pp. 148–49.

24. Hacker and Pierson, *Winner-Take-All Politics*, p. 25.

25. Ibid., p. 18.

26. G. William Domhoff, "Wealth, Income, and Power," updated November 2011, http://sociology.ucsc.edu/whorulesamerica/power/wealth.html (accessed March 6, 2011).

27. Gallup Poll, April 5, 2010, "Tea Partiers Are Fairly Mainstream in Their Demographics," http://www.gallup.com/poll/127181/tea-partiers-fairly-mainstream-demographics.aspx#1 (accessed December 26, 2011).

28. Naomi Oreskes and Erik M. Conway, *Merchants of Doubt: How a Handful of Scientists Obscured the Truth on Issues from Tobacco Smoke to Global Warming* (New York: Bloomsbury Press, 2010).

29. Ibid., pp. 5–6.

30. Ibid., pp. 126–35.

31. Ibid., pp. 85–98.

32. William Nierenberg, Chairman, *Report of the Acid Rain Peer Review Panel, July 1984* (Washington, DC: White House Office of Science and Technology Policy, 1984), p. v.

33. Steve LaRue, "Early Action Urged in Fight on Acid Rain," *San Diego Union* (August 8, 1984): B2.

34. Oreskes and Conway, *Merchants of Doubt*, pp. 95–100.

35. S. Fred Singer, "On a Nuclear Winter," *Science* 227, no. 4685 (1985): 356.

36. Union of Concerned Scientists, "Smoke, Mirrors & Hot Air: How ExxonMobil Used Big Tobacco Tactics to Manufacture Uncertainty on Climate Science," January 2007, http://www.ucsusa.org/assets/documents/global_warming/exxon_report.pdf (accessed March 10, 2011).

37. Michael Berkman and Eric Plutzer, *Evolution, Creationism, and the Battle to Control America's Classrooms* (Cambridge, UK; New York: Cambridge University Press).

38. For a nice, short review showing all the pertinent data, see John Cook, "The Scientific Guide to Global Warming Skepticism," *Skeptical Science*, December 2010, http://www.skepticalscience.com/docs/Guide_to_Skepticism.pdf (accessed July 2, 2011).

39. Peter T. Doran and Maggie Kendall Zimmerman, "Climate Change," *EOS* 90, no. 3 (2009): 22, online at http://tigger.uic.edu/%7Epdoran/012009_Doran_final.pdf (accessed March 18, 2011).

40. Intergovernmental Panel on Climate Change, http://www.ipcc.ch/index.htm (accessed March 18, 2011).

41. Intergovernmental Panel on Climate Change, "Climate Change 2007: Synthesis Report," http://www.ipcc.ch/publications_and_data/ar4/syr/en/mains6-1.html (accessed March 17, 2011).

42. National Research Council, "Frontiers in Understanding Climate Change and Polar Ecosystems: Summary of a Workshop," http://www.nap.edu/catalog.php?record_id=13132 (accessed March 29, 2011).

43. National Research Council, *America's Climate Choices* (Washington, DC: National Academies Press, 2011). Free download at http://www.nap.edu/catalog.php?record_id=12781 (accessed May 17, 2011).

44. Freeman Dyson, "Heretical Thoughts about Science and Reality," August 8, 2007, http://www.edge.org/3rd_culture/dysonf07/dysonf07_index.html (accessed March 8, 2011).

45. Freeman Dyson, "The Question of Global Warming," *New York Review of Books* (June 12, 2008).

46. Pew Research Center, "Religious Groups' Views on Global Warming," April 16, 2009, http://pewresearch.org/pubs/1194/global-warming-belief-by-religion (accessed March 9, 2011).

47. Lauri Lebo, "Creationism and Global Warming Denial: Anti-Science's Kissing Cousins?" *Religion Dispatches* (March 17, 2010), online at http://www.religiondispatches.org/archive/politics/2374/_creationism_and_global_warming_denial%3A_anti-science's_kissing_cousins (accessed March 18, 2011).

48. Leslie Kaufman, "Darwin Foes Add Warming to Targets," *New York Times* (March 3, 2010).

49. See my discussion in *The New Atheism: Taking a Stand for Science and Reason* (Amherst, NY: Prometheus Books, 2009), pp. 53–57.

50. Bernard Schoenburg, "Shimkus's Food for Thought at Hearing Prompts Snickers," http://www.sj-r.com/opinions/x180621374/Bernard-Schoenburg-Shimkus-food-for-thought-at-hearing-prompts-snickers#video1 (accessed February 9, 2011).

51. Ali Frick, "Barton: We Shouldn't Regulate CO_2 because 'It's in Your Coca-Cola' and 'You Can't Regulate God,'" *Think Progress*, May 15, 2009, http://thinkprogress.org/2009/05/19/barton-carbon-god/ (accessed March 29, 2011).

52. James M. Inhofe, "The Science of Climate Change: Senate Floor Statement July 28, 2003," http://inhofe.senate.gov/pressreleases/climate.htm (accessed March 7, 2011).

53. NOAA National Climatic Data Center, "State of the Climate Global Analysis: Annual 2004," December 2004, http://www.ncdc.noaa.gov/sotc/global/2004/13 (accessed March 7, 2011).

54. Chris Casteel, "U.S. Senator Jim Inhofe's Trips to Africa Called a 'Jesus Thing,'" *The Oklahoman* (December 21, 2008), http://newsok.com/u.s.-senator-jim-inhofes-trips-to-africa-called-a-jesus-thing/article/3331838#ixzz1FwUmLVma (accessed March 7, 2011).

55. Cornwall Alliance for the Stewardship of Creation, "An Evangelical Declaration on Global Warming," http://www.cornwallalliance.org/articles/read/an-evangelical-declaration-on-global-warming/ (accessed March 9, 2011).

56. The list of prominent signers is at http://www.cornwallalliance.org/blog/item/prominent-signers-of-an-evangelical-declaration-on-global-warming/, updated January 14, 2010 (assessed May 20, 2011).

57. Cornwall Alliance for the Stewardship of Creation, "Resisting the Green Dragon: A Biblical Response to One of the Greatest Deceptions of Our Day," http://www.resistingthegreendragon.com/ (accessed May 20, 2011).

58. Kaufman, "Darwin Foes Add Warming to Targets."

59. Richard Warren, *The Purpose-Driven Life: What on Earth Am I Here For?* (Grand Rapids, MI: Zondervan, 2002).

60. Laurie Goodstein, "Evangelical Leaders Join Global Warming Initiative," *New York Times* (February 8, 2006).

61. John Vidal and Tom Kington, "Protect God's Creation: Vatican Issues New Green Message for World's Catholics," *Guardian* (April 27, 2007).

62. Michael Specter, *Denialism: How Irrational Thinking Hinders Scientific Progress, Harms the Planet, and Threatens Our Lives* (New York: Penguin, 2009).

63. James E. Hansen, *Storms of My Grandchildren: The Truth about the Coming Climate Catastrophe and Our Last Chance to Save Humanity* (New York: Bloomsbury USA, 2009).

BIBLIOGRAPHY

Adam, T., N. Agafonova, A. Aleksandrov et al. "Measurement of the Neutrino Velocity with the Opera Detector in the CNGS Beam." *Journal of High Energy Physics*, preprint dated November 17, 2011. http://xxx.lanl.gov/pdf/1109.4897v2 (accessed December 22, 2011).

Aguirre, Anthony N., and Steven Gratton. "Inflation without a Beginning: A Null Boundary Proposal." *Physical Review D* 67 (2003): 083516.

Alcock, James E., "Back From the Future: Parapsychology and the Bem Affair." *Skeptical Inquirer* 35, no. 2 (2011): 31–39.

———. "Psychology and Near-Death Experiences." *Skeptical Inquirer* 3, no. 3 (1978): 25–41.

al-Khalili, Jim. *The House of Wisdom: How Arabic Science Saved Ancient Knowledge and Gave Us the Renaissance.* New York: Penguin, 2011.

Anderson, Elizabeth. "If God Is Dead, Is Everything Permitted?" In *Philosophers without Gods: Meditations on Atheism and the Secular Life*, edited by Louise M. Antony, 215–30. Oxford, UK; New York: Oxford University Press, 2007.

Anderson, P. W. "More Is Different." *Science* 177, no. 4047 (1972): 393–96.

Andrews, Edgar. *Who Made God: Searching for a Theory of Everything.* Darlington, UK: EP Books, 2009.

Angel, Leonard. *"Reincarnation and Biology* (Book Review)." *Skeptic* 9, no. 3 (2002): 86–90.

Armstrong, Karen. *The Great Transformation: The Beginning of Our Religious Traditions.* New York: Knopf, 2006.

Aspect, Alain, Phillipe Grangier, and Roger Gerard. "Experimental Realization of the Einstein-Podolosky-Rosen *Gedankenexperiment*: A New Violation of Bell's Inequalities." *Physical Review Letters* 49 (1982): 91.

Atran, Scott. *In Gods We Trust: The Evolutionary Landscape of Religion*, Evolution and Cognition Series. Oxford, UK; New York: Oxford University Press, 2002.

Attridge, Harold W., ed. *The Religion and Science Debate: Why Does It Continue?* New Haven: Yale University Press, 2009.

Augustine, Keith. "Does Paranormal Perception Occur in Near-Death Experiences?" *Journal of Near-Death Studies* 26, no. 4 (2007): 203–36.

———. "Halluncinatory Near-Death Experiences", http://www.infidels.org/library/modern/keith_augustine/HNDEs.html#veridical (accessed January 26, 2011).

———. "Near-Death Experiences with Hallucinatory Features." *Journal of Near-Death Studies* 26, no. 1 (2007): 3–31.

———. "Psychophysiological and Cultural Correlates Undermining a Survivalist Interpretation of Near-Death Experiences." *Journal of Near-Death Studies* 26, no. 2 (2007): 89–95.

Avalos, Hector. *Fighting Words: The Origins of Religious Violence*. Amherst, NY: Prometheus Books, 2005.

———. *Slavery, Abolitionism, and the Ethics of Biblical Scholarship*. Sheffield, UK: Sheffield Phoenix Press, 2011.

Aviles, J. M. et al. "Intercessory Prayer and Cardiovascular Disease Progression in a Coronary Care Unit Population: A Randomized Controlled Trial." *Mayo Clinic Proceedings* 76, no. 12 (2001): 1192–98.

Baars, Bernard J. "Can Physics Provide a Theory of Consciousness? A Review of *Shadows of the Mind* by Roger Penrose." *Psyche* 2, no. 8 (1995).

Ball, Philip. *Critical Mass: How One Thing Leads to Another*. New York: Farrar, Straus, and Giroux, 2004.

———. *The Self-Made Tapestry: Pattern Formation in Nature*. Oxford, UK; New York: Oxford University Press, 1999.

Barbour, Ian G. *Religion and Science: Historical and Contemporary Issues*. San Francisco, CA: HarperSanFrancisco, 1997.

Barker, A. T., R. Jalinous, and R. L. Freeston. "Non-Invasive Magnetic Stimulation of Human Motor Cortex." *Lancet* 1, no. 8437 (1985): 1106–1107.

Barker, Dan. *Godless: How an Evangelical Preacher Became One of America's Leading Atheists*. Berkeley, CA: Ulysses Press, 2008.

———. *The Good Atheist: Living a Purpose-Filled Life without God*. Berkeley, CA: Ulysses Press, 2010.

———. *Losing Faith in Faith: From Preacher to Atheist*. Madison, WI: Freedom From Religion Foundation, 2006.

Barrett, Justin. "Exploring the Natural Foundations of Religion." *Trends in Cognitive Science* 4 (2000): 29–34.

Barrow, John D. *The Book of Nothing: Vacuums, Voids, and the Latest Ideas about the Origins of the Universe*. New York: Pantheon Books, 2000.

———. *The Constants of Nature: From Alpha to Omega*. London: Jonathan Cape, 2002.

Barrow, John D., and Frank J. Tipler. *The Anthropic Cosmological Principle*. Oxford, UK; New York: Oxford University Press, 1986.

Bartusiak, Marcia. *The Day We Found the Universe*. New York: Pantheon Books, 2009.

Behe, Michael J. *Darwin's Black Box: The Biochemical Challenge to Evolution*. New York: Free Press, 1996.

Bem, Daryl J. "Feeling the Future: Experimental Evidence for Anomalous Retroactive Influences on Cognition and Alert." *Journal of Personality and Social Psychology* 100, no. 3 (2011): 407–25.

Benen, Steve. "From Genesis to Dominion: Fat-Cat Theocrat Funds Creationism Crusade." *Church & State*, July/August, 2000.

Benson, H. J. et al. "Study of the Therapeutic Effects of Intercessory Prayer (STEP) in Cardiac

Bypass Patients: A Multicenter Randomized Trial of Uncertainty and Certainty of Receiving Intercessory Prayer." *American Heart Journal* 154, no. 4 (2007): 934–42.

Bentley, Alex, ed. *The Edge of Reason? Science and Religion in Modern Society.* London; New York: Continuum, 2008.

Berkman, Michael B., and Eric Plutzer. *Evolution, Creationism, and the Battle to Control America's Classrooms.* New York: Cambridge University Press, 2010.

Bernstein, Morey. *The Search for Bridey Murphy.* Garden City, NY: Doubleday, 1956.

Birx, H. James. *Interpreting Evolution: Darwin & Teilhard de Chardin.* Amherst, NY: Prometheus Books, 1991.

Bishop, Jeffrey P., and Victor J. Stenger. "Retroactive Prayer: Lots of History, Not Much Mystery, No Science." *British Medical Journal* 329 (2004): 1444–46.

Blackmore, Susan J. *Dying to Live: Near-Death Experiences.* Amherst, NY: Prometheus Books, 1993.

Blackstone Group. "Our Businesses." http://www.blackstone.com/cps/rde/xchg/bxcom/hs/businesses.htm (accessed March 6, 2011).

Blaker, Kimberly, ed. *The Fundamentals of Extremism: The Christian Right in America.* New Boston, MI: New Boston Books, 2003.

———. "The Social Implications of Armageddon." In *The Fundamentals of Extremism: The Christian Right in America,* edited by Kimberley Blaker, 114–53. New Boston, MI: New Boston Books, 2003.

Blom, Philipp. *A Wicked Company: The Forgotten Radicalism of the European Enlightenment.* New York: Basic Books, 2010.

Bohm, David. "A Suggested Interpretation of Quantum Theory in Terms of 'Hidden Variables,' I and II." *Physical Review* 85 (1952): 166.

———. *Wholeness and the Implicate Order.* London: Routledge & Kegan Paul, 1981.

Bohm, David, and B. J. Hiley. *The Undivided Universe: An Ontological Interpretation of Quantum Theory.* London; New York: Routledge, 1993.

Borde, Arvind, Alan H. Guth, and Alexander Vilenkin. "Inflationary Spacetimes Are Not Past-Complete." *Physical Review Letters* 90 (2003): 151301.

Boyer, Pascal. *Religion Explained: The Evolutionary Origins of Religious Thought.* New York: Basic Books, 2001.

Brinkley, Dannion, and Paul Perry. *Saved By the Light: The True Story of a Man Who Died Twice and the Profound Revelations He Received.* New York: Villard Books, 1994.

Britt, Robert Roy. "Startling Discovery: The First Human Ritual." November 30, 2006. http://www.livescience.com/history/061130_oldest_ritual.html (accessed November 4, 2011).

Brooke, John Hedley. *Science and Religion: Some Historical Perspectives.* Cambridge Studies in the History of Science. Cambridge, UK; New York: Cambridge University Press, 1991.

Bruce, Steve. *God Is Dead: Secularization in the West.* Religion and Spirituality in the Modern World. Malden, MA: Blackwell, 2002.

Bruno, Giordano. *On the Infinite Universe and Worlds*. Venice: 1584.

Bullough, Vern L. "Science and Religion in Historical Perspective." In *Science and Religion: Are They Compatible?* edited by Paul Kurtz. Amherst, NY: Prometheus Books, 2003, 129–38.

Buosso, Raphael. "The Holographic Principle." *Reviews of Modern Physics* 74 (2002): 825–75.

Byers, Nina, "E. Noether's Discovery of the Deep Connection between Symmetries and Conservation Laws," 1999. http://www.physics.ucla.edu/~cwp/articles/noether.asg/noether.html (accessed December 26, 2011).

Byrne, Rhonda. *The Secret*. New York: Atria Books; and Hillsboro, OR: Beyond Words, 2006.

Capra, Fritjof. *The Tao of Physics: An Exploration of the Parallels between Modern Physics and Eastern Mysticism*. Berkeley [New York]: Shambhala [distributed in the United States by Random House], 1975.

Carr, Bernard, ed. *Universe or Multiverse?* Cambridge, UK: Cambridge University Press, 2007.

Carrier, Richard. "Attitudes Toward the Natural Philosopher in the Early Roman Empire (100 BC to 313 AD)." PhD dissertation, Columbia University, 2008.

———. "Christianity Was Not Responsible for Modern Science." In *The Christian Delusion: Why Faith Fails*, edited by John Loftus, 396–419. Amherst, NY: Prometheus Books, 2010.

Carroll, Sean M. *From Eternity to Here: The Quest for the Ultimate Theory of Time*. New York: Dutton, 2010.

Carter, Brandon. "Large Number Coincidences and the Anthropic Principle in Cosmology." Paper presented at the IAU Symposium 63: Confrontation of Cosmological Theories with Observational Data. Dordrecht, Germany, 1974.

Casteel, Chris. "U.S. Senator Jim Inhofe's Trips to Africa Called a 'Jesus Thing.'" *Oklahoman*, December 21, 2008.

Catechism of the Catholic Church. Rev. ed. London: Geoffrey Chapman, 1999.

Center for Responsive Politics. "Top Industries: Senator James M. Inhofe 2003–2008." http://www.opensecrets.org/politicians/industries.php?cycle=2008&cid=N00005582&type=I (accessed February 7, 2008).

Center for the Renewal of Science and Culture. "The Wedge Strategy." http://www.antievolution.org/features/wedge.html (accessed January 6, 2011).

Chalmers, David J. "Can Physics Provide a Theory of Consciousness? A Review of *Shadows of the Mind* by Roger Penrose." *Psyche* 2, no. 9 (1995).

Chaves, Mark. *American Religion: Contemporary Trends*. Princeton, NJ: Princeton University Press, 2011.

Chopra, Deepak. *Ageless Body, Timeless Mind: The Quantum Alternative to Growing Old*. New York: Harmony Books, 1993.

———. *Life after Death: The Burden of Proof*. New York: Harmony Books, 2006.

———. *Quantum Healing: Exploring the Frontiers of Mind/Body Medicine*. New York: Bantam Books, 1989.

Chown, Marcus. *The Never-Ending Days of Being Dead: Dispatches from the Frontline of Science*. London: Faber and Faber, 2007.

Churchland, Patricia Smith. *Neurophilosophy: Toward a Unified Science of the Mind-Brain*. Cambridge, MA: MIT Press, 1986.

Churchland, Paul M., and Patricia Smith Churchland, eds. *On the Contrary: Critical Essays, 1987–1997*. Cambridge, MA: MIT Press, 1998.

Clark, Kimberly. "Clinical Interventions with NDEs." In *The Near-Death Experience: Problems, Prospects, Perspectives*, edited by Bruce Greyson and Charles P. Flynn. Springfield, IL: C. C. Thomas, 1984.

Clayton, Philip. "Emergence: Us From It." In *Science and Ultimate Reality: Quantum Theory, Cosmology, and Complexity*, edited by John D. Barrow, Paul C. W. Davies, and Charles L. Harper Jr., 577–606. Cambridge: Cambridge University Press, 2004.

———. "Unsolved Dilemmas: The Concept of Matter in the History of Philosophy and in Contemporary Physics." In *Information and the Nature of Reality: From Physics to Metaphysics*, edited by Paul Davies and Niels Henrik Gregersen, 38–62. Cambridge, UK; New York: Cambridge University Press, 2010.

Clayton, Philip, and Paul C. W. Davies, eds. *The Re-Emergence of Emergence: The Emergentist Hypothesis from Science to Religion*. Oxford, UK; New York: Oxford University Press, 2006.

Clayton, Philip, Kirk Wegter-McNelly, and John Polkinghorne, eds. *Quantum Mechanics: Scientific Perspectives on Divine Action*. Vatican City, Rome, Italy: Vatican Observatory Publications, 2001.

Cleland, Carol E., and Christopher F. Chyba. "Defining 'Life.'" *Origins of Life and Evolution of the Biosphere* 32, no. 4 (2002): 387–93.

Collini, Elisabetta, et al. "Coherently Wired Light-Harvesting in Photosynthetic Marine Algae at Ambient Temperature." *Nature* 463 (2010): 644–47.

Collins, Francis S. *The Language of God: A Scientist Presents Evidence for Belief*. New York: Free Press, 2006.

Comings, David E. *Did Man Create God? Is Your Spiritual Brain at Peace with Your Thinking Brain?* Duarte, CA: Hope Press, 2008.

Conway Morris, Simon. *Life's Solution: Inevitable Humans in a Lonely Universe*. Cambridge, UK; New York: Cambridge University Press, 2003.

Cook, John. "The Scientific Guide to Global Warming Skepticism," December 2010. http://www.skepticalscience.com/docs/Guide_to_Skepticism.pdf (accessed July 2, 2011).

Cornwall Alliance for the Stewardship of Creation. "An Evangelical Declaration on Global Warming." http://www.cornwallalliance.org/articles/read/an-evangelical-declaration-on-global-warming/ (accessed March 9, 2011).

———. "Resisting the Green Dragon: A Biblical Response to One of the Greatest Deceptions of Our Day." http://www.resistingthegreendragon.com/ (accessed May 20, 2011).

Coyne, Jerry A. "Science and Religion Aren't Friends." *USA Today*, October 11, 2010.

———. *Why Evolution Is True.* Oxford, UK; New York: Oxford University Press, 2009. http://whyevolutionistrue.wordpress.com/ (accessed December 26, 2011).

Craddock, Travis John, and Jack A. Tuszinski. "A Critical Assessment of the Information Processing Capabilities of Neuronal Microtubules using Coherent Excitations." *Journal of Biological Physics* 26 (2010): 53–70.

Craig, William Lane. "Debates," 2006. http://www.leaderu.com/offices/billcraig/menus/debates.html (accessed February 3).

———. *The Kalâm Cosmological Argument.* London: Macmillan, 1979.

———. "Philosophical and Scientific Pointers to Creatio Ex Nihilo." *Journal of the American Scientific Affiliation* 32, no. 1 (1980): 5–13.

Craig, William Lane, and James Porter Moreland. *The Blackwell Companion to Natural Theology.* Chichester, UK; Malden, MA: Wiley-Blackwell, 2009.

Craig, William Lane, and Massimo Pigliucci. "The Craig-Pigliucci Debate: Does God Exist?" 1995. http://www.leaderu.com/offices/billcraig/docs/craig-pigliucci0.html (accessed February 3, 2011).

Craig, William Lane, and James D. Sinclair. "The Kalâm Cosmological Argument." In *The Blackwell Companion to Natural Theology*, edited by William Lane Craig and James Porter Moreland, 101–201. Chichester, UK; Malden, MA: Wiley-Blackwell, 2009.

Crean, Thomas. *God Is No Delusion: A Refutation of Richard Dawkins.* San Francisco, CA: Ignatius Press, 2007.

Crowe, Michael J. *Theories of the World from Antiquity to the Copernican Revolution.* 2nd rev. ed. Mineola, NY: Dover, 2001.

Cunningham, George C. *Decoding the Language of God: Can a Scientist Really Be a Believer? A Geneticist Responds to Francis Collins.* Amherst, NY: Prometheus Books, 2010.

Cushing, James T. *Philosophical Concepts in Physics: The Historical Relation between Philosophy and Scientific Theories.* Cambridge, UK; New York: Cambridge University Press, 1998.

D'Aquili, Eugene G., and Andrew B. Newberg. *The Mystical Mind: Probing the Biology of Religious Experience.* Minneapolis, MN: Fortress Press, 1999.

Darwin, Charles. *The Descent of Man, and Selection in Relation to Sex.* New York: D. Appleton & Co., 1871.

———. *On the Origin of Species by Means of Natural Selection.* London: J. Murray, 1859.

Davies, Paul C. W. *The Goldilocks Enigma: Why Is the Universe Just Right for Life?* London: Allen Lane, 2006.

———. *The Physics of Time Asymmetry.* Berkeley, CA: University of California Press, 1974.

Davies, Paul C. W., and Niels Henrik Gregersen. *Information and the Nature of Reality: From Physics to Metaphysics.* Cambridge, UK; New York: Cambridge University Press, 2010.

Davies, Paul C. W., and John R. Gribbin. *The Matter Myth: Dramatic Discoveries That Challenge Our Understanding of Physical Reality.* New York: Simon & Schuster, 1992.

Davis, Hank. *Caveman Logic: The Persistence of Primitive Thinking in a Modern World*. Amherst, NY: Prometheus Books, 2009.

Dawkins, Richard. *The God Delusion*. Boston, MA: Houghton Mifflin, 2006.

———. *The Greatest Show on Earth: The Evidence for Evolution*. New York: Free Press, 2010.

———. *The Selfish Gene*. Oxford, UK: Oxford University Press, 1976.

———. *The Selfish Gene*. New ed. Oxford, UK; New York: Oxford University Press, 1989.

Day, Vox. *The Irrational Atheist: Dissecting the Unholy Trinity of Dawkins, Harris, and Hitchens*. Dallas, TX: BenBella Books, 2008.

Deem, Rich. "Evidence for the Fine Tuning of the Universe." http://www.godandscience.org/apologetics/designun.html (accessed January 10, 2011).

Deisler, Paul F. Jr. "How Did Life Begin? A Perspective on the Nature and Origin of Life." *Skeptic* 16, no. 2 (2011): 34–40.

De la Mettrie, Julien O. *L'Homme Machine*. Lyde: 1748.

Dembski, William A. *Intelligent Design: The Bridge between Science and Theology*. Downers Grove, IL: InterVarsity Press, 1999.

Dennett, Daniel C. *Breaking the Spell: Religion as a Natural Phenomenon*. New York: Viking, 2006.

Dennett, Daniel C., and Alvin Plantinga. *Science and Religion: Are They Compatible?* New York: Oxford University Press, 2011.

Deprit, A. "Monsignor Georges Lemaître." In *The Big Bang and Georges Lemaître*, edited by A. Barger, 370. Dordrecht, Germany: Reidel, 1984.

De Waal, Frans B. M. "Food-Sharing and Reciprocal Obligations in Chimpanzees." *Journal of Human Evolution* 18 (1989): 433–59.

Dewdney, A. K. "The Hodge-Podge Machine Makes Waves." *Scientific American* 259, no. 2, 104–107.

Dewhurst, K., and A. W. Beard. "Sudden Religious Conversions in Temporal Lobe Epilepsy." *British Journal of Psychiatry* 117 (1970): 497–507.

d'Holbach, Paul-Henri Thiry. *The System of Nature*. New York: Garland, 1984.

Diamond, Jared. "Vengence Is Ours." *New Yorker*, April 21, 2008, 74–87.

Dirac, P. A. M. *The Principles of Quantum Mechanics*. Oxford, UK: Clarendon Press, 1930.

Dixon, Thomas. *Science and Religion: A Very Short Introduction*. New York: Oxford University Press, 2008.

Dixon, Thomas, G. N. Cantor, and Stephen Pumfrey. *Science and Religion: New Historical Perspectives*. Cambridge, UK; New York: Cambridge University Press, 2010.

Dobson, James C., Gary Lee Bauer, and Richard Rohan. *Children at Risk: The Battle for the Hearts and Minds of Our Kids*. Prince Frederick, MD: Recorded Books, 2003.

Domhoff, G. William. "Wealth, Income, and Power." updated November 2011. http://sociology.ucsc.edu/whorulesamerica/power/wealth.html (accessed December 26, 2011).

Donaldson, Dwight M. *Studies in Muslim Ethics*. London: SPCK, 1953.

Doran, Peter T., and Maggie Kendall Zimmerman. "Climate Change." *EOS* 90, no. 3 (2009): 22.

Dossey, Larry. *Recovering the Soul: A Scientific and Spiritual Search.* New York: Bantam Books, 1989.

Dostoevsky, Fyodor. *The Brothers Karamazov.* London: Heinemann, 1912.

Dowd, Michael. *Thank God for Evolution! How the Marriage of Science and Religion Will Transform Your Life and Our World.* Tulsa, OK: Council Oak Books, 2007.

Draper, John William. *History of the Conflict between Religion and Science.* 7th ed. London: Harry S. King, 1876.

Drori, Gili S., et al. *Science in the Modern World Polity: Institutionalization and Globalization.* Stanford, CA: Stanford University Press, 2003.

Drum, Kevin. "Plutocracy Now: What Wisconsin Is Really About." *Mother Jones,* March/April 2011.

Drummond, Henry. *The Lowell Lectures on the Ascent of Man.* London: Hodder & Stoughton, 1894.

D'Souza, Dinesh. *Life after Death: The Evidence.* Washington, DC: Regnery, 2009.

———. *What's So Great about Christianity?* Washington, DC: Regnery, 2007.

Durant, Will. *The Story of Philosophy: The Lives and Opinions of the Greater Philosophers.* New York: Simon & Schuster, 1953.

Dyson, Freeman. "Heretical Thoughts about Science and Reality," August 2008. http://www .edge.org/3rd_culture/dysonf07/dysonf07_index.html (accessed March 8, 2011).

———. "The Question of Global Warming." *New York Review of Books,* June 12, 2008.

Ebbem, Hayden, et al. "Maria's Near-Death Experience: Waiting for the other Shoe to Drop." *Skeptical Inquirer* 20, no. 4 (1996): 27–33.

Eberhard, Phillipe H., and Ronald R. Ross. "Quantum Field Theory Cannot Provide Faster-Than-Light Communication." *Foundations of Physics Letters* 2 (1989): 127–79.

Eckard, James Read. "The Logical Relations of Religion and Natural Science." *Biblical Reporatory and Princeton Review* 32 (1860): 577–608.

Ecklund, Elaine Howard. *Science vs. Religion: What Scientists Really Think.* New York: Oxford University Press, 2010.

Ecklund, Elaine Howard, and C. P. Scheitle. "Religion among American Scientists: Distinctions, Disciplines, and Demographics." *Social Problems* 54, no. 2 (2007): 289–307.

Eddington, A. S. "On the Instability of Einstein's Spherical World." *Monthly Notices of the Royal Society* 90 (1930): 668–88.

Edis, Taner. *An Illusion of Harmony: Science and Religion in Islam.* Amherst, NY: Prometheus Books, 2007.

Edwards, Paul. *Reincarnation: A Critical Examination.* Amherst, NY: Prometheus Books, 2002.

Ehrman, Bart D. *Jesus, Interrupted: Revealing the Hidden Contradictions in the Bible (and Why We Don't Know about Them).* New York: HarperOne, 2009.

Einstein, A., B. Podolosky, and N. Rosen. "Can the Quantum Mechanical Description of Physical Reality Be Considered Complete?" *Physical Review* 47 (1935): 777.

Ellis, George F. R. "True Complexity and Its Asociated Ontology." In *Science and Ultimate Reality: Quantum Theory, Cosmology, and Complexity*, edited by John D. Barrow, Paul C. W. Davies, and Charles L. Harper Jr., 607–36. Cambridge, UK: Cambridge University Press, 2004.

Ellison, C. G., and J. S. Levin. "The Religion-Health Connection: Evidence, Theory, and Future Directions." *Health Education and Behavior* 24 (1998): 700–20.

Englert, F., and R. Brout. "Broken Symmetry and the Mass of Gauge Vector Bosons." *Physical Review Letters* 13 (1964): 321.

England, Philip C., Peter Molnar, and Frank M. Richter. "Kelvin, Perry, and the Age of the Earth." *American Scientist* 95 (2007): 342–49.

Everitt, Nicholas. "The Argument from Imperfection: A New Proof of the Nonexistence of God." *Philo* 9, no. 2 (2006): 113–30.

Feeney, Stephen, et al. "First Observational Tests of Eternal Inflation," December 2010. http://arxiv.org/pdf/1012.1995v1 (accessed January 14).

Feferman, Solomon. "Penrose's Gödelian Argument: A Review of *Shadows of the Mind* by Roger Penrose." *Psyche* 2, no. 7 (1995).

Ferngren, Gary B., Edward J. Larson, and Darrel W. Amundsen. *The History of Science and Religion in the Western Tradition: An Encyclopedia*. Garland Reference Library of the Humanities, vol. 1833. New York: Garland, 2000.

Feynman, Richard. "The Theory of Positrons." *Physical Review* 76 (1949): 749–59.

Fischer, C. M. "If There Were No Free Will." *Med Hypotheses* 56, no. 3 (2001): 364–66.

Fisher, William L. "Christian Responsibility for Government." In *Christian Coalition Leadership Manual*, edited by William L. Fisher, Ralph Reed Jr., and Richard L. Wienhold, sec. 2.5–2.8. Christian Coalition, 1990.

Forrest, Barbara, and Paul R. Gross. *Creationism's Trojan Horse: The Wedge of Intelligent Design.* Oxford, UK; New York: Oxford University Press, 2004.

Fox, Mark. *Religion, Spirituality, and the Near-Death Experience.* London; New York: Routledge, 2003.

Frazier, Kendrick. "Journal That Published Bem's Paper Rejects Attempt to Replicate It." *Skeptical Inquirer* 35, no. 4 (2011): 7.

Freedman, David H. "Lies, Damned Lies, and Medical Science." *Atlantic*, November 2010, 76–86.

Frick, Ali. "Barton: We Shouldn't Regulate CO_2 Because 'It's in Your Coca-Cola' and 'You Can't Regulate God,'" May 2009. http://thinkprogress.org/2009/05/19/barton-carbon-god/ (accessed March 29, 2011).

Frith, Christopher D. *Making Up the Mind: How the Brain Creates Our Mental World.* Malden, MA: Blackwell, 2007.

Gardner, Martin. *Fads and Fallacies in the Name of Science*. New York: Dover, 1957.

Garriga, Jaume, and Alexander Vilenkin. "Many Worlds in One." *Physical Review D* 64 (2001): 043511.

Garrison, Becky. *The New Atheist Crusaders and Their Unholy Grail: The Misguided Quest to Destroy Your Faith*. Nashville: Thomas Nelson, 2007.

Gay, Mara. "Florida Education Bill Revives Evolution Debate." *Huffington Post*, March 10, 2011. http://www.aolnews.com/2011/03/10/florida-education-bill-revives-evolution-debate/ (accessed December 26, 2011).

Genz, Henning. *Nothingness: The Science of Empty Space*. Reading, MS: Perseus Books, 1999.

Gibbon, Edward. *The History of the Decline and Fall of the Roman Empire*. London: W. Strahan, 1783.

Gingerich, Owen. *God's Universe*. Cambridge, MA: Harvard University Press, 2006.

Gleick, James. *Chaos: Making a New Science*. 20th anniversary ed. New York: Penguin, 2008.

Gödel, Kurt. "Some Basic Theorems on the Foundations of Mathematics and Their Implications." In *Collected Works: Kurt Gödel, vol. 3*, edited by Solomon Feferman, 304–23. Oxford, UK: Oxford University Press, 1995.

Goldberg, Michelle. *Kingdom Coming: The Rise of Christian Nationalism*. New York: W. W. Norton, 2006.

Good, I. J. "Where Has the Billion Trillion Gone?" *Nature* 389, no. 6653 (1997): 806–807.

Goodstein, Laurie. "Evangelical Leaders Join Global Warming Initiative." *New York Times*, February 8, 2006.

Gott III, Richard, et al. "Will the Universe Expand Forever?" *Scientific American* (March 1976): 65.

Gould, Stephen Jay. *Ever Since Darwin: Reflections in Natural History*. New York: W. W. Norton, 1977.

———. *Rocks of Ages: Science and Religion in the Fullness of Life*. Library of Contemporary Thought. New York: Ballantine, 1999.

Granqvist, Pehr, et al. "Sensed Presence and Mystical Experiences Are Predicted by Suggestibility, Not by the Application of Transcranial Weak Complex Magnetic Fields." *Neuroscience Letters* 379, no. 1 (2005): 1–6.

Grant, Edward. *Science and Religion, 400 BC to AD 1550: From Aristotle to Copernicus*. Greenwood Guides to Science and Religion. Westport, CT: Greenwood Press, 2004.

Grassie, William. *The New Sciences of Religion: Exploring Spirituality from the Outside In and Bottom Up*. New York: Palgrave Macmillan, 2010.

Gray, John. "The Atheist Delusion." *Guardian*, March 15, 2008.

Greenblatt, Stephen. "The Answer Man: An Ancient Poem Was Rediscovered and the World Swerved." *New Yorker*, August 8, 2011, 28–33.

———. *The Swerve: How the World Became Modern*. New York: W. W. Norton, 2011.

Gregersen, Niels Henrik. "God, Matter, and Information: Towards a Stoicizing Logos

Christianity." In *Information and the Nature of Reality: From Physics to Metaphysics*, edited by Paul C. W. Davies, and Niels Henrik Gregersen, 319–48. Cambridge, UK: Cambridge University Press, 2010.

Grush, Rick, and Patricia Smith Churchland. "Gaps in Penrose's Toilings." In *On the Contrary: Critical Essays, 1987–1997*, edited by Paul M. Churchland and Patricia S. Churchland, 205–30. Cambridge, MA: MIT Press, 1998.

Gu, Mile, et al. "More Really Is Different." *Physica D* 238 (2009): 835–39.

Gunn, E. "Death by Prayer." *Freethought Today*, September 2008, 6–7.

Gurainik, G. S., C. R. Hagen, and T. W. Kibble. "Global Conservation Laws and Massless Particles." *Physical Review Letters* 13 (1964): 585.

Guth, Alan H. "Inflationary Universe: A Possible Solution to the Horizon and Flatness Problems." *Physcial Review D* 23, no. 2 (1981): 347–56.

Hacker, Jacob S., and Paul Pierson. *Winner-Take-All Politics: How Washington Made the Rich Richer—and Turned Its Back on the Middle Class*. New York: Simon & Schuster, 2010.

Hagan, S., S. Hameroff, and J. Tuszinski. "Quantum Computation in Brain Microtubules? Decoherence and Biological Feasibility." *Physical Review E* 65 (2002): 061901.

Hagerty, Barbara. "Nun Excommunicated for Allowing Abortion." National Public Radio, May 19, 2010.

Hallett, Mark. "Transcranial Magnetic Stimulation and the Human Brain." *Nature* 406 (2000): 147–50.

Halvorson, Hans, and Helge S. Kragh. "Theism and Physical Cosmology." In *Routledge Companion to Theism*, edited by Stewart Goetz, Victoria Harrison, and Charles Taliaferro (scheduled for release June 12, 2012).

Hamilton, W. D. "Altruism and Related Phenomena, Mainly in the Social Insects." *Annual Review of Ecology and Systematics* 3 (1972): 193–232.

Hannam, James. *God's Philosophers: How the Medieval World Laid the Foundations of Modern Science*. London: Icon, 2009.

Hansen, James E. *Storms of My Grandchildren: The Truth about the Coming Climate Catastrophe and Our Last Chance to Save Humanity*. New York: Bloomsbury USA, 2009.

Harris, Sam. *The End of Faith: Religion, Terror, and the Future of Reason*. New York: W. W. Norton, 2004.

———. *Letter to a Christian Nation*. 1st Vintage Books ed. New York: Vintage Books, 2008.

———. *The Moral Landscape: How Science Can Determine Human Values*. New York: Free Press, 2010.

Hartle, James B, and Stephen W. Hawking. "Wave Function of the Universe." *Physical Review D* 28 (1983): 2960–75.

Haught, James A. *Fading Faith: The Rise of the Secular Age*. Cranford, NJ: Gustav Broukal Press, 2010.

Haught, John F. "Information, Theology, and the Universe." In *Information and the Nature of Reality: From Physics to Metaphysics*, edited by Paul C. W. Davies and Niels Henrik Gregersen, 301–38. New York: Cambridge University Press, 2010.

Hawking, Stephen W. *A Brief History of Time: From the Big Bang to Black Holes*. Toronto; New York: Bantam Books, 1988.

Hawking, Stephen W., and Leonard Mlodinow. *The Grand Design*. New York: Bantam Books, 2010.

Hawking, Stephen W., and Roger Penrose. "The Singularities of Gravitational Collapse." *Proceedings of the Royal Society of London* Series A 314 (1970): 529–48.

Hedges, Chris. *American Fascists: The Christian Right and the War on America*. New York: Free Press, 2007.

Heisenberg, Werner. *Physics and Beyond: Encounters and Conversations*. London: G. Allen & Unwin, 1971.

Higgs, Peter W. "Broken Symmetries and the Masses of Gauge Bosons." *Physical Review Letters* 13 (1964): 508.

Hilbert, David. "On the Infinite." In *Philosophy of Mathematics*, edited by Paul Benacerraf and Hillary Putman, 139–41. Englewood Cliffs, NJ: Prentice-Hall: 1964.

Hines, Terence, ed. *Pseudoscience and the Paranormal: A Critical Examination of the Evidence*. Amherst, NY: Prometheus Books, 1988.

Hitchens, Christopher. *God Is Not Great: How Religion Poisons Everything*. New York: Twelve, Hachette Book Group, 2007.

Holden, Janice Miner, Bruce Greyson, and Debbie James, eds. *The Handbook of Near-Death Experiences: Thirty Years of Investigation*. Santa Barbara, CA: Praeger, 2009.

Holmyard, Eric John. *Makers of Chemistry*. Oxford, UK: Clarendon Press, 1931.

Hoyle, F., et al. "A State in C12 Predicted from Astronomical Evidence." *Physical Review Letters* 92 (1953): 1095.

Hume, David. *Dialogues Concerning Natural Religion*, edited by Henry David Aiken. New York: Hafner, 1972.

Hutchinson, Robert. *The Politically Incorrect Guide to the Bible*. Washington, DC: Regnery, 2007.

Huxley, Thomas Henry. *Evolution and Ethics*. London and New York: Macmillan, 1893.

Inhofe, James M. "The Science of Climate Change: Senate Floor Statement July 28, 2003." http://inhofe.senate.gov/pressreleases/climate.htm (accessed March 7, 2011).

Intergovernmental Panel on Climate Change (IPCC). "Climate Change 2007: Synthesis Report." http://www.ipcc.ch/publications_and_data/ar4/syr/en/mains6-1.html (accessed March 17, 2011).

Internet Encyclopedia of Philosophy. "Aristotle (384–322 BCE)." http://www.iep.utm.edu/aristotl/ (accessed January 30, 2011).

Isaak, Mark. "Index to Creationism Claims," 2006. http://www.talkorigins.org/indexcc/list.html (accessed April 1, 2011).

James, Craig A. *The Religion Virus: Why We Believe in God: An Evolutionist Explains Religion's Incredible Hold on Humanity*. O Books, 2010.

Jasnow, Richard. *A Late Period Hieratic Wisdom Text: P. Brooklyn P. 47.218.135*. Chicago: University of Chicago Press, 1992.

Jastrow, Robert. *God and the Astronomers*. 2nd ed. New York: W. W. Norton, 1992.

John Templeton Foundation. "Mission," 2011. http://www.templeton.org/who-we-are/about-the-foundation/mission (accessed February 7, 2011).

Johnson, Phillip E. *Defeating Darwinism by Opening Minds*. Downers Grove, IL: InterVarsity Press, 1997.

———. *Reason in the Balance: The Case against Naturalism in Science, Law & Education*. Downers Grove, IL: InterVarsity Press, 1995.

———. *The Wedge of Truth: Splitting the Foundations of Naturalism*. Downers Grove, IL: InterVarsity Press, 2000.

Jones, Richard H. *Piercing the Veil: Comparing Science and Mysticism as Ways of Knowing Reality*. New York: Jackson Square Books, 2010.

Jones, Robert P., and Daniel Cox. "Religion and the Tea Party in the 2010 Elections: An Analysis of the Third Biennial American Values Survey." http://www.publicreligion.org/objects/uploads/fck/file/AVS%202010%20Report%20FINAL.pdf (accessed August 26, 2011).

Jones III, John E. "*Kitzmiller v. Dover Area School District, et al.*, Case No. 04cv2688," 2005. http://www.pamd.uscourts.gov/kitzmiller/kitzmiller_342.pdf (accessed January 5, 2011).

Kaplan, Robert. *The Nothing That Is: A Natural History of Zero*. Oxford, UK; New York: Oxford University Press, 2000.

Katz, Leonard D. *Evolutionary Origins of Morality: Cross-Disciplinary Perspectives*. Thorverton, UK; Bowling Green, OH: Imprint Academic, 2000.

Kauffman, Stuart A. *Reinventing the Sacred: A New View of Science, Reason, and Religion*. New York: Basic Books, 2008.

Kaufman, Leslie. "Darwin Foes Add Warming to Targets." *New York Times*, March 3, 2010.

Kazanas, Demos. "Dynamics of the Universe and Spontaneous Symmetry Breaking." *Astrophysical Journal* 241 (1980): L59–L63.

Kazakov, D. I. "Beyond the Standard Model (In Search of Supersymmetry)." Lectures given at the European School for High Energy Physics, Caramulo, Portugal, August/September 2000. Cornell University Library, High Energy Physics—Phenomenology. http://arxiv.org/abs/hep-ph/0012288 (accessed December 26, 2011).

Keyes, Scott. "Jim DeMint's Theory of Relativity: 'The Bigger Government Gets, the Smaller God Gets,'" 2001. http://thinkprogress.org/2011/03/15/demint-big-govt/ (accessed March 29, 2011).

Kiel, K. A., et al. "Limbic Abnormalities in Affective Processing by Criminal Psychopaths as Revealed by Functional Magnetic Resonance Imaging." *Biological Psychiatry* 50, no. 9 (2001): 677–84.

Kirsch, Jonathan. *God against the Gods: The History of the War between Monotheism and Polytheism*. New York: Viking Compass, 2004.

Klein, Stanley A. "Is Quantum Mechanics Relevant to Understanding Consciousness? A Review of *Shadows of the Mind* by Roger Penrose." *Psyche* 2, no. 3 (1995).

Koenig, Harold G. *The Healing Power of Faith: Science Explores Medicine's Last Great Frontier*. New York: Simon & Schuster, 1999.

———. "What Religion Can Do for Your Health," 2006. http://www.beliefnet.com/Health/2006/05/What-Religion-Can-Do-For-Your-Health.aspx (accessed February 11, 2011).

Koenig, Harold G., Dana E. King, and Verna Benner Carson. *Handbook of Religion and Health*. 2nd ed. Oxford, UK; New York: Oxford University Press, 2011.

Kosmin, Barry A., and Ariela Keysar. "American Religious Identification Survey (ARIS) 2008." http://commons.trincoll.edu/aris/ (accessed February 21, 2010).

Kragh, Helge S. *Entropic Creation: Religious Contexts of Thermodynamics and Cosmology*. University of Aarhus, Demark, Science, Technology, and Culture Series, 1700–1945. Aldershot, Hampshire, England; Burlington, VT: Ashgate, 2008.

Krakauer, Jon. *Under the Banner of Heaven: A Story of Violent Faith*. New York: Doubleday, 2003.

Kristof, Nicholas D. "Our Banana Republic." *New York Times*, November 10, 2010.

Krucoff, M. W., S. W. Crater, and D. Gallup. "Music, Imagery, Touch, and Prayer as Adjuncts to Interventional Cardiac Care: The Monitoring and Actualisation of Noetic Trainings (Mantra) II Randomised Study." *Lancet* 366 (2005): 211–17.

Kurtz, Paul. *The Transcendental Temptation: A Critique of Religion and the Paranormal*. Amherst, NY: Prometheus Books, 1986.

LaCroix, Richard R. "Unjustified Evil and God's Choice." *Sophia* 13, no. 1 (1974): 20–28.

Lakoff, George, and Mark Johnson. *Philosophy in the Flesh: The Embodied Mind and Its Challenge to Western Thought*. New York: Basic Books, 1999.

Lamont, Corliss. *The Illusion of Immortality*. 4th ed. New York: F. Ungar, 1965.

Larson, D. B., J. P. Swyers, and M. E. McCullough eds. *Scientific Research on Spirituality and Health: A Report Based on the Scientific Progress in Spirituality Conferences*. Rockville, MD: National Institutes for Healthcare Research, 1998.

Larson, Edward J. "Leading Scientists Still Reject God." *Nature* 294, no. 6691 (1998): 313.

———. *Summer for the Gods: The Scopes Trial and America's Continuing Debate over Science and Religion*. New York: BasicBooks, 1997.

LaRue, Steve. "Early Action Urged in Fight on Acid Rain." *San Diego Union*, August 8, 1984, B2.

Lebo, Lauri. "Creationism and Global Warming Denial: Anti-Science's Kissing Cousins?" *Religion Dispatches*, March 17, 2010.

Lemaître, Georges. "Un Univers homogène de masse constante et de rayon croissant rendant compte de la vitesse radiale des nébuleuses extragalactiques (A homogeneous Universe of constant mass and growing radius accounting for the radial velocity of extragalactic nebulae)." *Annales de la Société Scientifique de Bruxelles* (Annals of the Scientific Society of Brussels) 47 (April 1927): 49.

Lennox, John C. *God's Undertaker: Has Science Buried God?* Oxford, UK: Lion Hudson, 2007.

Lewis, C. S. *Mere Christianity*. Westwood: Barbour, 1952.

Lewontin, R. C. "In the Beginning Was the Word." *Science* 291, no. 5507 (2001): 1263–64.

Libet, B., et al. "Time of Conscious Intention to Act in Relation to Onset of Cerebral Activity (Readiness-Potential). The Unconscious Initiation of a Freely Voluntary Act." *Brain* 106 (1983): 623–42.

Lindberg, David C. *The Beginnings of Western Science: The European Scientific Tradition in Philosophical, Religious, and Institutional Context, Prehistory to AD 1450.* 2nd ed. Chicago: University of Chicago Press, 2007.

Lindberg, David C., and Ronald L. Numbers. *God and Nature: Historical Essays on the Encounter between Christianity and Science.* Berkeley, CA: University of California Press, 1986.

Linde, Andrei D. "Eternally Existing Self-Reproducing Chaotic Inflationary Universe." *Physics Letters B* 175, no. 4 (1986): 395–400.

———. "A New Inflationary Universe Scenario: A Possible Solution of the Horizon, Flatness, Homogeneity, Isotropy, and Primordial Monopole Problems." *Physics Letters B* 108 (1982): 389.

Linder, Eileen W., ed. *Yearbook of American & Canadian Churches.* National Council of Churches, 2009.

Linker, Damon. *The Theocons: Secular America under Siege.* New York: Doubleday, 2006.

Livio, M., et al. "The Anthropic Significance of the Existence of an Excited State of C12." *Nature* 340 (1989): 281–84.

Lloyd, Seth. "The Computational Universe." In *Information and the Nature of Reality: From Physics to Metaphysics*, edited by Paul C. W. Davies and Niels Henrik Gregersen, 92–103. New York: Cambridge University Press, 2010.

———. *Programming the Universe: A Quantum Computer Scientist Takes on the Cosmos.* New York: Knopf, 2006.

Loftus, John W., ed. *The Christian Delusion: Why Faith Fails.* Amherst, NY: Prometheus Books, 2010.

———, ed. *The End of Christianity.* Amherst, NY: Prometheus Books, 2011.

———. *Why I Became an Atheist: A Former Preacher Rejects Christianity.* Prometheus Books, 2008.

Long, Jeffrey, and Paul Perry. *Evidence of the Afterlife: The Science of Near-Death Experiences.* New York: HarperOne, 2010.

Lorentz, H. A., et al. *The Principle of Relativity: A Collection of Original Memoirs on the Special and General Theory of Relativity.* London: Methuen & Co., 1923.

Lucretius, Titus. *De Rerum Natura (On the Nature of Things).* Cambridge, MA: Harvard University Press, 1992.

Lugo, Luis, et al. "U.S. Religious Landscape Survey," 2008. http://religions.pewforum.org/pdf/report-religious-landscape-study-full.pdf (accessed March 8, 2011).

Lutz, Diana. "Physicist Disputes 'Quantum Mind' in Debate Hosted by Deepak Chopra." *Skeptical Inquirer* 35, no. 3 (2011): 8.

Lyell, Charles, and G. P. Deshayes. *Principles of Geology; Being an Attempt to Explain the Former Changes of the Earth's Surface, by Reference to Causes Now in Operation.* London: J. Murray, 1830.

Matsubara, Hiroshi. "Western Eyes Blind to Spirituality in Japan." *Japan Times*, January 1, 2002.

Maudlin, Tim. "Between the Motion and the Act: A Review of *Shadows of the Mind* by Roger Penrose." *Psyche* 2, no. 2 (1995).

Maynard Smith, John. *On Evolution.* Edinburgh, UK: Edinburgh University Press, 1972.

McCarthy, John. "Awareness and Understanding in Computer Programs: A Review of *Shadows of the Mind* by Roger Penrose." *Psyche* 2, no. 11 (1995).

McCullough, Daryl. "Can Humans Escape Gödel? A Review of *Shadows of the Mind* by Roger Penrose." *Psyche* 2, no. 4 (1995).

McDermott, Drew. "[Star] Penrose Is Wrong." *Psyche* 2, no. 17 (1995).

McMullin, Ernan. "From Matter to Materialism and (Almost) Back." In *Information and the Nature of Reality: From Physics to Metaphysics*, edited by Paul Davies and Henrik Gregersen, 13–37. Cambridge, UK; New York: Cambridge University Press, 2010.

Meyer, Stephen C. "The Demarcation of Science and Religion." In *The History of Science and Religion in the Western Tradition: An Encyclopedia*, edited by Gary B. Ferngren, 17–23. New York: Garland, 2000.

Miller, John D., Eugenie C. Scott, and Shinji Okamoto. "Public Acceptance of Evolution." *Science* 11 (2006): 765.

Miller, Kenneth R. "Darwin, God, and Dover: What the Collapse of 'Intelligent Design' Means for Science and Faith in America." In *The Religion and Science Debate: Why Does It Continue?* edited by Harold W. Attridge, 55–92. New Haven: Yale University Press, 2009.

———. *Finding Darwin's God: A Scientist's Search for Common Ground between God and Evolution.* New York: Cliff Street Books, 1999.

Montgomery, Peter. "Jesus Hates Taxes: Biblical Capitalism Created Fertile Anti-Union Soil." *Religion Dispatches*, March 14, 2011.

Monton, Bradley John. *Seeking God in Science: An Atheist Defends Intelligent Design.* Peterborough, ON: Broadview Press, 2009.

Moody, Raymond A. *Life after Life: The Investigation of a Phenomenon—Survival of Bodily Death.* New York: Bantam Books; HarperCollins, 1976.

Mooney, Chris. *The Republican War on Science.* New York: Basic Books, 2005.

Moore, A. "The Christian Doctrine of God." In *Lux Mundi*, edited by C. Gore, 41–81. London: Murray, 1891.

Moravec, Hans. "Roger Penrose's Gravitonic Brains: A Review of *Shadows of the Mind* by Roger Penrose." *Psyche* 2, no. 6 (1995).

Moyers, Bill, "Welcome to the Plutocracy." Speech at Boston University on October 29, 2010. http://www.truth-out.org/bill-moyers-money-fights-hard-and-it-fights-dirty64766 (accessed December 26, 2011).

Mulhauser, Gregory R. "On the End of a Quantum Mechanical Romance." *Psyche* 2, no. 5 (1995).

National Academy of Sciences, *Teaching about Evolution and the Nature of Science*. Washington, DC: National Academies Press, 1998.

National Oceanic and Atmospheric Administration (NOAA) National Climatic Data Center. "*State of the Climate Global Analysis: Annual 2004*." http://www.ncdc.noaa.gov/sotc/global/2004/13 (accessed March 7, 2011).

National Research Council. *America's Climate Choices*. Washington, DC: National Academies Press, 2011.

———. "Frontiers in Understanding Climate Change and Polar Ecosystems: Summary of a Workshop," 2011. http://www.nap.edu/catalog.php?record_id=13132 (accessed March 29, 2011).

Nelson, Kevin. *The Spiritual Doorway in the Brain: A Neurologist's Search for the God Experience*. New York: Dutton, 2011.

Netz, Reviel, and William Noel. *The Archimedes Codex: How a Medieval Prayer Book Is Revealing the True Genius of Antiquity's Greatest Scientist*. Philadelphia, PA: Da Capo Press, 2007.

Newberg, Andrew B. *Principles of Neurotheology*. Farnham, Surrey, England; Burlington, VT: Ashgate, 2010.

Newberg, Andrew B., Eugene G. D'Aquili, and Vince Rause. *Why God Won't Go Away: Brain Science and the Biology of Belief*. New York: Ballantine Books, 2001.

Newberg, Andrew B., and Mark Robert Waldman. *Born to Believe: God, Science, and the Origin of Ordinary and Extraordinary Beliefs*. New York: Free Press, 2007.

———. *How God Changes Your Brain: Breakthrough Findings from a Leading Neuroscientist*. New York: Ballantine Books, 2009.

———. *Why We Believe What We Believe: Uncovering Our Biological Need for Meaning, Spirituality, and Truth*. New York: Free Press, 2006.

Newport, Frank. "In U.S., Very Religious Americans Still Align More with GOP," 2011. http://www.gallup.com/poll/148274/Religious-Americans-Align-GOP.aspx (accessed June 28, 2011).

———. "More Than 9 in 10 Americans Continue to Believe in God," 2011. http://www.gallup.com/poll/147887/Americans-Continue-Believe-God.aspx (accessed June 8, 2011).

Nichols, Peter. *Evolution's Captain: The Dark Fate of the Man Who Sailed Charles Darwin around the World*. New York: HarperCollins, 2003.

Nickell, Joe. *Inquest on the Shroud of Turin: Latest Scientific Findings*. Amherst, NY: Prometheus Books, 1998.

Nierenberg, William, Chairman. *Report of the Acid Rain Peer Review Panel, July 1984*. Washington, DC: White House Office of Science and Technology Policy, 1984.

Norris, Pippa, and Ronald Inglehart. *Sacred and Secular: Religion and Politics Worldwide*. 2nd ed. Cambridge, UK: Cambridge University Press, 2011.

Nottale, Laurent. "Scale Relativity and Fractal Space-Time: Theory and Applications." *Foundations of Science* 15, no. 2 (2008): 101–52.

Numbers, Ronald L. "Aggressors, Victims, and Peacemakers: Historical Actors in the Drama of Science and Religion." In *The Religion and Science Debate: Why Does It Continue?* edited by Harold W. Attridge, 15–53. New Haven: Yale University Press, 2009.

———. "The Creationists." In *God and Nature: Historical Essays on the Encounter between Christianity and Science*, edited by David C. Lindberg and Ronald L. Numbers, 391–423. Berkeley, CA: University of California Press, 1986.

———. *The Creationists: From Scientific Creationism to Intelligent Design*. Expanded ed. Cambridge, MA: Harvard University Press, 2006.

Nunez, J. M., B. J. Casey, and K. Sigmund. "Intentional False Responding Shares Neural Substrates with Response Conflict and Cognitive Control." *Neuroimage* 25, no. 1 (2005): 267–77.

Nussbaumer, Harry, and Lydia Bieri. *Discovering the Expanding Universe*. Cambridge, UK; New York: Cambridge University Press, 2009.

O'Connor, Eugene, trans. *The Essential Epicurus: Letters, Principle Doctrines, Vatican Sayings, and Fragments*. Amherst, NY: Prometheus Books, 1993.

Oreskes, Naomi, and Erik M. Conway. *Merchants of Doubt: How a Handful of Scientists Obscured the Truth on Issues from Tobacco Smoke to Global Warming*. New York: Bloomsbury Press, 2010.

Page, Don. "Does God So Love the Multiverse?" In *Science and Religion in Dialogue*, edited by Melville Y. Stewart. Chichester, UK: Wiley-Blackwell, 2010, 380–95.

———. "Evidence against Fine Tuning for Life." http://arxiv.org/abs/1101.2444v1 (accessed January 20, 2011).

Paley, William. *Natural Theology*. London: Printed for R. Faulder by Wilks and Taylor, 1802.

Pascual-Leone, Alvaro, Vincent Walsh, and John Rothwell. "Transcranial Magnetic Stimulation in Cognitive Neuroscience—Virtual Lesion, Chronometry, and Functional Connectivity." *Current Opinion in Neurobiology* 10 (2000): 232–37.

Patterson, Elissa. "The Philosophy and Physics of Holistic Health Care: Spiritual Healing as a Workable Interpretation." *British Journal of Advanced Nursing* 27 (1998): 287–93.

Paul, Gregory S. "The Chronic Dependence of Popular Religiosity upon Dysfunctional Psychosociological Conditions." *Evolutionary Psychology* 7, no. 3 (2009): 398–441.

———. "Cross-National Correlations of Quantifiable Societal Health with Popular Religiosity and Secularism in the Prosperous Democracies: A First Look." *Journal of Religion & Society* 7 (2005).

Peacocke, Arthur. "The Sciences of Complexity: A New Theological Resource?" In *Information and the Nature of Reality: From Physics to Metaphysics*, edited by Paul C. W. Davies, and Niels Henrik Gregersen, 249–81. Cambridge; New York: Cambridge University Press, 2010.

Peck, Tom, and Jerome Taylor. "Scientist Imam Threatened over Darwinist Views." *The Independent*, March 5, 2011.

Penfield, William, and T. Rasmussen. *The Cerebral Cortex of Man: A Clinical Study of Localization of Function*. New York: MacMillan, 1950.

Penrose, Roger. "Beyond the Doubting of a Shadow: A Reply to Commentaries on *Shadows of the Mind*." *Psyche* 2, no. 23 (1995).

———. *The Emperor's New Mind: Concerning Computers, Minds, and the Laws of Physics*. Oxford, UK; New York: Oxford University Press, 1989.

———. *The Road to Reality: A Complete Guide to the Laws of the Universe*. London: Jonathan Cape, 2004.

———. *Shadows of the Mind: A Search for the Missing Science of Consciousness*. Oxford, UK; New York: Oxford University Press, 1994.

Persinger, M. A. "Religious and Mystical Experiences as Artifacts of Temporal Lobe Function: A General Hypothesis." *Perceptual and Motor Skills* 57 (1983): 1255–62.

Pew Forum on Religion and Public Life. "Religion among the Millennials," 2010. http://pew forum.org/Age/Religion-Among-the-Millennials.aspx (accessed March 22, 2011).

———. "Religion and Public Life: A Faith-Based Partisan Divide," 2005. http://pewforum. org/uploadedfiles/Topics/Issues/Politics_and_Elections/religion-and-politics-report.pdf (accessed March 17, 2011).

———. "Religious Groups' Views on Global Warming," 2009. http://pewresearch.org/pubs/1194/global-warming-belief-by-religion (accessed March 9, 2011).

———. "U.S. Religious Knowledge Survey," September 2010. http://pewforum.org/other -beliefs-and-practices/u-s-religious-knowledge-survey.aspx (accessed January 5, 2011).

Pew Research Center, "World Publics Welcome Global Trade—But Not Immigration: October 4, 2007." http://pewglobal.org/files/pdf/258.pdf (accessed 2011).

Phillips, Kevin. *American Theocracy: The Peril and Politics of Radical Religion, Oil, and Borrowed Money in the 21st Century*. New York: Viking, 2006.

Pigliucci, Massimo. *Denying Evolution: Creationism, Scientism, and the Nature of Science*. Sunderland, MA: Sinauer Associates, 2002.

Pinker, Steven. *The Blank Slate: The Modern Denial of Human Nature*. New York: Viking, 2002.

Plantinga, Alvin. "Science and Religion: Why Does the Debate Continue?" In *The Religion and Science Debate: Why Does It Continue?* edited by Harold W. Attridge. New Haven: Yale University Press, 2009.

Pope Benedict XVI, "Easter Vigil Homily: We Celebrate the First Day of the New Creation," April 2011. http://www.catholic.org/clife/lent/story.php?id=41157&page=1 (accessed May 17, 2011).

Poupard, Cardinal Paul, (ed.) *Galileo Galilei: Toward a Resolution of 350 Years of Debate, 1633–1983*. Pittsburgh, PA: Duquesne University Press, 1987.

Powell, L. L., L. Shahabi, and C. E. Thoresen. "Religion and Spirituality: Linkages to Physical Health." *American Psychologist* 58, no. 1 (January 2003): 36–52.

Price, George McCready. *The Fundamentals of Geology and Their Bearings on the Doctrine of a Literal Creation.* Mountain View, CA: Pacific Press, 1913.

———. *Illogical Geology: The Weakest Point in the Evolution Theory.* Los Angeles: Modern Heretic Company, 1906.

———. *The New Geology: A Textbook for Colleges, Normal Schools, and Training Schools; and for the General Reader.* Mountain View, CA: Pacific Press, 1923.

———. *Outlines of Modern Christianity and Modern Science.* Oakland, CA: Pacific Press, 1902.

Price, Huw. *Time's Arrow & Archimedes' Point: New Directions for the Physics of Time.* New York: Oxford University Press, 1996.

Principe, Lawrence M. *Science and Religion.* Philosophy & Intellectual History. Chantilly, VA: Teaching Company, 2006.

Programme for International Student Assessment. "Pisa 2009 Results." http://www.oecd.org/document/61/0,3746,en_32252351_46584327_46567613_1_1_1_1,00.html (accessed March 15, 2011).

Provine, William. *Origins Research* 16, no. 1 (1994): 9.

Radin, Dean I. *The Conscious Universe: The Scientific Truth of Psychic Phenomena.* New York: HarperEdge, 1997.

Ramachandran, V. S., and Sandra Blakeslee. *Phantoms in the Brain: Probing the Mysteries of the Human Mind.* New York: HarperCollins, 1998.

Ramey, David, ed. *Consumers' Guide to Alternate Therapies in the Horse.* New York: Howell Book House, 1999.

Ray, Darrel W. *The God Virus: How Religion Infects Our Lives and Culture.* Bonner Springs, KS: IPC Press, 2009.

Rees, Martin J. *Just Six Numbers: The Deep Forces That Shape the Universe.* New York: Basic Books, 2000.

Rhine, J. B. *Extra-Sensory Perception.* Boston, MA: Boston Society for Psychic Research, 1934.

Richardo, Alonzo, and Jack W. Szostak. "Life on Earth: Fresh Clues Hint at How the First Living Organisms Arose from Inanimate Matter." *Scientific American* (September 2009): 54–61.

Ridley, Matt. *The Origins of Virtue: Human Instincts and the Evolution of Cooperation.* New York: Viking, 1997.

Ring, Kenneth, and Sharon Cooper. *Mindsight: Near-Death and Out-of-Body Experiences in the Blind.* Palo Alto, CA: William James Center for Consciousness Studies, 1999.

Roberts, Tom, and Siegmar Schleif. "What Is the Experimental Basis of Special Relativity?" 2007. http://www.xs4all.nl/~johanw/PhysFAQ/Relativity/SR/experiments.html#Tests_of_the_poR (accessed December 21, 2010).

Robinson, B. A. "Roman Catholicism and Abortion Access: Pagan & Christian Beliefs 400 BCE–1980 CE," 2010. http://www.religioustolerance.org/abo_hist.htm (accessed March 15, 2011).

Ross, Hugh. "Big Bang Model Refined by Fire." In *Mere Creation: Science, Faith & Intelligent Design*, edited by William A. Dembski. Downers Grove, IL: Intervarsity Press, 1988, 363–83.

Ross, Hugh. *The Creator and the Cosmos: How the Greatest Scientific Discoveries of the Century Reveal God*. Colorado Springs, CO: NavPress, 1993.

Rothman, Tony. "A 'What You See Is What You Beget' Theory." *Discover* (May 1987): 99.

Rouder, Jeffrey N., and Richard D. Morey. "A Bayes Factor Meta-Analysis of Bem's ESP Claim," 2011. http://www.springerlink.com/content/q113m61462793241/ (accessed June 22, 2011).

Rundle, Bede. *Why There Is Something Rather Than Nothing*. Oxford, UK; New York: Oxford University Press, 2004.

Ruse, Michael, ed. *But Is It Science? The Philosophical Question in the Creation/Evolution Controversy*. Amherst, NY: Prometheus Books, 1988.

———. *Science and Spirituality: Making Room for Faith in the Age of Science*. Cambridge, UK; New York: Cambridge University Press, 2010.

Russell, Bertrand. *Has Religion Made Useful Contributions to Civilization? An Examination and a Criticism*. London: Watts & Co., 1930.

———. *A History of Western Philosophy and Its Connection with Political and Social Circumstances from the Earliest Times to the Present Day*. New York: Simon & Schuster, 1945.

———. *Religion and Science*. New York: Oxford University Press, 1997.

Russell, Colin A. "The Conflict of Science and Religion." In *The History of Science and Religion in the Western Tradition: An Encyclopedia*, edited by Gary B. Ferngren, 12–16. New York: Garland, 2000.

Rutherford, M. "The Evolution of Morality." *Groundings* 1 (2007).

Sagan, Carl. *Cosmos*. New York: Random House, 1980.

Sample, Ian. "Stephen Hawking: 'There's No Heaven; It's a Fairy Story.'" *Guardian*, May 16, 2011.

Saunders, Nicholas. *Divine Action and Modern Science*. Cambridge, UK; New York: Cambridge University Press, 2002.

Scheiber, Béla, and Carla Selby. *Therapeutic Touch*. Amherst, NY: Prometheus Books, 2000.

Scheurer, P. B., and G. Debrock. *Newton's Scientific and Philosophical Legacy*. Dordrecht, Netherlands; Boston, MA: Kluwer Academic [now Springer Science+Business Media] 1988.

Schmidt, Alvin J. *How Christianity Changed the World*. Grand Rapids, MI: Zondervan, 2004.

Schoenburg, Bernard. "Shimkus's Food for Thought at Hearing Prompts Snickers," April 2, 2009. http://www.sj-r.com/opinions/x180621374/Bernard-Schoenburg-Shimkus-food-for -thought-at-hearing-prompts-snickers#video1(accessed February 9, 2011).

Schweber, S. S. *QED and the Men Who Made It: Dyson, Feynman, Schwinger, and Tomonaga*. Princeton Series in Physics. Princeton, NJ: Princeton University Press, 1994.

Schweitzer, Jeff, and Giuseppe Notarbartolo-di-Sciara. *Beyond Cosmic Dice: Moral Life in a Random World*. Los Angeles: Jacquie Jordan, 2009.

Scott, S. P. *History of the Moorish Empire in Europe*. vol. 3. Philadelphia and London: J. B. Lippincott, 1904.

Scott, Stephen K. *Oscillations, Waves, and Chaos in Chemical Kinetics*. Oxford, UK; New York: Oxford University Press, 1994.

Sedley, D. N. *Creationism and Its Critics in Antiquity*. Berkeley, CA: University of California Press, 2007.

Segal, Alan F. *Life after Death: A History of the Afterlife in the Religions of the West*. New York: Doubleday, 2004.

Shannon, Claude Elwood, and Warren Weaver. *The Mathematical Theory of Communication*. Urbana: University of Illinois Press, 1949.

Sharlet, Jeff. *C Street: The Fundamentalist Threat to American Democracy*. New York: Little, Brown, 2010.

———. *The Family: The Secret Fundamentalism at the Heart of American Power*. New York: Harper Perennial, 2009.

———. "This Is Not a Religion Column: Biblical Capitalism," 2008. http://www.religiondis-patches.org/archive/politics/562/ (accessed April 1, 2011).

Shermer, Michael. *The Believing Brain: From Ghosts and Gods to Politics and Conspiracies—How We Construct Beliefs and Reinforce Them as Truths*. New York: Times Books, 2011.

———. *How We Believe: The Search for God in an Age of Science*. New York: W. H. Freeman, 2000.

———. *In Darwin's Shadow: The Life and Science of Alfred Russel Wallace: A Biographical Study on the Psychology of History*. Oxford, UK; New York: Oxford University Press, 2002.

———. *The Science of Good and Evil: Why People Cheat, Gossip, Care, Share, and Follow the Golden Rule*. New York: Times Books, 2004.

Silk, Joseph. *The Big Bang*. 3rd ed. New York: W. H. Freeman, 2001.

Silliman, Benjamin. "Address before the Association of American Geologists and Naturalists, Assembled in Boston, April 24, 1842." *American Journal of Science* 43 (1842): 217–50.

Singer, S. Fred. "On a Nuclear Winter." *Science* 227, no. 4685 (1985): 356.

Sloan, Richard P., and Emila Bagiella. "Claims about Religious Involvement in Health Outcomes." *Annals of Behavioral Medicine* 24, no. 1 (2002): 14–21.

Sober, Elliott, and David Sloan Wilson. *Unto Others: The Evolution and Psychology of Unselfish Behavior*. Cambridge, MA: Harvard University Press, 1998.

Soon, Chun Siong, et al. "Unconscious Determinants of Free Decisions in the Human Brain." *Nature Neuroscience* 11, no. 5 (2008): 545.

Solt, Frederick, Philip Habel, and J. Tobin Grant. "Economic Inequality, Relative Power, and Religiosity." *Social Science Quarterly* 92, no. 2 (2011): 447–65.

Specter, Michael. *Denialism: How Irrational Thinking Hinders Scientific Progress, Harms the Planet, and Threatens Our Lives*. New York: Penguin, 2009.

Spencer, Herbert. *The Atheist Delusion*. New York: D. Appleton & Co., 1880.

————. *First Principles*. London: Williams and Norgate, 1863.

————. *The Principles of Biology*. London: William and Norgate, 1864.

Spooner, W. A. "The Golden Rule." In *Encyclopedia of Religion and Ethics*, edited by James Hastings, 310–12. New York: Charles Scribner's Sons, 1914.

Sproul, Barbara C. *Primal Myths: Creating the World*. San Francisco: Harper & Row, 1979.

Stark, Rodney. *For the Glory of God: How Monotheism Led to Reformations, Science, Witch Hunts, and the End of Slavery*. Princeton, NJ: Princeton University Press, 2003.

St-Pierre, L. S., and M. A. Persinger. "Experimental Facilitation of the Sensed Presence Is Predicted by the Specific Patterns of the Applied Magnetic Fields, Not by Suggestibility: Re-Analyses of 19 Experiments." *International Journal of Neurosciences* 116, no. 9 (2006): 1079–96.

Stefanatos, Joanne. "Introduction to Bioenergetic Medicine." In *Complementary and Alternative Veterinary Medicine: Principles and Practice*, edited by Allen M. Schoen and Susan G. Wynn. St. Louis: Mosby Year Book [Mosby is now an Elsevier Health Sciences imprint], 1997.

Stenger, Victor J. "Bioenergetic Fields." *The Scientific Review of Alternative Medicine* 13, no. 1 (1999): 26–30.

————. *The Comprehensible Cosmos: Where Do the Laws of Physics Come From?* Amherst, NY: Prometheus Books, 2006.

————. "Energy Medicine." In *Consumers Guide to Alternative Therapies in the Horse*, edited by David Ramey, 55–66. New York: Howell Book House, 1999.

————. *The Fallacy of Fine-Tuning: How the Universe Is Not Designed for Us*. Amherst, NY: Prometheus Books, 2011.

————. *God: The Failed Hypothesis: How Science Shows That God Does Not Exist*. Amherst, NY: Prometheus Books, 2007.

————. *Has Science Found God? The Latest Results in the Search for Purpose in the Universe*. Amherst, NY: Prometheus Books, 2003.

————. "In the Name of the Omega Point Singularity." *Free Inquiry* 27, no. 5 (2007): 62.

————. "Life after Death: Examining the Evidence." In *The End of Christianity*, edited by John W. Loftus, 305–32. Amherst, NY: Prometheus Books, 2011.

————. *The New Atheism: Taking a Stand for Science and Reason*. Amherst, NY: Prometheus Books, 2009.

————. *Physics and Psychics: The Search for a World beyond the Senses*. Amherst, NY: Prometheus Books, 1990.

————. "The Pseudophysics of Therapeutic Touch." In *Therapeutic Touch*, edited by Béla Scheiber and Carla Selby, 303–11. Amherst, NY: Prometheus Books, 2000.

————. *Quantum Gods: Creation, Chaos, and the Search for Cosmic Consciousness*. Amherst, NY: Prometheus Books, 2009.

————. "Responses to Edgar Andrews," 2001. http://www.colorado.edu/philosophy/vstenger/Godless/Andrews.htm (accessed June 7, 2011).

———. "Scientist Nitwit Atheist Proves Existence of God." *Free Inquiry* 15, no. 2 (1995): 54–55.

———. *Timeless Reality: Symmetry, Simplicity, and Multiple Universes.* Amherst, NY: Prometheus Books, 2000.

———. *The Unconscious Quantum: Metaphysics in Modern Physics and Cosmology.* Amherst, NY: Prometheus Books, 1995.

Stevenson, Ian. *Reincarnation and Biology: A Contribution to the Etiology of Birthmarks and Birth Defects.* Westport, CT: Praeger, 1997.

———. *Twenty Cases Suggestive of Reincarnation.* 2nd ed. Charlottesville: University Press of Virginia, 1974.

Stokes, Douglas M. "The Shrinking Filedrawer: On the Validity of Statistical Meta-Analysis in Parapsychology." *Skeptical Inquirer* 35, no. 3 (2001): 22–25.

Suppe, Frederick. "Episemology." In *The History of Science and Religion in the Western Tradition: An Encyclopedia*, edited by Gary B. Ferngren, 24–30. New York: Garland, 2000.

Susskind, Leonard. *The Cosmic Landscape: String Theory and the Illusion of Intelligent Design.* New York: Little, Brown, 2005.

Swinburne, Richard. "Argument from the Fine-Tuning of the Universe." In *Modern Cosmology and Philosophy*, edited by John Leslie, 160–79. Amherst, NY: Prometheus Books.

———. *Is There a God?* Oxford; New York: Oxford University Press, 1996.

Tegmark, Max. "The Importance of Quantum Decoherence in Brain Processes." *Physical Review E* 61, no. 4 (2000): 4194–4206.

Teilhard de Chardin, Pierre. *The Phenomenon of Man.* New York: Harper, 1959.

Thomson, Keith. "Huxley, Wilberforce, and the Oxford Museum." *American Scientist* 88, no. 5 (2000): 210.

't Hooft, Gerard. "Dimensional Reduction in Quantum Gravity," revised March 2009. http://lanl.arxiv.org/abs/gr-qc/9310026 (accessed January 11, 2011).

Tipler, Frank J. *The Physics of Christianity.* New York: Doubleday, 2007.

———. *The Physics of Immortality: Modern Cosmology, God, and the Resurrection of the Dead.* New York: Anchor Books, 1994.

Torres, Phillip. *A Crisis of Faith: Atheism, Emerging Technologies, and the Future of Humanity.* Norfolk, UK: Dangerous Little Books, 2011.

Union of Concerned Scientists. "Smoke, Mirrors & Hot Air: How ExxonMobil Used Big Tobacco Tactics to 'Manufacture Uncertainty' on Climate Science," 2007. http://www.ucsusa.org/assets/documents/global_warming/exxon_report.pdf (accessed March 10, 2011).

Vidal, John, and Tom Kington. "Protect God's Creation: Vatican Issues New Green Message for World's Catholics." *Guardian*, April 27, 2007.

Vilenkin, Alexander. "Creation of Universes from Nothing." *Physics Letters B* 117B (1982): 25–28.

———. *Many Worlds in One: The Search for Other Universes.* New York: Hill and Wang, 2006.

Vitzthum, Richard C. *Materialism: An Affirmative History and Definition.* Amherst, NY: Prometheus Books, 1995.

Volkow, Nora D., et al. "Effects of Cell Phone Radiofrequency Signal Exposure on Brain Glucose Metabolism." *Journal of the American Medical Association* 305, no. 8 (2011): 808–12.

Wagenmakers, Erik-Jan, et al. "Why Psychologists Must Change the Way They Analyze Their Data: The Case of Psi: Comment on Bem." *Journal of Personality and Social Psychology* 100, no. 3 (2011): 426–32.

Walsh, Vincent, and Alan Cowey. "Transcranial Magnetic Stimulation and Cognitive Neuroscience." *Nature Reviews Neuroscience* 1 (2000): 73–78.

Walsh, Vincent, and Alvaro Pascual-Leone. *Transcranial Magnetic Stimulation: A Neurochronometrics of Mind.* Cambridge, MA: MIT Press, 2003.

Ward, Keith. "God as the Ultimate Informational Principle." In *Information and the Nature of Reality: From Physics to Metaphysics,* edited by Paul C. W. Davies and Niels Henrik Gregersen, 282–300. Cambridge, UK; New York: Cambridge University Press, 2010.

Warren, Richard. *The Purpose-Driven Life: What on Earth Am I Here For?* Grand Rapids, MI: Zondervan, 2002.

Wegner, Daniel M. *The Illusion of Conscious Will.* Cambridge, MA: MIT Press, 2002.

———. "Précis of the Illusion of Conscious Will." *Behavioral Brain Science* 27, no. 5 (2004): 659–92.

Weinberg, Steven. *Dreams of a Final Theory.* New York: Pantheon Books, 1992.

———. "Living in the Multiverse." In *Universe or Multiverse?* edited by Bernard Carr, 29–42. New York: Cambridge University Press, 2007.

Weiss, Mechal. "Largest Study of Third-Party Prayer Suggests Such Prayer Not Effective in Reducing Complications Following Heart Surgery," 2007. http://www.templeton.org/pdfs/press_releases/060407STEP.pdf (accessed February 12, 2011).

Wheeler, John A. "Information, Physics, Quantum: The Search for Links." In *Complexity, Entropy, and the Physics of Information,* edited by W. Zurek. Redwood City, CA: Addison-Wesley, 1990.

Whitcomb, John Clement, and Henry M. Morris. *The Genesis Flood: The Biblical Record and Its Scientific Implications.* Philadelphia, PA: Presbyterian and Reformed Pub. Co., 1961.

White, Andrew Dickson. *A History of the Warfare of Science with Theology in Christendom.* New York: D. Appleton & Co., 1896.

———. *A History of the Warfare of Science with Theology in Christendom: Two Volumes in One.* Amherst, NY: Prometheus Books, 1993.

Whitehead, Alfred North, David Ray Griffin, and Donald W. Sherburne. *Process and Reality: An Essay in Cosmology.* Corrected ed. Gifford lectures; 1927–28, New York: Free Press, 1978.

Wigner, E. P. "The Unreasonable Effectiveness of Mathematics in the Natural Sciences." *Communications in Pure and Applied Mathematics* 13, no. 1 (1960): 1–14.

Wilczek, Frank. "The Cosmic Asymmetry between Matter and Antimatter." *Scientific American* 243, no. 6 (1980): 82–90.

Will, Clifford M. *Was Einstein Right? Putting General Relativity to the Test*. 2nd ed. New York: Basic Books, 1993. [Also 1st ed. New York: Basic Books, 1986.]

Wilson, Edward O. *Sociobiology: The New Synthesis*. Cambridge, MA: Belknap Press of Harvard University Press, 1975.

Woerlee, G. M. *Mortal Minds: The Biology of Near-Death Experiences*. Amherst, NY: Prometheus Books, 2005.

Wöhler, Friedrich. "Über künstliche Bildung des Harnstoffs" ("On theArtificial Formation of Urea"). *Annalen der Physik und Chemie* 88, no. 2 (1828): 253–56.

Wolfram, Stephen. *A New Kind of Science*. Champaign, IL: Wolfram Media, 2002.

Wright, Robert. *The Evolution of God*. New York: Little, Brown, 2009.

Wright, Ned. "Errors in the Steady State and Quasi-SS Models." updated December 2010. http://www.astro.ucla.edu/~wright/stdystat.htm (accessed January 11, 2011).

Wuthnow, Robert. *After Heaven: Spirituality in America since the 1950s*. Berkeley: University of California Press, 1998.

Young, Matt, and Taner Edis, eds. *Why Intelligent Design Fails: A Scientific Critique of the New Creationism*. New Brunswick, NJ: Rutgers University Press, 2004.

Young, Paul. *The Nature of Information*. New York: Praeger, 1987.

Zeh, H. D. *The Physical Basis of the Direction of Time*. 5th ed. Berlin; New York: Springer, 2007.

Zimmer, Carl. *Soul Made Flesh: The Discovery of the Brain—and How It Changed the World*. London: Heinemann, 2004.

Zuckerman, Phil. *Society without God: What the Least Religious Nations Can Tell Us about Contentment*. New York: New York University Press, 2008.

INDEX

ABOUT THE AUTHOR

Victor J. Stenger grew up in a Catholic working-class neighborhood in Bayonne, New Jersey. His father was a Lithuanian immigrant, his mother the daughter of Hungarian immigrants. He attended public schools and received a bachelor of science degree in electrical engineering from Newark College of Engineering (now New Jersey Institute of Technology) in 1956. While at NCE, he was editor of the student newspaper and received several journalism awards.

Moving to Los Angeles on a Hughes Aircraft Company fellowship, Dr. Stenger received a master of science degree in physics from UCLA in 1959 and a doctorate in physics in 1963. He then took a position on the faculty of the University of Hawaii and retired to Colorado in 2000. He currently is adjunct professor of philosophy at the University of Colorado and emeritus professor of physics at the University of Hawaii. Dr. Stenger has also held visiting positions on the faculties of the University of Heidelberg in Germany and the University of Oxford in England, and he has been a visiting researcher at Rutherford Laboratory in England, the National Nuclear Physics Laboratory in Frascati, Italy, and the University of Florence in Italy.

His research career spanned the period of great progress in elementary particle physics that ultimately led to the current *standard model*. He participated in experiments that helped establish the properties of strange particles, quarks, gluons, and neutrinos. He also helped pioneer the emerging fields of very high-energy gamma ray and neutrino astronomy. In his last project before retiring, Dr. Stenger collaborated on the underground experiment in Japan that in 1998 showed for the first time that the neutrino has mass. The Japanese leader of this project, Masatoshi Koshiba, shared the 2002 Nobel Prize for this work.

Victor J. Stenger has had a parallel career as an author of critically well-received popular-level books that interface between physics and cosmology and philosophy, religion, and pseudoscience. His 2007 book, *God: The Failed Hypothesis*, made the *New York Times* bestseller list in March of that year.

Dr. Stenger and his wife, Phylliss, have been happily married since 1962 and have two children and four grandchildren. They will celebrate their golden wedding anniversary on October 6, 2012. They now live in Lafayette, Colorado, and travel the world as often as they can.

Dr. Stenger maintains a popular website where much of his writing can be found, at http://www.colorado.edu/philosophy/vstenger/. He also maintains an e-mail discussion list, avoid-L, "Atoms and the Void," where the topics range from his own writings to the whole gamut of intellectual discourse and politics.

OTHER BOOKS BY VICTOR J. STENGER

Not by Design: The Origin of the Universe (1988)

Physics and Psychics: The Search for a World beyond the Senses (1990)

The Unconscious Quantum: Metaphysics in Modern Physics and Cosmology (1995)

Timeless Reality: Symmetry, Simplicity, and Multiple Universes (2000)

Has Science Found God? The Latest Results in the Search for Purpose in the Universe (2003)

The Comprehensible Cosmos: Where Do the Laws of Physics Come From? (2006)

God: The Failed Hypothesis—How Science Shows That God Does Not Exist (2007)

Quantum Gods: Creation, Chaos, and the Search for Cosmic Consciousness (2009)

The New Atheism: Taking a Stand for Science and Reason (2009)

The Fallacy of Fine-Tuning: Why the Universe Is Not Designed for Us (2011)